T0396202

VOLUME TWO HUNDRED AND THIRTY ONE

ADVANCES IN IMAGING AND ELECTRON PHYSICS

Nanolithography and Surface Microscopy with Electron Beams

VOLUME TWO HUNDRED AND THIRTY ONE

ADVANCES IN IMAGING AND ELECTRON PHYSICS

Nanolithography and Surface Microscopy with Electron Beams

LORD BROERS

Edited by

MARTIN HŸTCH
CEMES-CNRS
Toulouse, France

PETER W. HAWKES
CEMES-CNRS
Toulouse, France

Cover photo credit: “The author in 1967 with his newly invented electron microscope gun” by Lord Broers

Academic Press is an imprint of Elsevier
125 London Wall, London EC2Y 5AS, United Kingdom
525 B Street, Suite 1650, San Diego, CA 92101, United States
50 Hampshire Street, 5th Floor, Cambridge, MA 02139, United States

ISBN: 978-0-443-31462-9
ISSN: 1076-5670

For information on all Academic Press publications
visit our website at https://www.elsevier.com/books-and-journals

Publisher: Zoe Kruze
Acquisitions Editor: Jason Mitchell
Editorial Project Manager: Palash Sharma
Production Project Manager: James Selvam
Cover Designer: Gopalakrishnan Venkatraman

Typeset by STRAIVE, India

Contents

Preface

Papers in scientific journals do not tell the whole story of how discoveries and research are made. We learn to make a line of investigation seem to follow a logical and inevitable sequence of events, to the point that the process appears highly impersonal. In reality, scientific advancement results from the choices scientists make, their passion for a certain subject, the people they meet, and their funding and career opportunities. In the *Advances* series, we like to give the opportunity to distinguished scientists to document their life's work, including elements of biographical context. A scientist's career can also involve management, from small teams to corporate departments or whole universities.

It is therefore with great pleasure that we present the current volume of *AIEP* by Lord Broers. He gives a fascinating insight into not only the early days of scanning electron microscopy but also the incredible growth of the semiconductor industry. For readers interested in the very beginnings of scanning electron microscopy in the Cambridge group, we can direct them to Volume 133 (2004) dedicated to Charles Oatley and the more recent Volume 221 (2022), "The Beginnings of Electron Microscopy: Part 2." It was a time when the developments in electron physics fed directly into industrial development through the use of electron lithography. Lord Broers was an insider to all these developments and gives his personal view of the past and future prospects.

Peter W. Hawkes and Martin Hÿtch
Toulouse

Preface

[illegible]

Author's preface

I want to thank Martin Hytch and Peter Hawkes for their kindness in granting me the privilege of being one of their authors and for their sustained encouragement and help in producing this account of the research I carried out with my colleagues at Cambridge University, IBM, and IMEC. This volume is not a textbook or a comprehensive treatise on nanolithography and surface microscopy with electron beams. I have not attempted to describe the vast amount of research and development that has brought these technologies to their present advanced state. I have limited myself to only using original material created for this volume and from publications where I was an author. In many cases, the research was led by my colleagues, as described throughout the volume, and none of it would have happened without them.

Something I do not mention is that I became obsessively interested in sailing when I was very young and skiing in my teens. I mention these pastimes as they have been wonderful family activities that Mary and I have enjoyed with our two sons, Mark and Christopher, who developed expertise beyond our own in these sports. They have been important in balancing my life, especially when I found myself in stressful situations.

It is rare for research specializations in modern science and technology to remain center stage for a long time and many PhD graduates in science and engineering find themselves working on something different soon after graduating.

I was lucky. My project was to use a scanning electron microscope (SEM) to study the way metals were etched with ions. This was of interest to my sponsors, the UK Atomic Weapons Research Establishment because ion bombardment was damaging the walls of nuclear reactors. Serendipitously, it also proved possible to use the SEM and ion etching to make microstructures similar to the transistors in integrated electronic circuits, what we now call silicon chips. So in the early 1960s, I became involved with the miniaturization of electronic devices, just after the integrated circuit was invented. The benefits of crowding devices onto a single piece of silicon had already been recognized, but nobody imagined that miniaturization would reduce the cost of a transistor 10 billion times, nor that silicon chips would transform information and communication and change the way we live.

My involvement in the miniaturization of devices lasted for about 35 years. After completing my Ph.D. in Cambridge, I worked at IBM in the USA for about 20 years. I then returned to Cambridge in 1984 and continued exploring the limits of miniaturization until 1996 when I became a full-time administrator as Vice-Chancellor of the University. This was an exciting time in the evolution of microelectronics and many outstanding engineers and scientists were involved in solving the wide-ranging set of problems it presented. It was a privilege to be a member of that community.

I describe in this volume most of the research with which I was personally involved. As is the custom in this book series, I also describe how my research evolved and why I changed where I worked several times. My early life and the path that led me to my interest in electronics, and eventually to completing a PhD at the University of Cambridge using a scanning electron for microscopy and microfabrication, is described in Chapter 1.

I arrived in Cambridge in 1960 and completed the final year of the University's engineering undergraduate degree before starting my PhD in October 1961. My supervisor was Professor Charles Oatley whose research group developed the modern surface SEM. The history of the modern SEM is described in detail in Advances in Electronics and Electron Physics Volume 133, *Sir Charles Oatley and the Scanning Electron Microscope* edited by Peter Hawkes. Oatley's SEM used an efficient secondary electron detector developed at his suggestion by Tom Everhart and Richard Thornley. It produced noise-free images of the surface of bulk samples with a resolution of 10–20 nm. The narrow beam of the SEM produced images with a larger depth of focus than optical microscopes. These images were easy to interpret and immediately gained attention.

The resolution was not as high as that of the transmission electron microscope (TEM), but TEMs could not directly examine the surface of bulk samples. They could examine a thin replica of a surface that was transparent to electrons, but it was not possible to follow a specific area on a surface as it was changing which was possible with the SEM.

There were no commercial SEMs at that time. The microscope I inherited had originally been built by Professor Oatley and had subsequently been used by Garry Stewart, another of his research students. Stewart had designed and fitted an ion probe system that focused a beam of ions onto the SEM sample enabling the *in situ* study of ion-bombarded surfaces. Before starting my experiments, I replaced the electrostatic final lens of the SEM with a magnetic lens to improve the reliability of the instrument and improve its resolution. I also designed and fitted a magnetic deflection filter to the ion probe system to remove oxygen from the ion beam. Oxygen ions produced anomalous effects when etching the metal surfaces.

I describe my arrival in Cambridge, my completion of the final year of the engineering undergraduate degree, and the modifications to the SEM-ion beam system I made at the start of my PhD in Chapter 2. After characterizing its performance, I carried out a series of experiments in which I studied the formation of cones that appear on ion-bombarded metal surfaces and the evolution of ridges on the surface of ion-bombarded surfaces. These experiments are described in Chapter 3.

Chapter 4 describes the microfabrication experiments carried out during my PhD research in 1963 and 1964. I first used the SEM electron beam to create contamination patterns on thin layers of metal and used ion etching to remove the unprotected metal. This produced wires with a minimum width of 40 nm. Bill Nixon, who became my supervisor at that time, suggested that I also try using a standard photoresist. I did this and it produced linewidths down to about 0.1 μm which was about 50 times narrower than the lines in integrated circuits at that time.

I reported some of these results at the 1st International Electron and Ion Beam Science and Technology Conference in Toronto in 1964. There was considerable interest that these very small structures could be made, but I was told that the real challenge was to make these techniques practicable for the manufacture of devices.

I also learned that others had been working on the use of electron beams to make small structures including Buck and Shoulders in 1955, Mollenstedt and Spiedel at Tubingen University in 1960, Loeffler at Berlin University in 1964, and others. This early work is reviewed in Chapter 4.

Research on miniaturization was given impetus in December 1959 by Richard P. Feynman when he delivered his famous lecture entitled "Plenty of Room at the Bottom" to the American Physical Society in Pasadena, in which he explored the immense possibilities afforded by miniaturization.

Chapter 5 describes how in 1965, at the end of my Ph.D. studies, I moved to the IBM Research Laboratory in Yorktown Height, New York, to join a group managed by Alan Brown that was working on memory systems that used electron beams to write on photographic film. Richard Thornley, who had developed the secondary electron detector for the SEM with Tom Everhart, brought the electron optical expertise from Cambridge to the IBM group in 1960.

I was assigned the task of finding a long-life cathode for the electron beam systems that were writing data on photographic film. They were using the tungsten wire hairpin cathodes that were universally used in electron microscopes and demountable electron beam equipment at the time. I found several refractory metal cathodes that would last 200 h. I also found that a rod cathode made of lanthanum hexaboride lasted more than a thousand hours and could produce much higher electron brightness than the metal cathodes. This work on cathodes is described in Chapter 5.

The higher brightness of the lanthanum hexaboride cathode offered the possibility of building an electron probe for scanning microscopy and microfabrication with higher resolution than had previously been possible. I went ahead and did this and completed the new electron probe in 1967. Its design and the results it produced as a surface scanning microscope, and as an electron beam lithography exposure system, are described in Chapter 6. The beam diameter of the new probe was close to the theoretical minimum of 27 Å as determined by the aberrations of the relatively long focal length final lens needed for secondary electron surface microscopy and diffraction.

As a SEM, the higher beam current allowed noise-free images to be obtained at higher magnifications and I used it in collaboration with biologists, and physicians to examine blood cells, human marrow, and elephant's teeth.

As a lithography tool, the LaB_6 cathode probe could focus 1 nA into a beam diameter of 15 nm and 1 μA into a beam diameter of 200 nm. Michael Hatzakis and I used it to make surface wave transducers with finger widths of 0.15 μm that operated at 3.5 GHz which was the highest frequency achieved at the time. At this time Tom Sedgwick and I invented the window substrates that eliminated the back-scattering of electrons from the substrate and made it possible to examine nanostructures with transmission microscopy. Most importantly, they made it possible to make electrical contact to these tiny fragile structures.

Chapter 7 describes how in 1970 I rebuilt this probe with a short focal length final lens and a 75 kV electron gun. Together these reduced the beam diameter from 27 to 5 Å and allowed us to explore the limits of surface scanning electron microscopy using low-loss high-energy electrons scattered from the sample surface, rather than secondary electrons, as proposed by Oliver Wells. It also made it possible to take full advantage of the thin window substrates in measuring the point spread function for PMMA resist.

Before designing the new probe, I analyzed the different column configurations available for electron probes of this type. In particular, I compared the conventional SEM column where the cross-over formed in a thermal cathode electron gun is imaged onto the sample, with critical and Kohler illumination systems where the source is an optimally illuminated physical aperture. The new column was designed so that it could be operated in the conventional cross-over demagnification mode or with Kohler illumination.

I considered thermal, Schottky, thermal field emission, and cold field emission sources for a wide range of beam currents and diameters. The alignment coils and apertures needed for Kohler illumination allowed the LaB_6 cathode electron gun to be set up for maximum brightness. All of these analyses, and the choice of, are contained in Chapter 7.

Chapter 8 contains the measurements of the performance of the short focal length probe. This is followed by a description of the energy filter designed to select the low-loss electrons scattered from the sample surface and its application to high-resolution low-loss surface microscopy.

The column could focus a few picoamps into a beam diameter that was close to the theoretical limit set by spherical aberration and diffraction of 0.5 nm. For applications that needed higher current levels, 1 nA could be focused into a beam diameter of 4 nm and 1 μA into a beam diameter of 40 nm.

This chapter also includes an extended study of Bacteriophages with Barbara Panessa and Joseph Gennaro at NYU. This was the first time these bacterial viruses were examined in a surface SEM. The resolution of conventional secondary electron scanning microscopy is not able to show the details of these very small viruses that are typically only 500 Å in size.

Chapter 9 describes the application of the short focal length probe to nanofabrication.

The first experiments with contamination resist and the thin window substrates produced gold wires that were 80 Å wide. It was also possible with PMMA resist and the lift-off invented by Michael Hatzakis process to produce linewidths of 20 nm. An array of 15 nm lines on 45 nm centers was fabricated with PMMA in the negative mode and ion etching. All of these dimensions were smaller than had been achieved before.

Following these initial experiments, and to provide the means to accurately determine the resolution of electron beam lithography using PMMA resist, the point spread function for the resist was measured using a beam diameter of 0.5 nm diameter and thin window substrates. The sigma of the point spread function, assuming it was Gaussian, was measured to be between 12 and 15 nm.

Having determined the point spread function, the resolution of electron beam lithography is compared with the resolution of optical and X-ray lithography.

Electron beam methods for directly patterning ionic crystals and Langmuir Blodgett films with Mel Pomerantz are described at the end of Chapter 9. These methods do not require a separate development process.

Chapter 10 describes nanoscale devices fabricated with the short focal length electron probe.

Working in close collaboration with Bob Laibowitz, Frank Mayadas, Richard Voss, Praveen Chaudhari, Cory Umbach, and others we fabricated and measured the characteristics of Josephson effect niobium nanobridges, Nanobridge SQUIDs and Aharonov Bohm rings. All of these devices had smaller dimensions than similar devices that had been previously measured.

In 1977, I became an IBM Fellow, a special position created to encourage engineers and scientists to pursue their research interests. Fellows were no longer part of the standard management structure and were free to determine for themselves what they did. I stepped down from management for several years and concentrated on making devices for scientific exploration and using high-resolution scanning electron microscopy to study biological samples as already described.

In 1982, I accepted a challenge from Holly Caswell to join him in IBM East Fishkill where the electronic components for IBM's computer processors were designed and manufactured. Caswell had been Director of Applied Research at IBM's Thomas J. Watson Research Center and had moved to East Fishkill to be Vice President and Director of the Fishkill Development Laboratory.

My appointment as an IBM Fellow and my move to East Fishkill are described at the beginning of Chapter 11. This is followed by a description of what went on at East Fishkill and how IBM developed and manufactured electronic components in the 1980s. In particular, I review the capabilities of optical, electron beam, and X-ray lithographies at the time and the strategy for their further development in the years to come.

In 1984, I decided that I wanted to return to my research and after spending some time in IBM's headquarters in Armonk New York as a member of the company's Corporate Technical Committee, I moved to Cambridge University to pursue my science-oriented interest in the ultimate resolution of what by then we were calling nanolithography. I describe the reasons I made this difficult decision at the beginning of Chapter 12. It was not a complete break from IBM as I became a member of IBM's Science Advisory Committee, and IBM generously allowed me to continue to use my Yorktown laboratory during vacations from Cambridge.

After a few months of settling into academic life and teaching in Cambridge, of which I had little or no experience, I slowly, with the help of colleagues found the funds and space to set up a nanolithography laboratory in the University's Engineering Department. The centerpiece for the laboratory was a JEOL 400 kV transmission electron microscope that I modified so that it could also be used as an electron probe for nanolithography. The microscope and all of the equipment we needed for thin film processing and lithography were housed in a new cleanroom. I was joined in this nanolithography research by David Allee, Andrew Hoole, Xiaodan Pan, Joe Ryan, and Shanhong Xia.

This all went well, and we used the 4000EX to re-measure the resolution of contamination lithography and PMMA with the 4000EX. Projection transmission microscopy proved to be a better way to examine our nanostructures than scanning transmission microscopy because the contamination rate and the risk of damaging resist structures were much lower. For nanolithography, the high accelerating voltage greatly reduced the forward scattering of the beam within resist layers and produced dramatically high aspect ratios in contamination resist. The latter allowed us to produce high-quality 10 nm wide gold wires on window substrates. The sigma (σ) of the point spread function for PMMA remained at 15 nm at 350 kV.

Having established the ability to use the standard fabrication processes we turned our attention to the direct exposure SiO_2 a process discovered by O'Keefe and Handy in 1967. This proved to have higher resolution than PMMA or contamination resist. We were able to produce periodic structures with center-to-center spacings below 10 nm. However, the ratio between the solubility of the exposed and unexposed areas was only 3–4 which made it difficult to produce high aspect ratio structures.

All of these activities in Cambridge are described in Chapter 12.

By 1989, we had characterized the capabilities of the 4000 EX and were ready to use it to make devices, and we joined a European consortium led by IMEC to apply our lithographic capabilities to the design and fabrication of MODFETs. I had earlier worked with Luc van den Hove and Marc van Rossum at IMEC and collaborated with Fabian Pease and David Allee at Stanford on MODFETs. Some of the results produced by the consortium are described in Chapter 13.

Despite my intention to concentrate mainly on research and teaching in Cambridge, I became increasingly involved in academic administration becoming Master of Churchill College in 1990, Head of the Engineering Department in 1993, and Vice Chancellor of the university in 1996. The Vice Chancellorship of the University had been a part-time position until 1992 but became full-time following an in-depth inquiry into the University's administration in the late 1980s. As Vice-Chancellor I had no time for experimental research although my involvement in science and engineering policy and industry continued.

I enjoyed working with the brilliant entrepreneurs that surrounded the university and who in the 1980s and 1990s built the Cambridge high technology cluster into one of the largest and most influential in Europe. They included Hermann Hauser, Andy Hooper, David Cleeveley, Roger Needham, Richard Friend, Andy Harter, Alan Munro, Robin Saxby, Warren East, Charles Cotton, and there were many others.

The Duke of Edinburgh was the Chancellor of the University from 1976 to 2011 so, as Vice-Chancellor, I, in effect reported to him. He visited the University several times a year and would occasionally stay with my wife and me in the Vice-Chancellor's Lodge. I learned a great deal from him. He had a sharp intellect, and like Prince Albert, consort to Queen Victoria, was intensely interested in engineering and technology. Our relationship continued after I left Cambridge as I was for 5 years the President of the Royal Academy of Engineering, the UK's national academy of engineering. He was the Academy's Senior Fellow, and in many ways the father of the Academy as he had played a pivotal role in its founding and had tirelessly supported its activities since its inception in 1976.

My term as Vice Chancellor ended in 2003 and I became a Life Peer in the House of Lords in 2004 where I chaired for several years the Select Committee for Science and Technology and became President of the Parliamentary and Scientific Committee. My interests inevitably expanded to include a range of science and industrial policy matters including those related to pandemics and aging as well as nuclear power and climate change. I retired as an active Peer in 2021 and returned to my interest in the semiconductor industry as the dimensions of the transistor elements on silicon chips approached the ultimate limits of fabrication and power dissipation, hence my sitting down to write this account.

LORD BROERS

CHAPTER ONE

Early life

Contents

I was born in Calcutta, now Kolkata, in Bengal, India in 1938. My parents were English. My father was born in London in 1899 and spent his working career as an insurance company executive. His father and mother were English, although his family's earlier roots on his father's side were Dutch. Our family name Broers is described in Wikipedia as "A Dutch patronymic surname meaning 'Broer's (son)'. Broer and Broeder mean 'brother' in Dutch."

My mother's parents were also English with the common English and Welsh surname of Cox. They decided, when she showed a talent for playing the piano at the age of 5, that she should be a concert pianist. From then on, she played the piano for many hours a day and went on to study at the Royal Academy of Music in London where the excellence of her performance led her into a career as a performer. Her career was relatively short because she married my father in 1931, at the age of 24, and went with him to India where the opportunities to perform were limited, but her interest in music persisted and influenced my life.

My father was, luckily, just too young to participate in the First World War, although he trained as a soldier when he left school. He was nearly killed when a trainee next to him pulled the trigger pin out of a hand grenade and then dropped it in panic rather than throwing it into a pit. Fortunately, the supervising sergeant saw what had happened and kicked the live grenade into the pit where it exploded harmlessly. My father would tell this story without drama as if it was the sort of thing that happened to everyone.

After the war, my father became a clerk in the London office of the Caledonian Insurance Company, one of the first insurance companies to insure motor cars. In 1924, when the company was looking for someone to go to India and open a branch office, he volunteered and found himself on his own on an ocean liner headed for Calcutta to do just that. It took more than 3 weeks for an ocean liner to travel the 8700 nautical miles to

Advances in Imaging and Electron Physics, Volume 231
ISSN 1076-5670
https://doi.org/10.1016/B978-0-443-31462-9.00001-0

Calcutta (20.2 days at 18 knots) and business between England and India was routinely conducted by sea mail post so it took more than a month to get a response from England. Telephone and telegraph communications were only for emergencies.

It must have taken much self-confidence and entrepreneurship to get a branch operating in such a remote and culturally different place, but my father seems to have met the challenges without difficulty. Disappointingly, I know little about how he accomplished this. He rarely talked about his early days in India, but the insurance business grew to be stable and profitable, and he enjoyed the independence that remoteness from London gave him. He was in effect running his own business, and throughout his life was never really comfortable when this was not the case. He was a gentle and calm man with a strict sense of responsibility.

He spent his working life entirely in insurance and ended up as Managing Director of the Guardian Group in Australia in the late 1950s after the Guardian took over the Caledonian. This take-over was initially difficult for him as he found himself reporting to a Guardian manager with no experience in Australia who was sent out from London. Thankfully, it was not long before the Guardian CEO visited Australia and saw that my father, as a senior respected figure in Australian insurance with deep knowledge of the business, was the more appropriate person to run the new combined Australian business. The Guardian manager was returned to London.

While my father was primarily a successful businessman, he became intensely interested in radio and photography. These were more than hobbies to him, and his expertise became close to that of a professional.

He was very proud of his Rolleiflex camera with its Zeiss lens that he had re-coated. He used it to win several awards including one for a picture of the Taj Mahal that had a special place on the wall of our home.

His interest in radio led him to become a columnist for the Calcutta Statesman. He wrote more than 60 articles about radio that ranged from general comments on radio as a medium of communication, to in-depth reviews of the technical performance of short-wave receivers. I was told by one of his friends that he was the first member of the public to receive the BBC by short-wave in India. This may well have been true as the BBC's first broadcast of any kind was in November of 1922.

The 1920s were the early days for radio. Signal strength was low, and to achieve higher noise-free gain and better selectivity, the superheterodyne circuit was invented in which the signal from the antenna is mixed with another signal from a local oscillator to produce a fixed intermediate frequency that can more easily be selectively amplified. Previous radio receivers had used a succession of amplifying stages all of which had to be tuned simultaneously to the radio station's frequency. The word superheterodyne was generally shortened to superhet. My father wrote under the pseudonym "Superhet."

Some 75 years later, when I was asked to deliver the BBC's Reith Lectures, named after Lord Reith, General Manager and Director-General of the BBC, I remembered vaguely that my father had said something about Lord Reith in one of his Statesman articles. I found the scrapbook full of his articles and opened it to find, by strange coincidence, that the page before me had an article about Lord Reith, dated June 26, 1938. He wrote twice about Reith, describing him as "building up the BBC from its beginnings to the mighty machine which today transmits music, entertainment, and information to no less than 8,600,000 homes in Great Britain", and later pointing out that he had behaved as a virtual dictator in his management style. This was in 1938 when Reith was about to leave the BBC to become Chairman of Imperial Airways. I included these remarks in my first Reith Lecture on the BBC and, amusingly, found several years later when editing this lecture before it was included in the book *Remarkable Minds: A Celebration of the Reith Lectures,* published by BBC Radio 4 (4, 2020) that the remark about his management style had been deleted from the official transcript of my lecture. I replaced it in the transcript of my lecture in Remarkable Minds.

My interest in things electrical coincided with my very first memories at the age of 3, about a year after we arrived in Sydney in 1940. I remember my father showed me that a little light bulb from a torch lit up if one held the bottom of the bulb on the top of a torch battery and used a wire to connect the side of the bulb to the bottom of the battery. At the age of three, I was transfixed.

We had moved to Australia because my father had contracted amoebic dysentery and was told that he would die if he did not leave India and find a cure. The year 1941 was not the time for international travel and a return to London was out of the question as the air raids over London were getting more and more serious. Getting to Australia was also high risk because we had to stop in Bangkok, where there were many Germans and my mother was worried that they might sabotage the plane, and Singapore where the threat of Japanese occupation was already apparent, but it was decided that we had to go to Australia. Singapore did not fall to the Japanese until about a year later in February 1942, so ironically, the urgency of my father's blight saved us from what might have been a worse fate.

The journey by flying boat from Calcutta to Sydney was graphically described by my mother in a letter dated December 3rd, 1940, to her parents in Kent, England. This journey, which would take 6h today, took 6 days. We had to stop in Bangkok, Singapore, Surabaya in East Java, Darwin, and Townsville before finally arriving in Sydney. Between each of the stops, the plane had to land and re-fuel every four and a half hours. Sunderland flying boats could only cruise at about 80 mph and at a maximum height of about 10,000 ft where the air was not as smooth as it is at 30,000–40,000 ft where modern jetliners cruise. Air sickness was a real problem especially when the plane had to descend through the rough tropical air to land.

The only saving grace was that the seats in the plane were luxurious by modern standards, more like today's business class seats. They reclined to be almost horizontal and there was a bunk for my father, which he did not use as it was needed by my mother and me when we felt too sick to sit upright. We had to be up every morning at about 4 a.m. so you can imagine what it was like for my mother and father who had always had nannies to look after my brother and me. Looking after two little boys at the ages of 2 and 3 in such circumstances must have been stressful, to say the least.

The members of the airline crew were evidently very kind and attentive, and the passengers stayed in first-class hotels at the stopovers, but my mother describes how "she was so done up and her nerves so on edge" by the time they reached Sydney that she nearly burst into tears of relief as they got into the launch to go ashore for the last time.

My father's health improved in Australia, and he managed to travel around New Zealand as well as Australia exploring the possibilities for expanding the Caledonian's fledgling business, but unfortunately, the doctors in Sydney failed to completely cure his dysentery and we had to leave. His travels, however, were to be relevant to the family's future.

As soon it became clear he was not going to recover without the specialist treatment that was available in London, he began exploring means for us to get to London. Flying back through India, the Gulf, and Europe was not possible and the Japanese navy was still patrolling the Pacific making it difficult to travel east by sea. In the end, despite the Japanese presence in the Pacific, he found us passage on a troop ship across the Pacific and through the Panama Canal to London. I recall my brother and I being severely reprimanded for chasing each other around this ship after dark and opening doors exposing light. The ship was blacked out to avoid being seen by Japanese submarines. In any case, we made it across the Pacific and through the Panama Canal and arrived back in London safely in 1943.

The doctors at the School of Tropical Medicine in London did manage to get my father's illness under control, if not cured, and he returned to full-time work and immediately set about persuading the Caledonian that they should expand their operations in the Far East, especially in Australia. He succeeded and in 1948 we set off back to Australia, this time enjoying the luxury of the Orient liner, Orontes. Our destination was

Melbourne, the business capital of Australia, where he became the "Eastern Superintendent" for the Caledonian's businesses in the Far East. In the end, this was mainly concerned with expanding the business in Australia.

My brother Richard and I enjoyed this trip because our fellow passengers on the Orontes included the whole of the Australian cricket team captained by Don Bradman that had just won the 1948 Ashes test series from England. The Australians won all 34 matches played on that tour, an unprecedented feat. I remember playing deck tennis with Ray Lindwall, perhaps the most famous of Australia's fast bowlers, and Keith Miller the all-rounder who was capable of winning a match almost singlehandedly.

Geelong Grammar School

On arriving in Australia my brother and I went to Geelong Grammar School, a well-known boarding school in the English tradition about 40 miles from Melbourne. I was only 10 years old but was already very interested in electronics and my first project of course was a crystal set, so I could listen to the radio in bed in the school dormitory. The school frowned upon the use of radios in the dormitories but rather than discouraging me this was an incentive. I hid the aerial in the ivy that covered the school buildings. I soon added a valve to amplify the tiny signal from the crystal set so that it could drive more than one pair of headphones. This let me connect others in the dormitory provided they were near a window so that the connecting wires could also be hidden in the ivy. None of us could afford to own a ready-made radio as they cost hundreds of dollars in today's money. I nearly electrocuted myself several times with the 200 volts needed for the valve, but fortunately, the human body has a higher impedance to DC currents than to AC, so I survived.

I was a troublesome student because the necessary rules of a boarding school rankled me and I was continuously getting into trouble for disobeying them. I was fortunate, however, to be among the first students to spend time at Timbertop, a branch of the school built in the bush 213 km from Melbourne. Here, the rules were relaxed, and we had more freedom to do what we liked.

Initially, there was no electricity in the buildings at Timbertop. Kerosene lamps with mantles provided light, and we burned wood for heat, but after about a month it was decided that a generator should be installed, and I was able to watch and play a small role in this exciting project. The motor and the generator were mounted on a concrete slab and initially were not aligned accurately enough and within a few days the bearings failed. The second time around, after many small adjustments, the alignment was corrected, and the generator was still running when I left Timbertop a few months later to return to the main school at Corio.

Knowing that it was possible to accurately align components remained with me and led me many years later to design the components of the electron beam columns of the electron microscopes I built so that they were aligned to within a few microns. This made it much easier to ensure that the electron beam traveled precisely along the axes of the lenses and the microscope reach its theoretical resolution.

Geelong Grammar School, and especially Timbertop, were much in the news in 1966, 11 years after I was there, when the Prince of Wales, now King Charles III, attended the school for 6 months. He later said that his time at Timbertop was the most enjoyable part of his whole education. It was certainly the happiest time for me at school, although I greatly enjoyed my time at both Melbourne and Cambridge universities.

Melbourne University

I left school when I was just 17, a year earlier than usual because I had passed most of the necessary exams and was looking forward to the freedom I had been told one had at university. University for me was Melbourne University as it was the only university in Victoria until Monash University was founded in 1960.

There was, however, a last-minute problem. In order to be admitted to a university the state of Victoria required that students had not only to obtain satisfactory results in their final school exams in the subjects that they were going to study, but had in the previous year passed the exams in English and a foreign language. My language had been French, and I had failed. I had hoped that nobody would notice this, but when it was discovered, my acceptance at Melbourne University was suspended. This was a crisis. I did not know what to do, but my parents stepped in and arranged for me to be tutored over the summer by John Bedggood, one of the teachers at Geelong Grammar. I knew John well because I had participated in many of the musical activities he had organized. He was a talented teacher and despite my lack of linguistic skills, I managed to scrape through the re-take of the exam that was allowed at the end of the summer, and all was saved. I went on to study science with physics as my major subject. Some 10 years later, John became one of Prince Charles's tutors when the Prince spent two terms at GGS.

During my Ph.D. research, I found that I needed to read about the research of several French and German engineers and scientists and was grateful that I had at least some understanding of French. I then took a course in "Science German," which proved valuable. As I got older, I appreciated more and more what an asset it was to know a language and when I became the Head of Engineering in Cambridge, I supported very strongly a successful program of language teaching set up by Anny King, Edith Esch, and Sarah Springman. About a third of the students voluntarily took this course.

Jumping even further forward, when I was a member of the House of Lords, I enthusiastically joined those who were defending the teaching of modern languages in schools when funding for schools was being threatened. In one speech I remember citing some Canadian research that showed that the development of dementia and Alzheimer's was delayed several years for people who spoke two languages regularly.

Returning to my interest in electronics, by the time I started at Melbourne University I had formed a small business making high-quality audio systems for rich farmers, and music lovers, and soon learned that I could make money, not only by making the valve amplifiers from scratch but by buying the other components wholesale and selling them on at full retail price. I worked with a skilled cabinet maker and our small enterprise went well, enabling us to buy the best loudspeakers and turntables to demonstrate the magic of high-fidelity sound to our customers. Much of the best equipment was made in England. I enjoyed this and the business thrived, but I was determined to learn more and to seek a career that involved the design and building of state-of-the-art electronic systems. So, when Melbourne University introduced a new final year in electronics in 1958, I enthusiastically took this course after completing my Physics degree.

A contributing factor to this decision was that I had failed to get an FM radio tuner to work and wanted to become expert enough to do this. I built the tuner when I learned that FM radios produced higher quality sound than AM radios and for the first time, experimental FM radio transmissions were being made in Australia. I had found the circuit in *Wireless World* and bought the components in a radio parts store in Melbourne called McGraths, but failed to make it work. I learned enough to make it work in my year studying electronics but by then the tuner had been destroyed.

In the 1950s in Australia, young men were required to complete National Service in the Australian armed forces. The National Service Act 1951 stipulated that Australian males turning 18 on or after 1 November 1950 would do 176 days of standard recruit training in the Navy, Army or Air Force to be followed by 5 years of follow-up in their respective Reserves. University students were allowed to split the 176 days into two periods of about 90 days in the summer holidays. After the first summer of basic training, I opted to serve in the Melbourne University Regiment Band as a tuba player. John Bedggood had persuaded me to learn to play the tuba while I was at Geelong Grammar School as players were needed for the school band. I am not a big person and was particularly small as a child so this was a strange choice as I was about the same size as the tuba. But I managed, and the tuba was not as heavy as a car battery, which I learned I would have to carry if I opted for my second choice in the Melbourne University Regiment, which was the Signal Corps.

Halfway through my second summer in the army, I was quite unexpectedly summoned to the Camp Commander's Office. My first thought was that the military police had discovered my little Austin A30 that I had hidden in the bush adjacent to the camp so that we could escape to a pub in the local town. This was forbidden, so I was frightened that I was going to be severely punished. When I arrived at the Commander's Office he greeted me sympathetically, which surprised me, and told me to sit down. He then told me that my house in Melbourne had burned to the ground and that I was to return there to join my mother who had been there alone but had escaped and was okay. I had what must be a common reaction to receiving such news and said that there must be a mistake and that I was Private Broers and such things did not happen to me. But it had and I returned to Melbourne by train and took another train to the suburb where we lived and walked up the hill to our house. It was nighttime by then and when I arrived, to my great relief, it appeared that the house was still there. The front wall was still standing, but when I looked behind the wall there was just a pile of ashes.

My mother was with the neighbors next door who had been kind and helpful and fed and clothed us as we only had the clothes we were wearing. Things were not that bad as my father, brother, and sister were at our beach house outside Melbourne and no one had been injured, but the fire was very intense and destroyed almost everything. The house stood on red-wood stumps so there was an airspace underneath it that supplied oxygen to the fire. Among the ashes, we found what remained of my mother's jewelry. In particular, we were looking for a platinum ring with a large diamond. We found the ring, but the diamond had burned. We learned that diamond burns at 850 °C and platinum melts at 1768 °C. I found the blades of my ice skates, but no sign of the hand-made boots. I also found a lump of melted aluminum. It was all that remained of my FM tuner. I still remember the smell of the ashes.

CHAPTER TWO

SEM/ion beam system for examining ion etched surfaces☆

Contents

Singing was an important part of my life at Geelong Grammar School, and my mother arranged for me to be professionally trained when I left school. While I was studying at Melbourne University, I was a member of the choir at St Paul's Cathedral in Melbourne. I was paid the princely sum of 12 shillings and nine pence a week (about A$20 in 2024) to be the paid tenor on Cantoris, the side occupied by the cantor. A cathedral choir is split into two. This obliged me to attend all of the choral services held at the cathedral, four during the week and two on Sundays. It sounds a lot, but I enjoyed it a great deal and it fitted my daily schedule. I attended lectures for my BSc degree at the university in the mornings and practical classes in the early afternoon, and then I would take one of Melbourne's famous trams to the other side of Melbourne, where I would ice-skate for an hour. The ice rink was just over the river Yarra from the cathedral, where I would sing Evensong at 5 pm. It was then time to take a train home, have dinner with my parents, and build hi-fi sets. Not a very social life but it kept me out of trouble.

☆University of Cambridge—Mechanical Science Final Year—Ph.D. research describing the modification of the SEM/Ion Beam system to improve the resolution and reliability of the SEM and to remove oxygen ions from the ion beam.

Advances in Imaging and Electron Physics, Volume 231
ISSN 1076-5670
https://doi.org/10.1016/B978-0-443-31462-9.00002-2

My musical activities were important as I wanted to go to the University of Cambridge to explore the opportunities for research in radio astronomy, and several of the University's colleges offered Choral Scholarships. Competition for these scholarships was relatively strong but most of those seeking them were younger than I was and did not have any formal training, so I was fairly sure that I could win one and did so at Gonville and Caius College. As I was studying engineering, I could not accept a scholarship at King's of St John's colleges, where the choirs were essentially professional, because their compulsory rehearsals were held in the afternoons and clashed with engineering laboratory classes.

My naive dream was to study for a PhD in which I would build radio telescopes with Martin Ryall. When I arrived, surprisingly considering his fame and importance—after all, he was to win the Nobel Prize with Tony Hewish in 1974—he agreed to meet me but told me that his group had only recently finished building a new interferometer and that research for the next few years would mainly be observational and theoretical. In reaction to my obvious disappointment, he suggested that I explore possibilities in the Engineering Department. I did this and was told that there were possibilities for me to study for a PhD, perhaps to work on the new scanning electron microscope, but was told that I should complete Part II of the Mechanical Sciences Tripos before starting research. It was still the era when Cambridge was reluctant to recognize degrees from other universities, let only those in the colonies! In any case, I was thrilled that I would be able to study in Cambridge and rushed to a library to find out what a scanning electron microscope was.

In retrospect, I am pleased that I was made to remain an undergraduate for a fifth year as I had already learned most of what was in the syllabus, and it allowed me to indulge in extracurricular activities, especially singing, sailing, skiing, tennis, reading novels, and drinking wine. Despite, or perhaps because of these, I completed the Part II of the Engineering Tripos and started as Professor Charles Oatley's research student in 1961. Instead of making radio telescopes, I was to make microscopes. I was to take over one of Charles Oatley's scanning electron microscope projects.

This was an exciting time in the development of the scanning electron microscope. Charles Oatley arrived in Cambridge in 1945 and for the next 20 years, he and his research students showed that the SEM was an instrument with significant potential for science, and especially for the development of a broad spectrum of high technologies. The story of Charles Oatley and his research group, and the subsequent development of the SEM, is comprehensively told in volume 133 of *Advances in Electronics and Electron Physics* by editor-in-chief Peter Hawkes, which was published in 2004 (Hawkes, 2004). This volume, which was entitled *Sir Charles Oatley and the Scanning Electron Microscope,* was originally suggested by Bernie

Breton, who worked for many years in the Cambridge Instrument Company and later in the Engineering Department at Cambridge University with Oatley. It is edited by Breton (2004), McMullan (2003) who was the first of Oatley's research students to work on the SEM starting in 1948, and Smith (1956) who started in 1952 and built the first SEM to be applied outside Cambridge

By the time I started my research in 1961, the first commercial SEM was about to be built by the Cambridge Instrument Company. A resolution of 200 Å had been demonstrated in several of the SEMs built in Oatley's group and the microscope's large depth of focus had caught everyone's attention because the images looked familiar to the human eye and were easy to interpret.

The SEM I inherited was originally designed by Oatley in 1958 for his own use, but he handed it over to Garry Stewart (Stewart, 1962) in 1958 for his PhD project, which was to study ion etched surfaces. Ion etching damaged the walls of nuclear reactors and was of interest to the UK Atomic Energy Research Establishment, AERE, at Harwell. Charles Oatley had been Director of the Royal Radar Establishment at Malvern at the end of the Second World War and knew senior people at all of the UK research establishments. This was valuable in gaining funds for research and Garry and my PhDs were funded by AERE.

The SEM used three electrostatic lenses to demagnify the cross-over formed in the electron gun. Garry Stewart brought the microscope into operation and went on to improve the electrode configuration of the SEM's final lens, He also fitted a new specimen chamber and specimen stage, and most importantly designed and built a system that focused an ion beam onto the specimen in the SEM. This allowed him to use the SEM to examine the same area on the specimen through many stages of ion etching without removing the sample from the vacuum. It was not possible to observe the surface while it was being etched, because the flood of secondary electrons created by the ions drowned out the secondary signal from the SEM electron beam, but this was of little importance.

Garry used this system to examine the structures that formed on the surface of metals as they were etched. He studied in detail the formation of spikes, or cones as I will call them, that form under particles as the surface is etched away. He also examined etched, and un-etched, animal teeth in a fascinating set of experiments in collaboration with Alan Boyde (Boyde, 2004) at London University. I continued these experiments with Alan Boyde years later after I built an SEM at IBM (Fig. 2.1).

Garry had an encyclopedic knowledge of physics, electronics, vacuum systems, and mechanical engineering, all of the disciplines needed to build and operate scanning electron microscopes. He was immensely helpful to me although soon after I started, he went off to the Cambridge Instrument

Fig. 2.1 ANB at the controls of the SEM/ion beam system shown in Fig. 2.7, which was used to carry out the experiments described in Chapter 3 and to make the nanostructures described in Chapter 4.

Company to guide the development of the Stereoscan, the first successful commercial SEM. I needed help as I was faced with six racks of electronics and the complicated microscope column with its electron gun, lenses, and vacuum pumps, and not an instruction manual in sight. Thank heavens I had been playing with electronics for a long time.

My first task, at Garry's suggestion, was to take stereo images in the SEM so that the height of features on a surface could be measured. I calibrated the specimen tilting mechanism and took a series of stereo pairs at different tilting angles. I was amazed that Garry could just place a pair of stereo images on a table and merge them without using a stereo viewer. In the end, I taught myself to do this. It was one of the many things he taught me. It was not surprising that the first SEM produced by the Cambridge Instrument Company was called the Stereoscan, nor that it was a success as it drew upon 15 years of research in the university. It was an outstanding example of how industry and universities can collaborate. Before the first Stereoscan was completed, I took micrographs for potential customers and have always been pleased that I managed to play a small role in this important project.

Before a year had passed, I had to move the SEM/Ion beam system because the building it was in was to be demolished. This was a traumatic learning experience. The system had to be pulled apart completely and many of the multi-wire cables that connected the six electronic racks to the microscope column and the operating console had to be replaced.

I had a lot of help from the expert technicians in doing this, but they enjoyed practical jokes and could not resist the opportunity to frighten me. I was in the middle of the lengthy process of carefully labeling every single wire and cable so that I could be sure everything could be correctly re-connected, when I returned from lunch one day to find one of the technicians with a huge pair of metal shears, ostentatiously slicing through two large multi-wire cables. None of the wires were labeled, or even colored. I collapsed in despair in a chair with my head in my hands, only to have them, with great laughter and amusement, reveal that these were dummy cables they had buried amongst the other cables and were not part of the system. They and I have never forgotten this.

Having survived the drama of moving the system and getting it working again, I decided to modify the microscope to improve its resolution and ease of operation. The most important change was to replace the electrostatic final lens with a magnetic lens. The microscope was unreliable mainly because it was difficult to keep the electrodes of this final electrostatic lens clean. Small pieces of insulating dust, and the layer of carbonaceous contamination formed by scattered electrons, had to be removed regularly. To gain access to the lens, the sample chamber had to be removed, and after taking the lens apart and cleaning it, it had to be re-assembled and its electrodes re-aligned. This was done on a jig borer in a machine shop, where there was every likelihood of gathering fresh dust. Magnetic lenses do not have such problems because their pole pieces can be electrostatically shielded from the electron beam by a non-magnetic metal tube.

Charles Oatley had opted for electrostatic lenses because the stability required of the high voltage power supply was eased if the supply was used to drive both the electron gun and the lenses. However, power supplies with adequate stability had become available and I decided that the cleaning problem was very serious, and I would go ahead and replace the electrostatic lens with a magnetic lens. I retained the first two electrostatic lenses as their performance was relatively unimportant. I was reassured in making this decision because Fabian Pease (Pease, 1963), who worked in the room next to my own, had just built the highest resolution SEM in the world with a beam diameter of 50 Å, and it used magnetic lenses.

The methods I used for designing the magnetic lens were based on the publications of Liebmann (Liebmann, 1951, 1955; Mulvey, 1959). They have been widely used and I subsequently used them to design several lenses including the final lenses of the two high-resolution probe systems I built at IBM and for other microfabrication systems at IBM including the system used by my close colleague Hatzakis (Broers, 1970). I discuss them in some detail as a guide for others working with electron probe systems.

The final lens of an SEM has to be a pin-hole lens with the second bore smaller than the first to reduce the magnetic field at the sample to about 3 G. The field on the axis was estimated using Liebmann's data. At this level, it will not interfere with the collection of secondary electrons from the specimen surface. The focal position was also limited by the position at which it was possible to bring in the ion probe. For in situ experiments it was desirable to have the ion beam bombard the sample in the middle of the area examined by the SEM.

The minimum beam diameter for a given current depends crucially on the aberration coefficients of the final lens as discussed in the next section. Spherical aberration is the most important but chromatic aberration can also be important in the high-current probes used for microfabrication where the Boersh effect (Boersch, 1954) can increase the energy spread in the beam by more than an order of magnitude. I will discuss this later.

The aim of the design is, therefore, to minimize the spherical aberration coefficient C_S and the chromatic aberration coefficient C_C at a given working distance L. This can be done using Mulvey curves of L/C vs excitation $V/(NI)^2$ for various values of S/D, where S is the gap between the pole pieces, $D=(R_1+R_2)$ is the sum of the radii of the two pole pieces, V is the accelerating voltage, and NI the ampere-turns. L/C_S goes through a maximum as $V/(NI)^2$ is varied and becomes more favorable as S/D is reduced. The minimum gap is set by saturation of the pole pieces.

The mechanical tolerances required for the pole pieces were obtained from Archard's (Archard, 1953) interpretation of Sturrock's (Sturrock, 1951) calculations. To achieve the theoretical performance, the bores of the pole pieces had to be honed and polished until they showed less than 0.2 μm eccentricity and 2.5 μm face roughness. This requires great skill and we found that final finishing with a wooden lap was needed to reach these values. The quality of the steel used is also critical. It had to be free of non-magnetic inclusions and pores, and have a high saturation level.

The major consideration in the mechanical design of the new lens was compactness. Magnetic lenses necessarily have more bulk than electrostatic lenses and it was desirable to replace the final lens with as little alteration to the existing microscope as possible. The maximum overall diameter of the lens was limited by the position of the final lens of the ion probe forming system. To help reduce the overall diameter, the volume occupied by the coil was reduced to the minimum acceptable and the coil was water-cooled.

Mild steel was used to form the magnetic circuit and special high-purity Sandvik iron for the pole pieces. The maximum flux density in the mild steel, corresponding to the lens focusing a 28 kV electron beam (this was the maximum potential that the first two electrostatic lenses could tolerate, and therefore the maximum voltage at which the electron beam could operate), is less than 6000G. To calculate this, the total flux across the lens gap, and the leakage flux across the slot for the winding, were graphically integrated.

The total power dissipated in the lens winding was 16W. The water-cooled annulus conducted away the heat preventing any thermal expansion of the lens components that might have led to distortion of the pole piece configuration. The final aperture carrier was designed to hold and locate three apertures, as shown in the inset in Fig. 2.2.

The lens was designed to be accurately aligned with the rest of the column, but electromagnetic alignment coils were also in case the magnetic axis of the lens did not coincide with its mechanical axis. These coils proved to be unnecessary.

A cross-sectional drawing of the new lens is shown in Fig. 2.2, and the lens itself in Fig. 2.3.

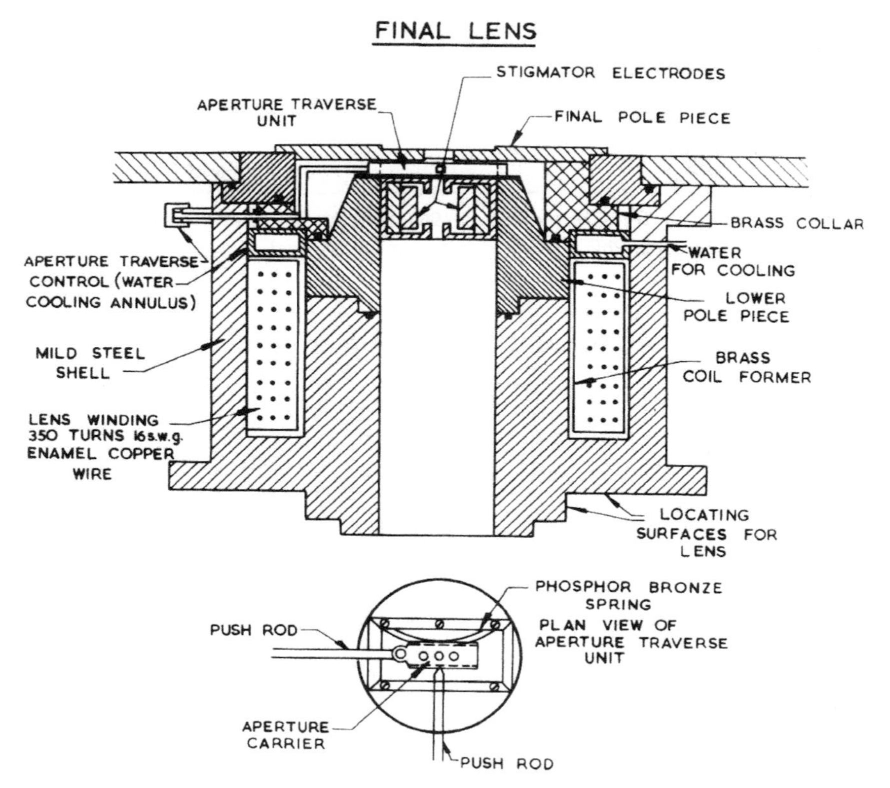

Fig. 2.2 Cross-section drawing of new final lens for SEM. *(Broers, A., 1965. Selective ion beam etching in the scanning electron microscope. PhD Thesis, University of Cambridge).*

Fig. 2.3 New final lens for SEM. *(Broers, A., 1965. Selective ion beam etching in the scanning electron microscope. PhD Thesis, University of Cambridge).*

Summary of the final lens specifications

R_1 = radius of the first pole-piece = 2.0 cm
R_2 = radius of second pole-piece = 0.5 cm
L = working distance from the inside surface of the final pole piece = 1.4 cm
S = distance between the pole pieces = 0.5 cm
f = focal length = 1.8 cm
C_S = 3.6 cm, C_C = 1.44 cm
(NI) = 1200 A-turns
H_0 = 740 G
V_T = 28 keV
$k^2 \sim 0.5$

Theoretical estimation of beam diameter

The minimum useful beam diameter in an SEM depends on the beam current needed to form a reasonably noise-free image. This current has generally been assumed to be about 10^{-12} A which corresponds to about 1 million electrons/s. For a satisfactorily noise-free image, the signal-to-noise ratio should be about 10, which will require about 100 electrons per pixel, so with a beam current of 1 pA it will take about 100 s to record a 1000-line, 1 million-pixel image. Digital imaging did not exist in those days. Images were recorded on silver halide photographic film using a slow frame scan. The camera recording the image on the cathode ray tube had a magazine that held several meters of 35 mm film. Before I finished my Ph.D. I had taken several thousand micrographs and spent a lot of time in the darkroom developing the film and printing the micrographs.

The minimum beam diameter for a given current depends on the electron beam brightness, the aberration coefficients of the final lens, and ultimately for the highest brightness electron guns, diffraction. It can be estimated using the formulae of K. C. A. Smith who assumed that the square of the diameter of a focused electron probe was equal to the sum of the squares of the diameters of the disks of confusion due to spherical aberration, chromatic aberration, astigmatism and diffraction and the nominal demagnified image of the gun cross-over. Smith combined the calculated probe diameter with the brightness formula of Langmuir to arrive at the following expressions for the optimum operating aperture α_{OPT} and the relationship between the probe diameter (d) and the probe current I. The diameter of the electron probe is defined as the distance between opposing points where the current density is one-fifth of the maximum value.

$$d^2 = P/\alpha^2 + C_S\,\alpha^6 + Q\,\alpha^2$$

$$\alpha_{OPT}{}^4 = \left((Q^2 + 12CP)^{1/2}Q/6C\right.$$

where

$$P = i/B + (1.22\lambda)^2,$$

$$B = 5.65 J_C V \times 10^3/T,$$

$$C = (0.5\,C_S)^2,$$

$$Q = (C_C \delta V/V)^2 + Z_\alpha$$

λ is the wavelength of the electrons, V is the accelerating potential, T is the cathode temperature, J_C is the cathode emission density, δV is the electron velocity spread, and Z_α is the distance between the two line foci when there is residual astigmatism.

At the optimum aperture, the disk of confusion created by the aberrations is roughly the same as the demagnified image of the gun cross-over. The beam convergence angle is set by the physical aperture in the final lens which generally has a diameter of about a fifth of a millimeter. Increasing the diameter of the aperture increases the current quadratically, but the growth in the diameter of the disk of confusion due to spherical aberration is cubic. Chromatic aberration is generally not important for the beam currents needed for surface microscopy but can be significant in the high current probes used for microfabrication when the Boersh Effect can increase the energy spread in the beam by more than an order of magnitude.

For the highest resolution the operating conditions $i = 10^{-12}$ A, $V = 28$ kV, $\alpha_{OPT} = 5 \times 10^{-3}$ rad, and $d_{min} = 50$ Å. Normally, to ensure noise-free images, a current of $2\text{–}3 \times 10^{-12}$ A was used.

Much to my amazement and relief, as I had never designed anything like it before, the new lens operated as it was meant to and after a couple of weeks of tuning, the microscope beam diameter was reduced to 100 Å. The microscope also seemed to be more reliable.

Encouraged by these improvements I went on to fit a new electron gun that could operate at higher voltage, and replaced the eight-pole astigmatism corrector that had been located down the column with one inside the first pole piece of the new lens where it could be easily removed for cleaning. I also fitted new lining tubes and spray apertures throughout the column to reduce sensitivity to dust and contamination and make cleaning easier.

To improve the vacuum level and reduce the time it took to change the specimen, I fitted a larger diffusion pump and increased the diameter of the vacuum pipes to the microscope. A new specimen stage was also needed to eliminate differential vibration that was limiting resolution when the beam size was reduced to 100 Å. This stage rested on the final pole piece on three pads, so it could not rock, and there was a small amount of play between the shafts of the two micrometers and the stage itself so that they could be retracted when the specimen was in the desired position.

The SEM was now ready to use but I still needed to design and fit a mass filter for the ion beam system to eliminate oxygen in the ion beam. Selective oxidation by oxygen ions in the beam had been obscuring the fine ridges formed by sputtering with inert Argon ions and the aim was to study the mechanisms of ion beam sputtering rather than oxidation. This proved to be a much larger task than I anticipated.

Mass filter for the ion beam

The purpose of the mass filter was to remove oxygen from the beam of inert ions that bombarded the surface of the SEM specimen. Unfortunately, the beam produced by the high-frequency ion source inevitably contained small amounts of impurities including oxygen. The argon gas leaked into the ion source was of commercial grade (0.01%

impurities) and any gas remaining in the system due to the imperfect vacuum added to these impurities. The background pressure in the system was about 10^{-6} mmHg and the operating pressure inside the ion source was about 10^{-3} mmHg. The background level of impurities was therefore 0.1%.

The resolution of the filter could be low compared with the resolution of many mass spectrometers because there was no need to separate isotopes. It would have been a disadvantage to improve the resolution as the current density of the ion beam emerging from the filter would have been reduced. The current density of the ion beam at the specimen was as high as possible so that the ratio of etching ions, to atoms from the residual vacuum arriving at the specimen surface, was a maximum. For an argon beam, a resolution of 15 was sufficient to eliminate ions of the closest adjacent element on the atomic scale, chlorine, but a resolution of about 10 would be sufficient to ensure that unwanted impurities were eliminated.

To counteract the divergence of the ion beam due to space charge and therefore maintain the current density in the beam, electrostatic lenses were used to focus the beam before and after the filter. The general arrangement is shown in Fig. 2.7. Under normal conditions, only the second lens of the double electrostatic lens unit following the filter was used. The distance the ions travel in the filter was kept at a minimum to minimize the divergence of the beam due to space charge.

As the purpose of the ion optical unit is to produce a flood beam at the specimen and not a fine probe, the ion optics of the system did not need to be accurately determined. An exact estimate was not possible in any case because the precise characteristics of the ion beam produced by the RF ion source could not be predicted. An approximate estimate of the current density in the final beam was made, however, to ensure that it would exceed 0.5 mA/cm^2. The current density had to be greater than 0.5 mA/cm^2 if the ratio of ions to residual atoms arriving at the specimen was to be greater than 10 when the pressure in the specimen chamber was about 10^{-6} mmHg.

The RF ion source produced a beam of 5 keV argon ions 0.8 mm in diameter carrying 300 μA and it was estimated, using the approximate data of Field et al. quoted by Klemperer for the divergence of charged particle beams, that 20 cm was the maximum permissible ion path length in the filter.

Two types of mass filter were considered, electrostatic quadrupole and magnetic deflection.

The main advantage of the quadrupole filter was that it would have been relatively straightforward to install because there was no deflection of the beam. There would also have been no question about it interfering with the SEM. Its disadvantages were that its resolution, while barely adequate, its design complex, and its drive electronics expensive.

The magnetic deflection filter, although it required more extensive changes to the rest of the ion beam system, had higher resolution and its operation did not involve any complex electronic equipment, so it would cost less. In the final assessment, I chose the magnetic filter although it ended up being a much larger major part of my PhD research than I anticipated.

Electrostatic quadrupole filter

This was to be an adapted form of a spectrometer developed by Woodward and Crawford (Woodward, 1963). The original work on this type of filter was carried out by Paul (1953, 1955).

In the quadrupole filter, the ion beam passes along the axis of an electrostatic quadrupole lens. Opposite electrodes of the quadrupole are interconnected, and DC and RF voltages are simultaneously applied between the two electrodes. The equations for the resulting transverse ion motion are Mathieu equations. Their solutions (Mathieu functions) are either stable or unstable depending on the RF frequency, the magnitude of the RF and DC voltages, the physical dimensions of the quadrupoles, and the charge-to-mass ratio of the ions. By suitable choice of these parameters, the quadrupole can be designed to only pass ions of a single charge-to-mass ratio. Other ions will be ejected from the sides of the filter. However, for this particular application, especially because of the length constraint imposed by the space charge, the resolution would only have been 3.7.

Magnetic deflection filter

Two factors governed the design of the magnetic filter:

(a) The distance the ions traveled in the filter had to be kept at a minimum (i.e., less than about 20 cm) for the required resolution of about 10. This meant maximizing the magnetic field strength.

(b) The external magnetic field of the filter had to be below that at which it would interfere with the operation of the SEM. To achieve this, the magnetic circuit connecting the pole pieces had to enclose the pole pieces. This increased the distance the ions had to travel so the design was a compromise between ion path length and resolution.

12,000 G was taken as the maximum deflection field that could be readily obtained. Saturation would not have occurred with high-grade iron until 22,000 G, but the maximum attainable field in the gap was limited to a lower level by losses in the connecting magnetic circuit. The magnetic circuit enclosed the activating coil so its length depended on the size of the coil. The space occupied by the coil was set by the amount of heat it produced and to maximize this it was water-cooled.

Shape of the magnetic deflecting field

A sector-shaped field with the ion beam entering and leaving the field at right angles has a focusing action in the plane of the pole pieces only and introduces astigmatism into the beam. It is possible, however, by adjusting the shape of the sector field so that the ions enter it at an angle other than 90°, to use the fringing field to focus the beam in the plane normal to the plane of the pole piece faces. Camac had calculated the condition for the two focusing effects to be equal.

Such a field will give anastigmatic deflection of the ion beam and is desirable in this application as higher current density will be available in the final ion probe. The condition for double-focusing is complicated by the electrostatic lens which precedes the filter. The source distance as treated by Camac becomes negative. A calculation using Camac's equations, however, showed that with the beam entering the field at 8° to the normal the desired condition for double-focusing was satisfied. This calculation and the rest of the design details can be found in Broers (1965).

Performance of the magnetic deflection filter

Figs. 2.4 and 2.5 show plan and section drawings of the filter, and Fig. 2.6 shows a photograph of the filter components.

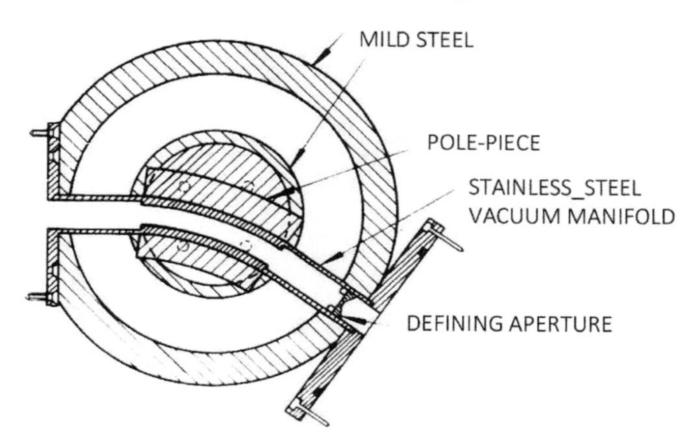

Fig. 2.4 Plan view of the magnetic filter. *(Broers, A., 1965. Selective ion beam etching in the scanning electron microscope. PhD Thesis, University of Cambridge).*

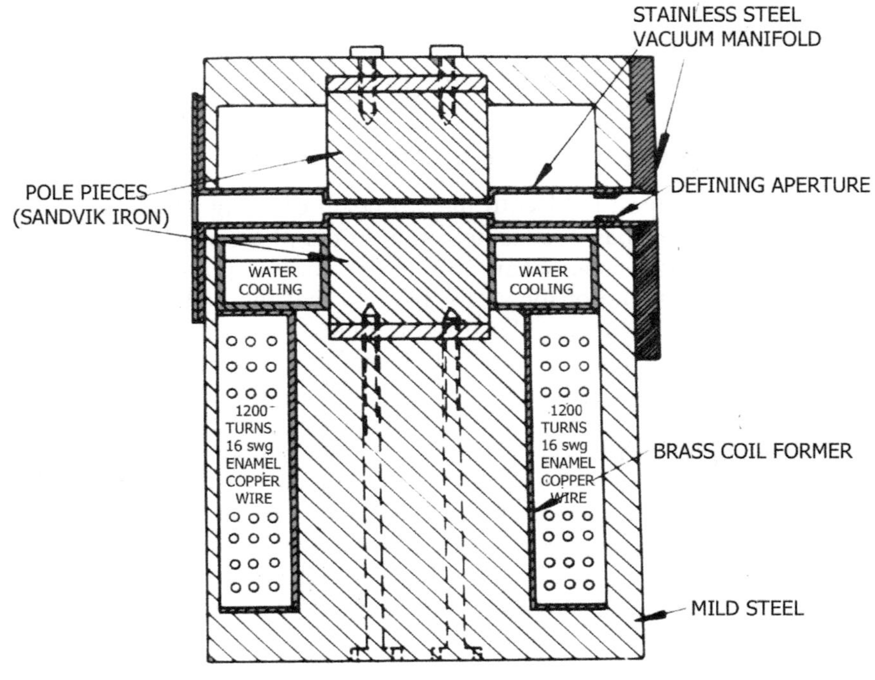

Fig. 2.5 Cross-section view of the magnetic filter. *(Broers, A., 1965. Selective ion beam etching in the scanning electron microscope. PhD Thesis, University of Cambridge).*

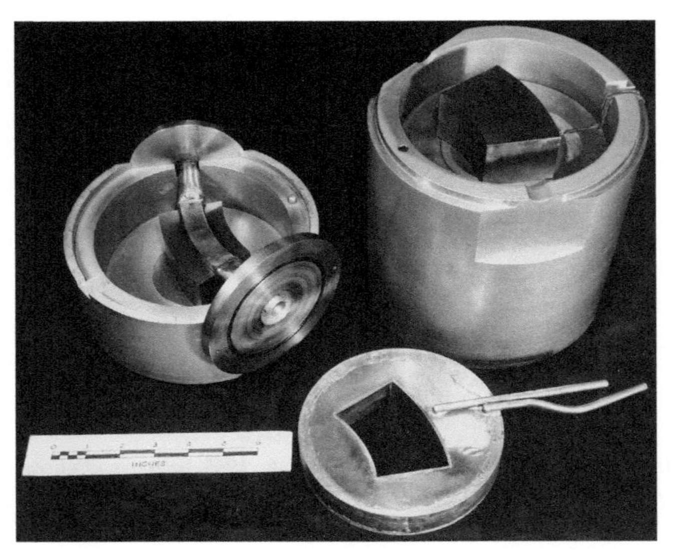

Fig. 2.6 Photograph of the components of the magnetic filter. *(Broers, A., 1965. Selective ion beam etching in the scanning electron microscope. PhD Thesis, University of Cambridge).*

The maximum field attained in the 6 mm gap between the filter's pole pieces was 12,200 G. This agreed with the calculation that considered the losses in the magnetic circuit and assumed a maximum current of 5.16 A in the 1200 turn, 3.77 Ω, coil. This meant that the power dissipated in the coil was 100 W which could be handled easily by the water cooling. 12,200 G allowed 8 keV Xenon ions to pass through the filter with its radius of curvature of 12.3 cm. 8 keV Xenon ions were thought to be the ions with the highest energy that the filter would have to pass. The ion current to the specimen under normal operating conditions was about 20 μA and the beam diameter, which was generally limited by the final aperture placed immediately before the specimen, was about 1.5 mm. This corresponded to an average current density in the beam of 0.7 mA/cm^2. It was sufficient to know that the current density was greater than 0.5 mA/cm^2 to ensure that the ratio of ions to residual atoms arriving at the sample surface was greater than 10, as discussed above. This was certainly the case because the area studied was at the center of the bombarded area.

The measured current to the specimen under normal conditions was in fact higher than 20 μA because of the secondary electrons created by the ions. A Faraday cage, with the secondary electrons suppressed, was used to monitor the current to the beam.

The resolution of the filter was about 10. This was measured by the fractional change in the current through the magnetizing coil of the filter required to vary the ion current to the specimen from one side of a single peak to the other. The 10% of maximum points on either side of the ion current peak are used for this measurement. The theoretical resolution of the filter was calculated to be 11.2.

Fig. 2.7 is a diagrammatic view of the SEM-ion beam systems after the addition of the new magnetic final lens and the magnetic filter (Broers, 2004).

I realize now, much more than I did then, how lucky I was in the advice and help I had from everyone in the engineering department, especially those in the workshop. I remember the first time I took the microscope apart to clean the final lens I broke the intricate glass tube that carried the high voltage wire to the lens. I thought that my research career was probably over, I had broken this magnificent piece of equipment and at the least, Garry Stewart would give up helping me. But they took pity on me in the workshop and made a new one in a couple of days and when I finally dared to tell Garry he laughed and said he knew that would happen. He had broken it a couple of times himself.

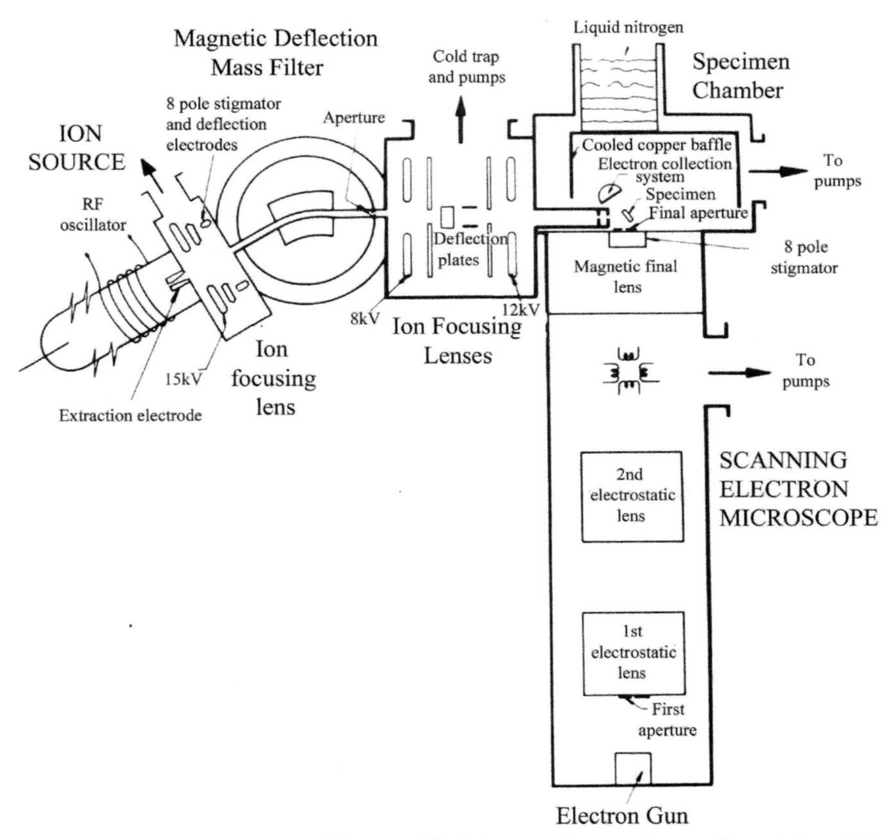

Fig. 2.7 Diagrammatic view of the modified SEM-ion beam system. *(Broers, A., 1965. Selective ion beam etching in the scanning electron microscope. PhD Thesis, University of Cambridge. (Hawkes, 2004).*

References

Archard, G., 1953. Magnetic electron lens aberrations due to mechanical defects. J. Sci. Instrum. 30 (10), 352.

Boersch, H., 1954. Z. Phys. 139, 115.

Boyde, A., 2004. A. D. G. Stewart and an early biological application. Adv. Imaging Electron Phys. 133, 165–175.

Breton, B.C., 2004. My life with the Stereoscan. Adv. Imaging Electron Phys. 133, 439–465.

Broers, A., 1965. Selective Ion Beam Etching in the Scanning Electron Microscope. PhD thesis, University of Cambridge.

Broers, A.a., 1970. Microcircuits made through microscopes. Ind. Res. 1970, 56–58.

Broers, A., 2004. Chapter 210. The application of the scanning electron microscope to microfabrication and nanofabrication. In: Hawkes, P. (Ed.), Sir Charles Oatley and the Scanning Electron Microscope, Advances in Imaging and Electron Physics. vol. 133. Elsevier, pp. 207–226.
Hawkes, P.E., 2004. In: Hawkes, P.W. (Ed.), Advances in Imaging and Electron Physics. vol. 133. Elsevier Academic Press.
Liebmann, G.a., 1951. Proc. Phys. Soc. 64B, 960.
Liebmann, G., 1955. Proc. Phys. Soc. *68B*, 682–737.
McMullan, D., 2003. The development of the first Cambridge scanning electron microscope 1948-1953. Adv. Imaging Electron Phys. 133, 39–57.
Mulvey, T., 1959. J. Sci. Instr. 36, 3509.
Paul, W.a., 1953. Z. Natutforsch. *8a*, 448.
Paul, W.a., 1955. Z. Phys. 140, 262.
Pease, R., 1963. Thesis. Cambridge University.
Smith, K., 1956. PhD thesis, University of Cambridge.
Stewart, A., 1962. Fifth International Congress on Electron Microscopy., p. D12.
Sturrock, P., 1951. Phil. Trans. R. Soc. A 243, 387.
Woodward, C.a., 1963. Technical Report 176, Lab. of Insulation Res. MIT.

CHAPTER THREE

SEM examination of ion-etched metal and semiconductor crystals

Contents

Introduction

Of the many phenomena that arise when ions strike a surface, the ejection of microscopic particles of solid material is the most interesting. It is commonly referred to as sputtering. The sponsors for my Ph.D., the UK's Atomic Energy Research Establishment, AERE, were interested in sputtering because it was damaging the walls of nuclear reactors at

Advances in Imaging and Electron Physics, Volume 231
ISSN 1076-5670
https://doi.org/10.1016/B978-0-443-31462-9.00003-4

Aldermaston, and they were concerned about what happened to the bombarded material.

Sputtering can also damage spacecraft as it occurs naturally in outer space. However, there are many useful applications of sputtering. For example, it has been used for coating samples with thin films since the early 1800s, and by the late 1800s, sputtered films dominated the optical-coating market. Today, it is used widely both as a source of material for thin films and for removing material selectively, for example, in the fabrication of semiconductor chips.

A comprehensive review of the mechanisms of sputtering was published by Wehner in 1955 (Wehner, 1955a,b) and this was followed by reviews by Koedam (Koedam, 1961), Thomson (Thomson, 1962), Kay (Kay, 1962), and Behrisch (Behrisch, 1964).

At the time of my research in 1962 two theoretical models had been proposed for sputtering: the vaporization theory, which assumed that incident ions caused a local hot spot in the solid from which vaporization occurred, and the momentum transfer theory, which assumed that elastic impacts occurred between incident ions and atoms in the solid, with a transfer of momentum to the atoms creating collision cascades that ejected atoms from the surface.

There is now adequate experimental evidence to suggest that the evaporation theory has limited relevance, at least for crystallographic materials. This is primarily because it has been shown that atoms are largely ejected along a few principal crystallographic directions. This was discovered originally by Wehner (1956, 1957) and many others have observed that the paths of the sputtered atoms are aligned with closely packed crystallographic planes.

According to H. Tsuge and S. Esho (Tsuge, 1981), atoms sputtered from Au, Al, Pt, and NiFe targets are ejected preferentially in the direction of their close packing in the lattice. These results are explained in terms of the crystal structure of the target surface during and after ion bombardment. The surface structure of the metal targets remains essentially unchanged. Sputtered-atom ejection normally increases with increasing incident-ion energy. Even at oblique incidence, heavy target atoms such as Au are ejected preferentially in the close-packed directions.

All of my results supported the momentum transfer theory, although no ridges appeared on the surface of sputtered silicon crystals.

In this chapter, I describe examples of the experiments in which I observed, at high magnification, the same area of the surface of metal crystals through many stages of etching with 5-10 keV Argon ions (Broers, 1965, 1985). The results presented are only a small fraction of those obtained, but they include examples of most of the significant features that characterize the etched surfaces.

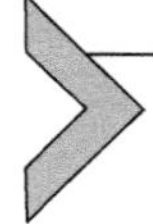

Formation of cones underneath particles of shielding material on the surface of ion-etched samples

Introduction

The formation of cones underneath shielding particles on the surface of ion-etched samples was first reported by Wehner (1955a,b) who observed cones underneath particles of polishing dust when ion-etching a silver rod. He and others studying the surface of ion-etched surfaces could only examine their specimens with optical microscopes or by making replicas of the specimen surface and examining the replica in a transmission electron microscope. They could not follow specific areas on a sample with a resolution of 200 Å as was possible in the system that Stewart and I used (Stewart, 2004). Stewart also showed that inclusions with a slower etch rate than the bulk material could give rise to cones. I examined these phenomena on many electro-polished crystals of aluminum, silver, and silicon through extended etching mostly with a filtered beam of argon ions.

Results

The eight micrographs in Fig. 3.1 illustrate the cone formation process. Fig. 3.1A shows a piece of nickel dust on the surface of an aluminum crystal before ion bombardment, and Fig. 3.1B–H show the same area of the surface after successive stages of ion etching with a beam of 5 kV argon ions with a current density of 1 mA/cm^2. The times of bombardment are shown beneath each micrograph.

Fig. 3.2 shows three cones formed after etching a silver crystal with a beam of 5 kV argon ions with a current density of 0.6 mA/cm^2 for an hour. Cone (1) still has the shielding particles on top of it, but the particles under which cones (2) and (3) formed have themselves sputtered away. It should be noted that when the shielding particle has been sputtered away the cones become pointed [see Fig. 3.2 cones (2) and (3) and also Fig. 3.1D] and after further etching the cones become reduced in size [see Fig. 3.2 cone (3) and also Fig. 3.1E] and eventually disappear or end up as pits [see Fig. 3.1G and H].

When a large cone (i.e., several tens of microns high) is formed underneath a large shielding particle, or underneath a piece of material that sputters very much slower than the bulk material, its walls often develop facets before the cone itself is finally etched away. This is evident in the case of the cone in Fig. 3.3A which is, even at this advanced stage of etching, 80 μm high, and it is still more evident in the case of the remains of this large cone in Fig. 3.3B, and in the case of the base of the cone shown in Fig. 3.3G.

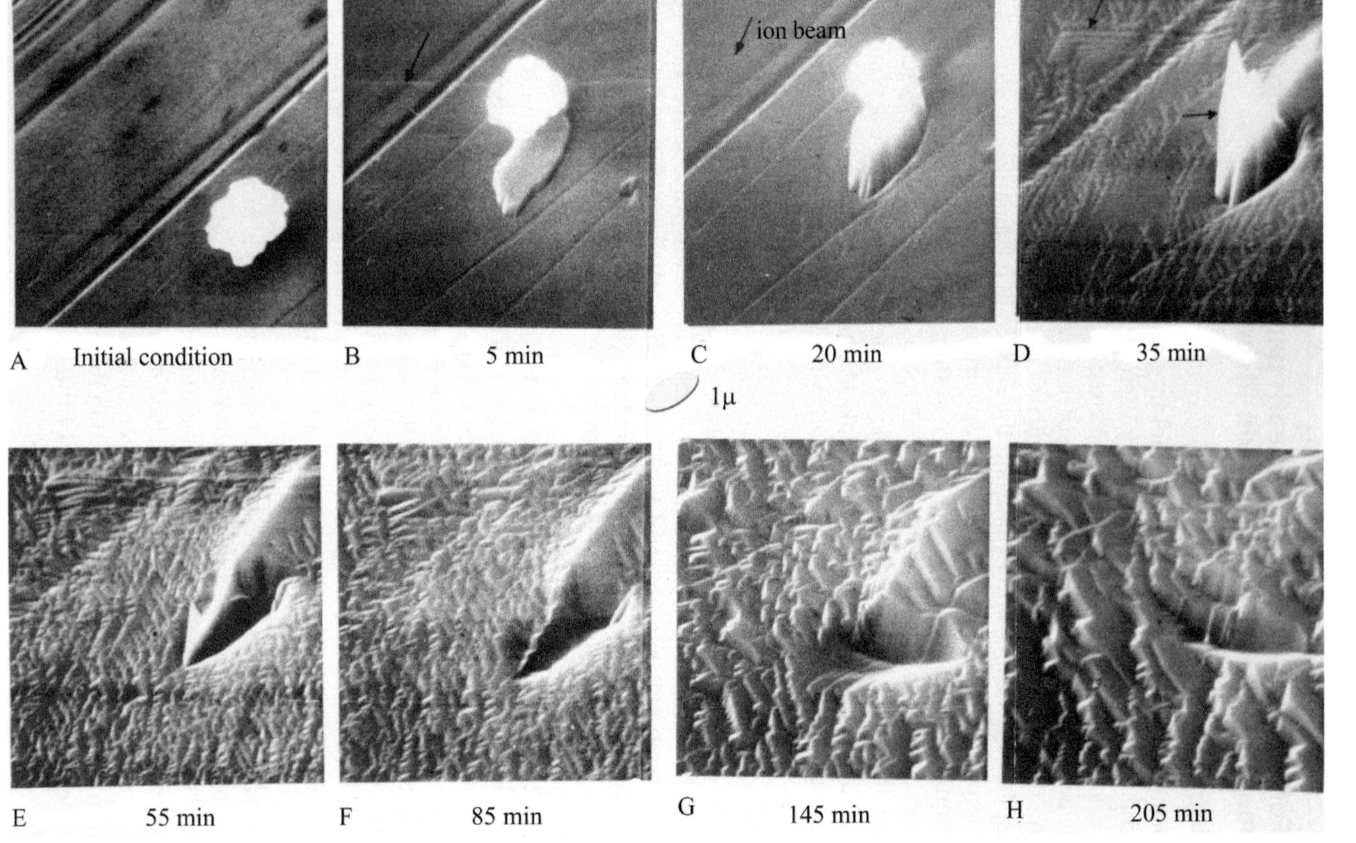

Fig. 3.1 Etching of an aluminum crystal with 5 keV Argon ions at a current density of $0.5\,mA/cm^2$. *(Used with permission of John Wiley & Sons—Books from Broers, A., 1985. High-resolution electron beam fabrication: a brief review of experimental studies that began in Cambridge University in 1962. J. Microscopy. 139, 139–152, conveyed through Copyright Clearance Center, Inc.; Broers A., 1965. Selective Ion Beam Etching in the Scanning Electron Microscope, PhD Thesis University of Cambridge).*

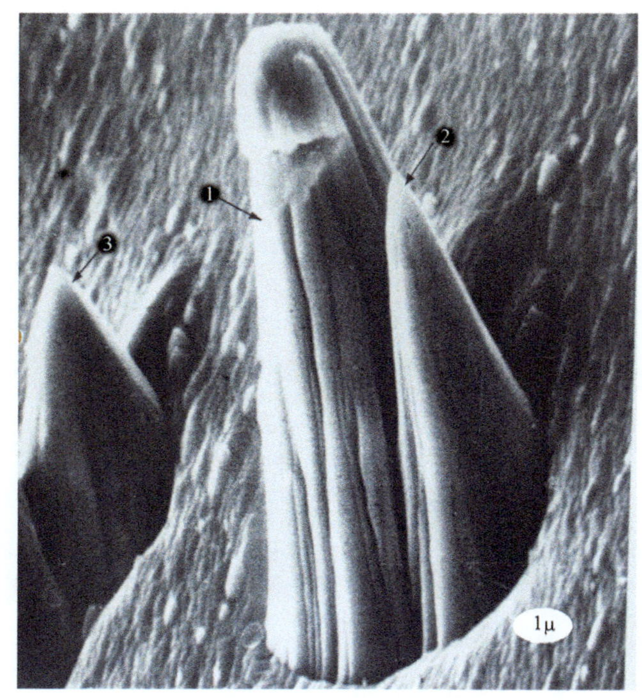

Fig. 3.2 Cones on a silver crystal etched with 5 keV Argon ions at a current density of 1.0 mA/cm^2. *(Broers A., 1965. Selective Ion Beam Etching in the Scanning Electron Microscope, PhD Thesis, University of Cambridge).*

When the surface is bombarded with a beam of ions at a low angle of incidence, the cones are inclined to the normal to the surface. A ridge appears that leads out from the base of the cones in the direction of the incoming ion beam (see Fig. 3.4). These "up-stream tails" were first observed by Stewart who explained them as being the result of material sputtered from the cones back onto the surface. The tails are usually thickest at the base of the cones and gradually taper away until they disappear at a distance from the cone that depends on, the height of the cone, the crystallographic orientation of the surface being sputtered, and the angle of incidence of the bombarding beam. When the surface being sputtered exhibits a fine structure, the "upstream tails" exhibit a similar structure. Fig. 3.4A shows transverse ridges on both the crystal surface and the tail of a cone formed on an aluminum crystal that was bombarded for about 1 h with 5 keV argon ions at a current density of 1 mA/cm^2. Fig. 3.4B shows longitudinal ridges on both the crystal surface and the tail of the cone formed on a silver specimen bombarded under the same conditions. Fig. 3.4C shows a silicon crystal bombarded under the same conditions but at a lower angle of incidence.

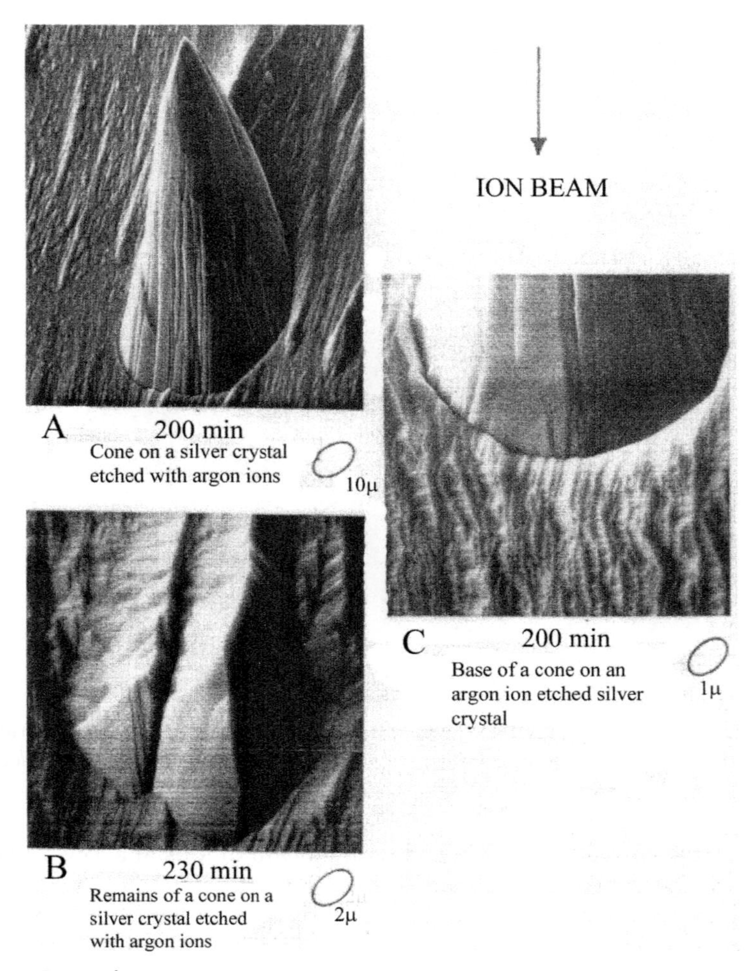

ION BEAM
A 200 min
Cone on a silver crystal etched with argon ions
10μ
B 230 min
Remains of a cone on a silver crystal etched with argon ions
2μ
C 200 min
Base of a cone on an argon ion etched silver crystal
1μ

Fig. 3.3 Faceted cones.

Interpretation of results

The first point of interest is that the shielded cones are cones and not parallel-sided columns with cross-sections determined by the shape of the shielding particles. The broadening of the shielded columns into cones as surface removal continues was first explained by Wehner as the result of material sputtering from the surface around the columns onto the flanks of the columns. This explanation has been treated in more detail by Stewart (2004).

There is another possible explanation which may be complementary to Wehner's explanation. The ions that contribute to etching the surface at

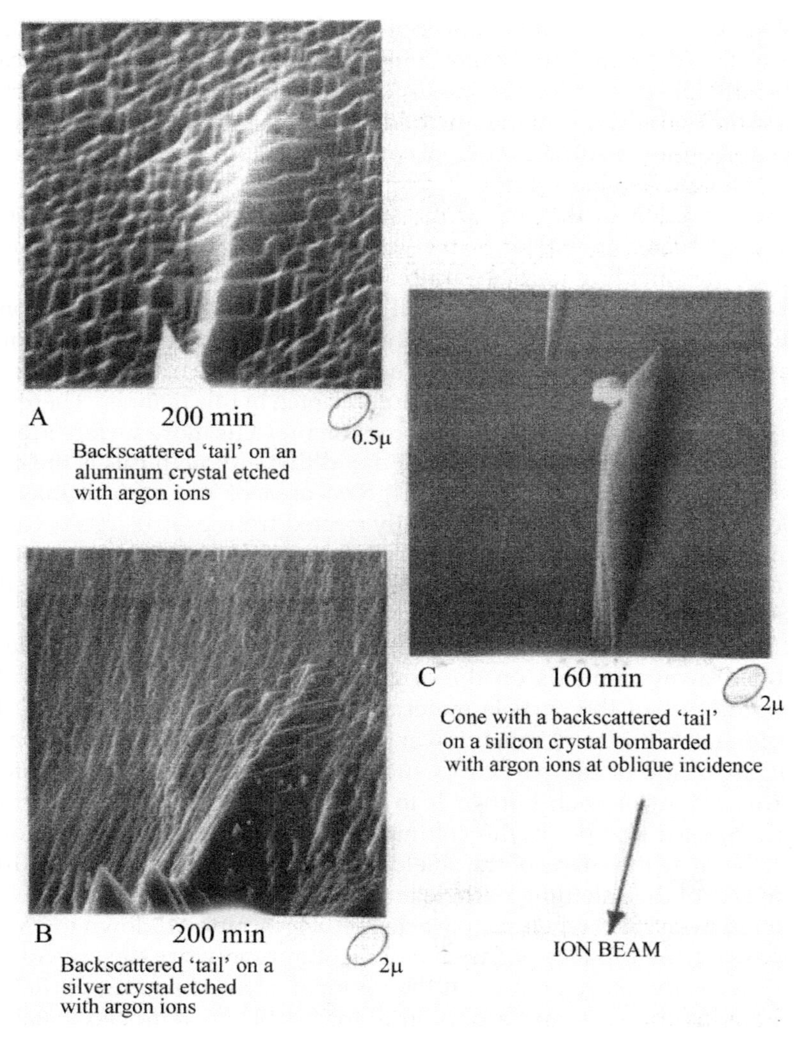

A
200 min
0.5μ
Backscattered 'tail' on an
aluminum crystal etched
with argon ions
B
200 min
2μ
Backscattered 'tail' on a
silver crystal etched
with argon ions
C
160 min
2μ
Cone with a backscattered 'tail'
on a silicon crystal bombarded
with argon ions at oblique incidence
ION BEAM

Fig. 3.4 Cone "Tails" formed by material backscattered onto the surface. *(Broers A., 1965. Selective Ion Beam Etching in the Scanning Electron Microscope, PhD Thesis. University of Cambridge).*

the edge of the shadow cast by the shielding particle only come from one side of the shadow, and the sputtering rate at the edge is therefore reduced. In other words, if it is assumed that an atom on the surface is removed by the action of an energetic ion striking the surface within a certain distance of the atom, a distance determined by the energy of the ion and the momentum transfer process in the specimen, then only half the number of ions that usually contribute to the removal of surface atoms, contribute to the removal of atoms lying along the edge of the shadow cast by the shielding particle. The continual reduction in the sputtering rate around the bottom edge of the shielded column makes the base of the cone wider and wider as surface removal continues and finally produces a cone and not a parallel side column.

The reduction in the size of the shielding particle as it is sputtered away accentuates the slope of the column faces, provided the outside edges of the shielding particle sputter away first.

It can be assumed, because in all cases observed the cones become smaller and eventually disappear as the etching continued, that they sputter faster than the surrounding surface of the specimen. All protrusions on the surface of bulk specimens were found to etch in this manner. There are two possible reasons for this: (1) a protrusion presents more surface area to the bombarding ion beam (it must be noted that although this is the case the total number of ions striking a given area of the surface may be reduced); (2) atoms may be more easily ejected from protrusions because the incoming momentum of the bombarding ions does not always have to be transferred through the large angles necessary to eject atoms from flat surfaces.

The shape of the shielding column immediately after the particle is sputtered away depends on the original shape of the particle and the sputtering rate of the particle material. When the sputtering rate of the particle material is very much lower than that of the base material, and when the shape of the particle is such that its thickness, and therefore the time taken for each part of it to sputter away, becomes uniformly greater toward one point, the column has the shape of a pointed cone. This is because the shape of the shielded column is adjusted to the reduction in size of the shielding particle more quickly than the particle itself is sputtered away, and when the particle has been sputtered down to a very small speck, the top face of the shielded column only has the area of the speck and is therefore pointed. If the removal of the shielding particle is not a gradual process, and a portion of the particle is removed instantaneously, the shielded column takes on a lop-sided shape before its equilibrium shape is attained (see the arrow in Fig. 3.1D).

The faceting of the flanks of the shielded column is likely to be due to the influence of the crystallographic orientation of the specimen. Material will be ejected from the different parts of the column at different rates depending on their crystallographic orientation.

If the "upstream tails" described in Results above are caused by material sputtered back onto the surface from the cones, then the fact that the ridge structure observed on the tails is similar to that of the surrounding surface indicates that this back-sputtered material is deposited epitaxially.

The ridge structure that appears on the surface of crystalline specimens after ion etching is characteristic of the crystallographic orientation of the surface, as discussed above. The material that makes up the tail is being continually sputtered and deposited and yet retains the same ridge structure as the surrounding surface, further supporting the conclusion that it must be deposited epitaxially.

Ion etching of mechanically distorted aluminum crystals

Introduction

Experiments had been carried out by several authors to determine whether crystallographic faults such as stacking faults and screw dislocations could be revealed by ion etching.

Wehner (1958) observed a row of pits along a lineage boundary after etching a germanium single crystal with 150 eV Hg ions. He also observed that specimens were etched more along the grain boundaries and that these developed into furrows, just as in a chemical etching. Ogilvie reports the appearance of steps between grains of different crystallographic orientations of polycrystalline specimens. Stewart, using this apparatus in its unmodified state, carried out a detailed investigation of etch effects at grain boundaries. He observed irregularities along the grain boundaries of polycrystalline aluminum specimens and explained the irregularities as being caused by preferential oxidation of these areas of the surface.

Yurasova (1957) reported that bombardment with 2 keV neon ions revealed slip lines in cadmium and zinc and the same effect was noted by Cunningham (1963) for 8 keV argon ions on zinc. Wehner (1958) ion etched and chemically etched a germanium single crystal and found no correlation between the pits developed by the two forms of etching. A 6° tilt boundary however developed into a sharply indented furrow for both ion and chemical etching. Cunningham reports the lack of any preferential etching along slip lines introduced in an aluminum crystal by bending the specimen around a cylinder with a radius of 5 cm. The experiments in which single crystals of silicon were etched with ions describer later failed to reveal crystallographic faults.

The following section describes experiments in which slip lines, introduced into the grains of polycrystalline aluminum specimens by bending the crystals, were revealed by ion etching, sometimes as rows of pointed cones, sometimes as rows of etch pits, and sometimes as grooves or walls on the surface, depending on which grain was examined.

Fig. 3.5 shows the general configuration of the cones that appear along slip lines on a polycrystalline aluminum sample.

Fig. 3.5 General configuration of the cones that appear along slip-lines on a polycrystalline aluminum sample. *(Broers A., 1965. Selective Ion Beam Etching in the Scanning Electron Microscope, PhD Thesis. University of Cambridge).*

Experimental results

To examine how distorting aluminum crystals affected the way they etched, the following procedure was used: (1) the sample was ion etched with 5 keV argon ions at a current density of about $1\,\mathrm{mA/cm^2}$, (2) it was examined with the SEM, (3) it was mechanically distorted by bending it through an angle of about 15°, (4) it was re-examined with the SEM, (5) it was ion etched as in (1), (6) examined with the SEM.

A high purity (the exact purity was not known) polycrystalline aluminum specimen was lightly polished on a series of alumina dust wheels, electro-polished in a perchloric acid-alcohol-water solution, and wiped with a Kleenex tissue soaked in Analar acetone. The Kleenex tissue was used to lightly scratch the surface so that it would be possible to distinguish between the etching of scratches (surface topography) and slip lines (crystallographic faults introduced by the mechanical distortion that penetrated into the crystal). The slip lines appeared as steps on the specimen

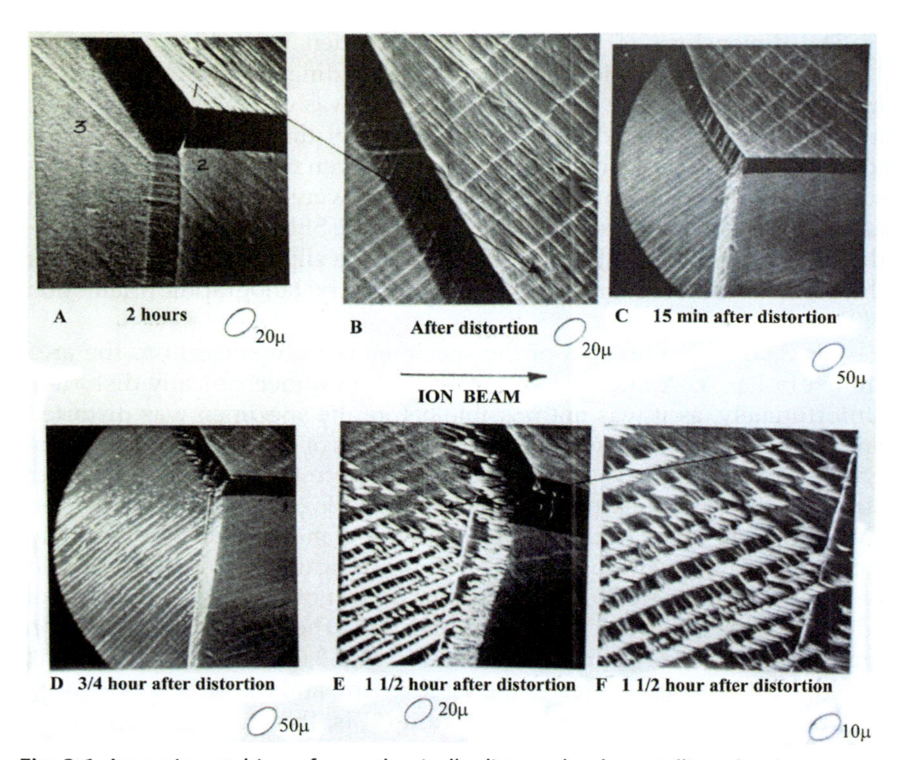

Fig. 3.6 Argon ion etching of a mechanically distorted polycrystalline aluminum sample. *(Broers A., 1965. Selective Ion Beam Etching in the Scanning Electron Microscope, PhD Thesis. University of Cambridge).*

surface and, therefore affected the surface topography as well as the crystal lattice. The remains of the surface scratches after the initial ion etching can be seen in Fig. 3.6A running diagonally upwards across the micrograph from right to left. The surface was carefully wiped in one direction only with the Kleenex tissue so that all the scratches would run in the same direction on the specimen surface and could, therefore, be easily identified.

Fig. 3.6A shows the aluminum surface after the initial ion etching at the intersection of three grains. Because the grains had different crystallographic orientations, they sputtered at different rates, and steps appeared between adjacent grains. The relative depths to which the grains have been etched were estimated by the alteration in the final lens focusing current required to bring the SEM into focus on each grain. Grain (1) is the most deeply etched, grain (3) the east etched, and grain (2) etched to a depth between the other two.

The dimensions of the aluminum specimen were 15 mm × 8 mm × 1.5 mm. It was mechanically distorted by bending it through about 15°. In all cases, the surface that was ion etched was convex and, therefore, in tensile stress. After distortion, the grains of the specimen were seen to be striped with slip lines that were visible even under an optical microscope at a magnification of 20 times. The slips were, therefore, gross on an atomic scale and were seen in some cases by the SEM as steps several hundred Å high on the surface of the crystal. The slip lines ran in different directions on different grains according to the crystallographic orientation of the grains.

Fig. 3.6B shows an area of the specimen surface adjacent to the area shown in Fig. 3.6A after the specimen had been mechanically distorted. Unfortunately, as it was not possible before the specimen was distorted and etched again, to know which were the areas of greatest interest, micrographs were not always obtained of these interesting areas at the earlier stages of the experiment. The micrographs in Figs. 3.6 and 3.7 however, show adjacent areas of the surface, and the common topographical features shown in the various micrographs are indicated. Careful note should be taken of the magnifications in each case as not all the micrographs of adjacent areas are at the same magnification. The fine structure seen on grain (3) in Fig. 3.6A is not so evident in Fig. 3.6B because in Fig. 3.6B, the contrast is not expanded to the same extent, and the SEM was focused on grain (1) rather than grain (3). Grain (3) is, therefore, slightly out of focus. The slip lines can be seen running across the surface approximately at right angles to the scratches.

Fig. 3.6C shows the same area as Fig. 3.6C but at lower magnification and after the specimen had been mechanically distorted and lightly etched. It can be seen that the slip lines are etching differently from the rest of the surface, particularly on the steep face between grains (1) and (3). Fig. 3.6D–F show the surface after considerably more etching. The slip lines on grain (3) are now seen as rows of pointed cones of the type shown in Fig. 3.5.

A central area of grain (3) is shown in Fig. 3.7. Figs. 3.7A–C show the surface after light ion etching and Fig. 3.7D after heavy ion etching. Figs. 3.7C and D is at the same magnification and demonstrate clearly how the cones grow as the ion etching continues. A more widely spaced row of cones on another aluminum specimen is shown in Fig. 3.8A and B. This specimen had an identical history to the specimen discussed above and was again polycrystalline.

Rows of cones have appeared along slip lines on 4 out of 15 aluminum grains that had been distorted, ion etched, and examined with the SEM. In one case, the slip lines after ion etching appeared as walls (see Fig. 3.9A), but in most cases, they were revealed as rows of etch pits. Fig. 3.9B and C

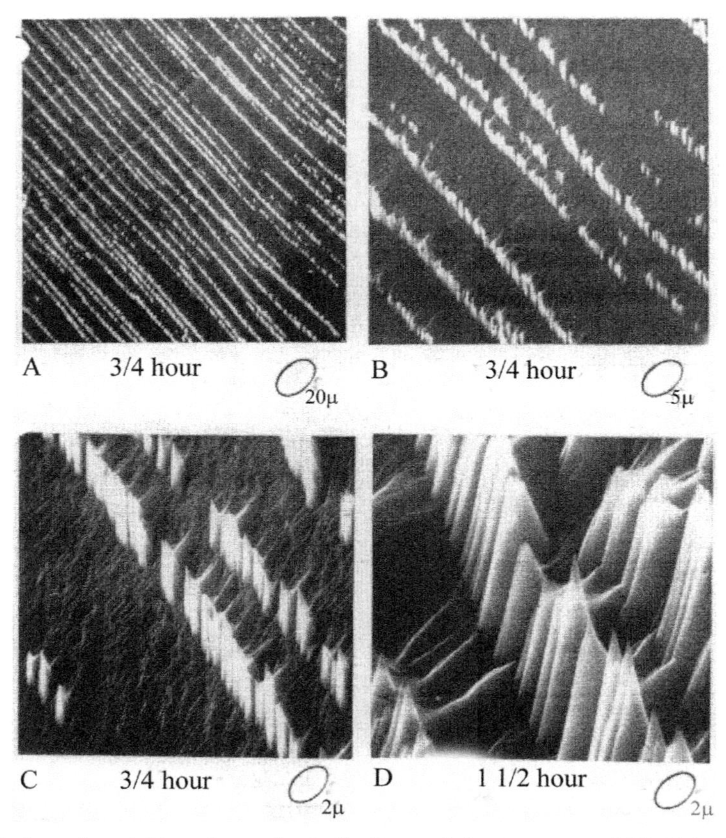

Fig. 3.7 Argon ion etching of a mechanically distorted aluminum crystal (5 keV, 1 mA/cm^2). *(Broers A., 1965. Selective Ion Beam Etching in the Scanning Electron Microscope, PhD Thesis. University of Cambridge).*

show one of these rows of etch pits. In all cases, the etch pits on a particular grain had the same shape.

Summary and interpretation of results

The intersections of slip planes with the surfaces of aluminum crystals have been found to etch differently from the rest of the crystal surfaces. The slip planes were produced by bending the crystals through angles of about 15° so that the surface of the crystal being etched was slightly curved. In all cases, the surface that was ion etched was convex and, therefore, in tensile stress. The slips were gross on an atomic scale sometimes producing steps several hundred Å high on the surface of the crystals. After ion etching, slip lines were revealed as rows of pointed cones, rows

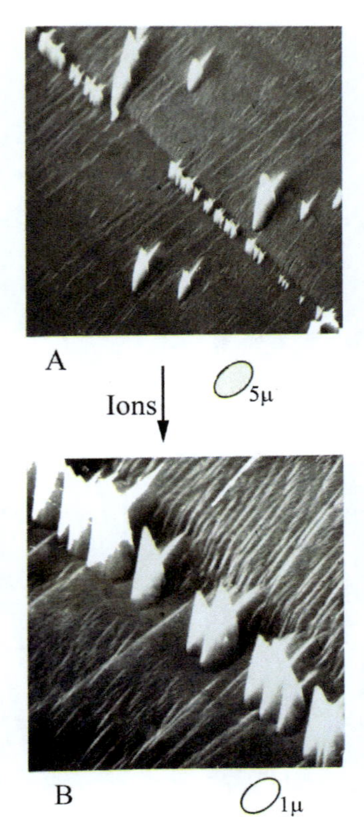

Fig. 3.8 Row of cones along a slip-line on an argon ion etched aluminum crystal (5 keV, 1 mA/cm^2). *(Broers A., 1965. Selective Ion Beam Etching in the Scanning Electron Microscope, PhD Thesis. University of Cambridge).*

of etch pits, or continuous grooves or ridges, depending on the crystal grain examined. In all cases where there were discrete cones or pits their shape was always the same on a particular grain. Two possible explanations for these etch phenomena will be considered:

(A) The gross disturbance of the crystal lattice in the vicinity of the slip lines disturbs the sputtering mechanism by interfering with the momentum transfer process in the crystal lattice, and by altering the binding energy of the surface atoms. Grooves or etch pits will appear after ion etching if the sputtering rate is increased in the vicinity of slip planes, and ridges or cones will be formed if the sputtering rate is decreased.

Individual pits and cones could be the result of individual stacking faults or dislocations. Cones would be expected at points of low sputtering rate for the same reasons that cones formed underneath

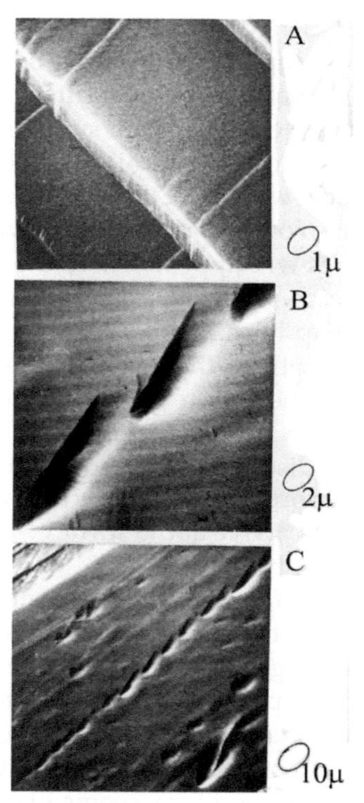

Fig. 3.9 Ridges and pits along slip-lines on an argon ion etched aluminum crystal (5 keV, 1 mA/cm^2). *(Broers A., 1965. Selective Ion Beam Etching in the Scanning Electron Microscope, PhD Thesis. University of Cambridge).*

shielding particles as discussed above. The continued growth of the cones as the etching proceeds is expected because the crystallographic faults extend into the crystal. The cones are pointed indicating that the shielding area on the surface is small. This also agrees with the supposition that they are the sites on the surface of single crystallographic faults, such as stacking faults and screw dislocations. The regular shape of the etch features on a particular grain would be expected because the crystallographic faults are associated with the crystal lattice which is a repetitive structure.

All the experimental evidence supports the idea that the etch phenomena are due to the effect on the sputtering mechanism of the disturbance to the crystal lattice in the vicinity of the slip lines. The second explanation that will be considered does not account for the observed experimental phenomena so completely.

(B) The disturbance of the crystal lattice in the vicinity of the slip planes affects the growth of oxide on the specimen surface. The important experiments were all carried out with aluminum specimens on which thin oxide layers grow rapidly. Aluminum oxide has been shown by Cunningham (1963) to have a slower sputtering rate than aluminum, and the growth of oxide on aluminum surfaces depends on the crystallographic orientation of the aluminum substrate (Kubaschewski and Hopkins (Kubaschewski, 1962)). If the oxide grows preferentially along the slip lines, its slower sputtering rate will give rise to ridges or perhaps rows of cones will appear where the slip lines intercept the surface. Grooves or etch pits will appear if oxide growth is restricted in the region of the slip lines.

The continued growth of the cones as the etching proceeds can only be explained if the oxide grows faster than the material is removed from the surface by the sputtering. This is not likely to be the case because, as I discuss at the end of the next section on ridge formation, 150 times more atoms are removed from the specimen surface than oxygen atoms arrive at the surface. The mass filter removes all oxygen from the ion beam.

The oxide layer that forms on the specimen before the surface is ion etched would affect the topography of the surface after etching, but only in the initial stages of etching. As a result, small surface perturbations would be expected, but as the etching continues these would disappear when the oxide has been removed. The cones observed in practice continued to grow indefinitely, and some 20 μm high were seen. It, therefore, seems likely that the initial oxide film could not have given rise to the etching phenomena observed, and it can be generally concluded that the etch effects are not caused by preferential oxidation of the surface.

The way slip lines etch depends on the crystallographic orientation of the crystal etched, which is different for different crystal grains. Therefore, the etch phenomena are likely to depend on the incident direction of the ion beam as well. This dependence on the particular conditions of bombardment could explain the conflicting evidence obtained by the workers mentioned in the Introduction to this section on the etching of distorted crystals.

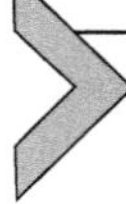

Formation of ridges on the surface of ion-etched crystalline metal specimens

Introduction

When an electro-polished single crystal metal specimen is etched with ions, the surface of the specimen often becomes ridged. Haymann had observed this (Haymann, 1962a,b), Wehner (Wehner, 1955a,b, 1956),

Cunningham (Cunningham, 1960, 1963), Fluit (Fluit, 1964), and others starting in the mid-1950s, long before I carried out my experiments, but none of them could observe surfaces at high magnification through many stages of etching.

Wehner (1958) discovered when etching a smooth silver crystal with 150 eV Hg ions that very soon (110) planes became exposed microscopically all over the crystal's surface.

Haymann (1962) made a detailed study of the ridge structure developed on the surface of Hg crystals bombarded with argon ions at 10 keV at varying angles of incidence. The ridge structure depended on the angle of incidence of the ion beam and the crystallographic orientation of the surface bombarded. Ridges formed along (110) and (100) directions and revealed (110), (111) and (110) planes. The surfaces were examined by electron diffraction, which showed that the crystallographic orientation remained unperturbed, and by electron microscopy. A double replica technique was used for the electron microscopy with a resolution of 50–100 Å.

A statistical analysis of the length L_2 (exp(-log $L_2)_2$) rather than a Gaussian distribution. This suggested that the lengths were multiples of an elementary length L1 and the average distance of a focused collision sequence along the (110) direction in the silver lattice. This distance was estimated at 65 ± 5 Å. No specific mechanism was proposed.

The ridge structure was found to depend on the current density of the ion beam. The measured length of the ridges was found to be less for the lower current densities. No mention was made of the amount of material removed from the specimen for the various bombardment current densities. It was suggested that the dependence of the structure on the current density was connected with the interactions between collision cascades.

Fluit and Datz (Fluit, 1964) determined the orientation of the facets on the regular ridge-like structure that appeared on the (100) surface of a copper single crystal after it had been bombarded with an analyzed beam of 20 keV argon ions. The ridge structure depended on the angle of incidence of the ion beam. For incident angles corresponding to a small ion penetration depth, the structure was explained on the basis that there were two competing mechanisms: one favoring the development of facets normal to the incident ion direction and the other favoring the development of (110) surfaces. It was assumed that these surfaces would develop because they were the surfaces from which material was removed fastest. No justification for the assumption was given. Surfaces other than (110) surfaces were revealed when the ion beam was directed at 45° to the surface (corresponding to a (110) crystal direction). The difference in behavior was attributed to the large increase in mean penetration depth for the incident angle.

In all observations that had been made, the ridge structure appeared to depend on the crystallographic orientation of the surface being etched and the angle of the incidence of the ion beam. However, many other factors influence the ridge structure. These include oxidation, contamination of the surface, sputtered material re-deposited on the surface, specimen temperature, crystallographic faults in the specimen, the energy of the ion beam, the current density of the ion beam, the species of ion used, and the specimen material. The effect of many of these factors on ridge formation is difficult to assess, and, based on data available at the time of my research, it was impossible to arrive at any definite conclusions about the mechanism of ridge formation. All that I attempted to do was to report the mechanisms of ridge formation as far as they were experimentally observed and to put forward some tentative conclusions that might be drawn from the experimental evidence.

Experimental results

I carried out many experiments to investigate the formation of the ridges discussed above. The general trend in forming the ridges when they occurred was the same in all crystalline metal specimens and for all angles of incidence of the ion beam. The ridges always increased in spacing and depth as the etching continued and never attained an equilibrium size. This is illustrated by the series of micrographs in Figs. 3.10 and 3.11, which show aluminum crystals and a silver crystal after various stages of etching with a beam of analyzed 5 keV argon ions. Ridges did not form on silicon crystals (Fig. 3.17).

No matter how the ridges appeared in the early stages of etching, they always developed facets after prolonged etching. Figs. 3.10 and 3.11 are good examples of this, and Fig. 3.12 shows four examples of heavily etched metal crystals exhibiting facets.

The general form of the ridges was found to depend on five factors:

(1) The crystallographic orientation of the specimen surface. This is illustrated by Fig. 3.10, which shows an area of an aluminum specimen where three differently oriented crystals join.

(2) The direction of the ion beam with respect to the crystallographic orientation of the surface. Fig. 3.13A and B show the same surface of a silver crystal after similar amounts of silver have been removed by the ion etching with a beam of 5 keV argon ions from different directions. The angle of incidence of the ion beam was the same in both cases but the specimen was rotated 90° between the two cases. The scratches shown on the micrographs run in the same direction on the specimen surface and allow the two areas of the surfaces to be orientated with resspect to each other. The amount of material removed was measured by the height of the shielded cones that formed under

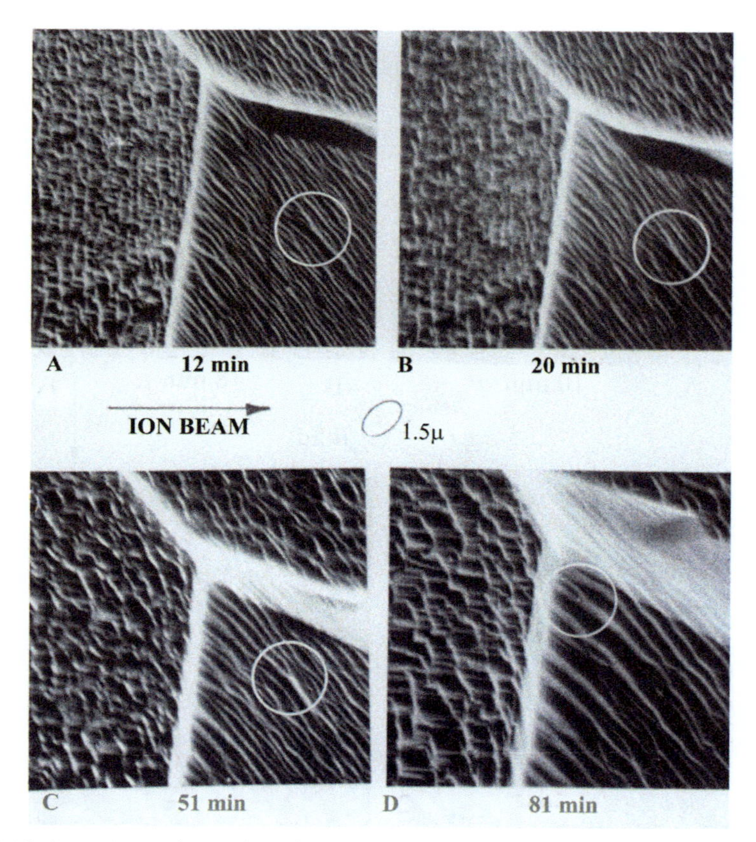

Fig. 3.10 Argon ion etching of a polycrystalline aluminum sample. (A) Shows the sample after 12 min of etching with 5 keV Argon ions at a current density of 2 mA/cm^2, (B) after 21 min, (C) after 51 min, and (D) after 81 min. *(Broers A., 1965. Selective Ion Beam Etching in the Scanning Electron Microscope, PhD Thesis. University of Cambridge).*

shielding particles, usually nickel or graphite dust, placed deliberately on the surface for this purpose.

(3) The species of ions used for etching. Fig. 3.14A and B show the same aluminum crystal etched with argon and oxygen ions. Again the energy, current density, and direction of the ion beam were the same, and similar amounts of aluminum were removed in each case. The amount of material removed from the surface was measured by the height of the cones that formed under shielding particles, usually nickel or graphite dust, placed deliberately on the surface for this purpose

The effect observed in the case of the oxygen beam was almost certainly due to surface oxidation and does not, therefore, reflect any change in the etched surface because of a difference in the mass of

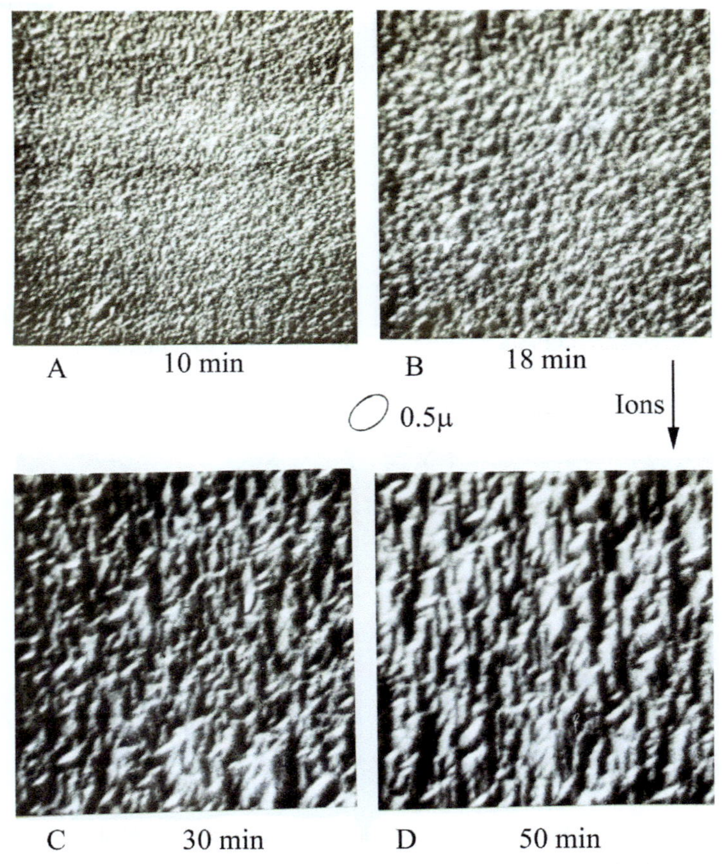

Fig. 3.11 Argon ion etching of a silver crystal with 5 keV Argon ions for increasing times. (A) for 10 min, (B) for 18 min, (C) for 30 min, and (D) for 50 min. *(Broers A., 1965. Selective Ion Beam Etching in the Scanning Electron Microscope, PhD Thesis. University of Cambridge).*

the ions. Experiments were planned using hydrogen ions but were not carried out due to a shortage of time.

(4) The specimen material. In many cases, the exact crystallographic orientation of the specimen was not known, and it was not possible to compare the ridges formed on identical crystallographic faces of different materials for the same etching conditions. The ridge structure, however, is almost certainly dependent upon the specimen material because no ridges formed in the case of silicon (Figs. 3.17 and 3.18).

(5) The amount of material etched from the specimen. The sequences of micrographs in Figs. 3.10 and 3.11 illustrate this.

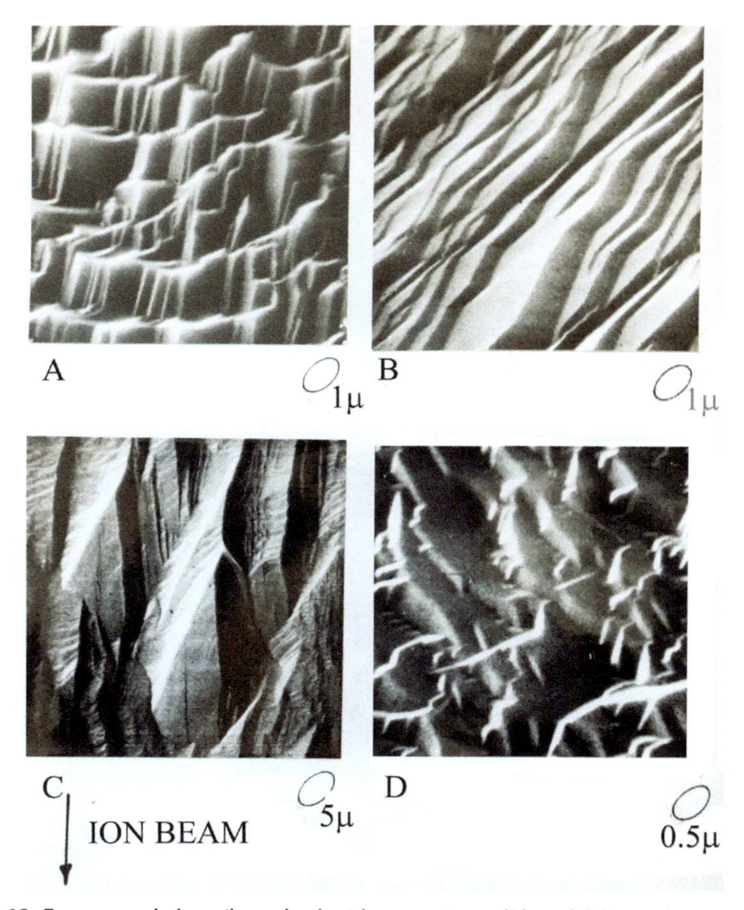

Fig. 3.12 Four crystals heavily etched with argon ions. (A) and (B) are aluminum and (C) and (D) are silver. *(Broers A., 1965. Selective Ion Beam Etching in the Scanning Electron Microscope, PhD Thesis. University of Cambridge).*

The general form of the ridges was found to be independent of the following two factors:

(1) The current density in the ion beam. Fig. 3.15A and B show different parts of the surface of the same silver crystal after equal amounts of silver had been removed by etching with a beam of 5 keV argon ions. The amount of silver removed was measured in the manner described in (2) above. The area shown in Fig. 3.15A was bombarded by a central portion of the ion beam for 5 min and Fig. 3.15B by the edge of the ion beam, where the current density was considerably lower, for 60 min. In both cases, the ridge structure is similar in form and magnitude.

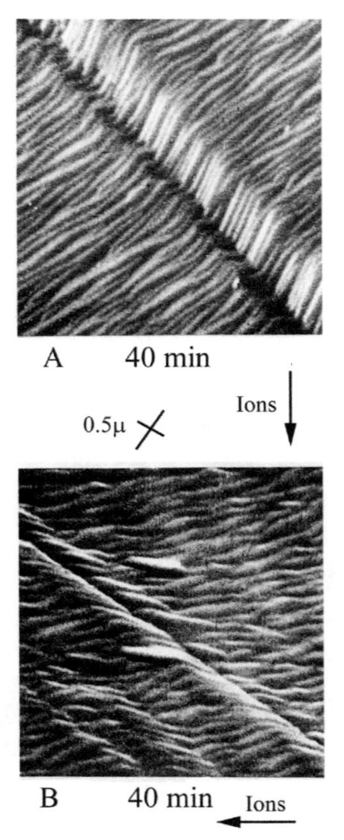

Fig. 3.13 Argon ion etched silver crystal in the vicinity of a scratch (5 keV, 1 mA/cm^2). *(Broers A., 1965. Selective Ion Beam Etching in the Scanning Electron Microscope, PhD Thesis. University of Cambridge).*

The current density in the center of the ion beam was about 1.5 mA/cm^2 and at about 0.15 mA/cm^2.

(2) The energy of the ion beam. No alteration in the form and magnitude of the ridge structure was found when the ion beam voltage was varied from 4 to 7 keV. This was the maximum voltage range over which the RF ion source would operate efficiently.

It was noticed when studying the formation of cones on the surface of ion-etched samples that ridges on the surface first appeared along the edges of scratches, i.e., surface perturbations (Fig. 3.1). To investigate this effect further, the microfabrication technique described in Chapter 4 was used to produce a narrow wall on the surface of a silver crystal. The electron beam was left scanning a single line for several

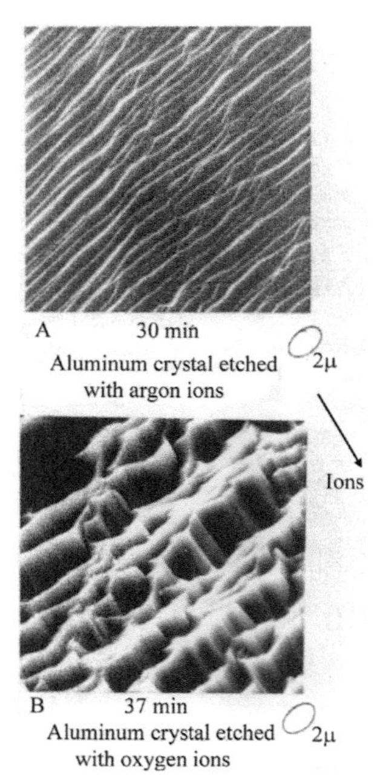

Fig. 3.14 Aluminum crystals etched with argon ion and oxygen ions to approximately the same depth. *(Broers A., 1965. Selective Ion Beam Etching in the Scanning Electron Microscope, PhD Thesis. University of Cambridge).*

minutes until a detectable line of contamination was laid down on the surface. The crystal was then etched with 5 keV argon ions. Because the sputtering rate of the contamination was low compared with that of the silver, a narrow wall was created on the surface and this wall nucleated ridges before they appeared on the rest of the surface, even though the ridges ran at right angles to the machined wall (Fig. 3.16).

To test whether the surface of the specimen became oxidized or contaminated between the doses of ion etching, an aluminum specimen was bombarded in two different areas with a beam of 5 keV argon ions in the following way. The first area was bombarded continuously for 60 min and the second area for 6 times 10 min with 4-min intervals between each of the 10-min bombardments. The ridges that appeared were similar in form in both cases, but in the second case, where the surface was etched in six separate steps, the ridges were smaller. The amount of material removed in this case was also less than when the surface was continuously etched. The amount of material removed was measured as described in (2) above.

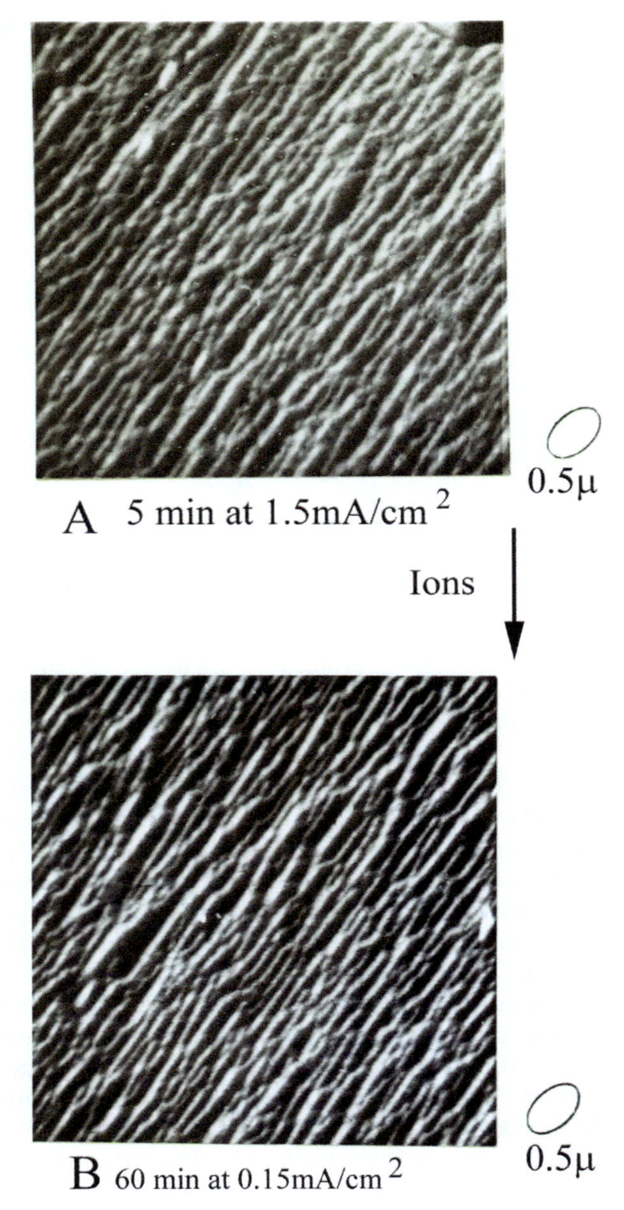

Fig. 3.15 Silver crystal etched with argon ions at different current densities. *(Broers A., 1965. Selective Ion Beam Etching in the Scanning Electron Microscope, PhD Thesis. University of Cambridge).*

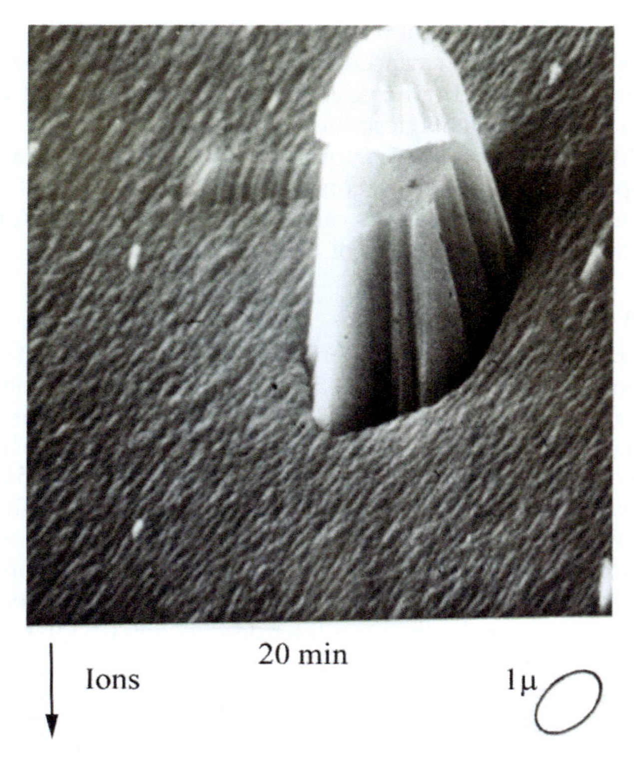

Fig. 3.16 Argon ion etched silver crystal showing ridges across a micro-machined line. *(Broers A., 1965. Selective Ion Beam Etching in the Scanning Electron Microscope, PhD Thesis. University of Cambridge).*

Summary and conclusions

All the experimental evidence indicated that the ridges that form on the surface of electro-polished crystalline metal specimens after ion etching have no equilibrium size and shape but continue to increase in spacing and depth as the etching continues.

It was possible by introducing a small topographical perturbation on a flat electro-polished surface to initiate the ridge formation earlier than it appears on a smooth surface. This could suggest that ridges only appear when the surface is either imperfectly polished, i.e., not perfectly smooth, or if the etching is non-uniform. Non-uniform etching could result from the surface contaminating or oxidizing selectively or because the ion beam was inhomogeneous.

Ridges always developed facets if etching was continued long enough. These facets appear to be connected with the orientation of the crystal and the incident direction of the ion beam. Their formation can be explained in the manner suggested by Fluit if it is assumed that, as etching continues, there is a tendency for the surfaces that sputter fastest to be revealed. No experiments were carried out to determine the orientation of the facets observed in the experiments described in this chapter.

It should be explained that factors such as oxidation, contamination, and back-sputtered material will complicate the mechanism of ridge formation. The effect of these factors cannot easily be assessed. A brief discussion of the possible influences of these factors follows.

Oxidation

The arrival rate of ions at the specimen surface, when the current density in the ion beam is $1\,\mathrm{mA/cm^2}$, is about 5×10^{15} ions/$\mathrm{cm^2}$/s, and an ambient impurity gas pressure of 10^{-6}mmHg is equivalent to about 4×10^{14}/$\mathrm{cm^2}$/s, arriving at the specimen surface. It is unlikely that >20% of the impurity gas arriving at the specimen surface will be oxygen, and each 5 keV argon ion will, for example, remove about three aluminum atoms. The ratio of atoms removed from an aluminum surface to oxygen atoms arriving at the surface from the vacuum is, therefore, about 150:1. The effect of oxidation during ion bombardment is, therefore, likely to be slight. Oxygen is eliminated from the ion beam by the mass filter described in Chapter 2. The surface may become oxidized, however, between successive bombardments, for example, when it is being examined with the SEM. This was checked in one case for an aluminum specimen. The experiment is described at the end of the experimental results described in the section entitled "Formation of ridges on the surface of ion-etched crystalline metal specimens." Because the surface, which was bombarded in discrete steps, sputtered more slowly than the continually etched surface, it can be assumed that the surface may have been oxidized, or contaminated while it was examined with the SEM. It was not possible from the results of this experiment to distinguish between the effects of oxidation and contamination. The shape of the ridges in both cases was the same but they were larger on the continually etched surface where more material has been removed. This, however, was only the usual behavior of the ridges as more material had been removed.

Contamination

There will have been oil vapor near the specimen surface, even with the liquid nitrogen baffle operating. The oil vapor will have condensed on the specimen surface and been polymerized by both the ion beam and the electron beam. The layer of polymerized oil vapor sputters slower than the metal surface as is shown in Chapter 4. For the same reason as those given for oxidation, the effect of contamination will not be serious while the specimen is being bombarded with ions, but a layer of oil vapor may condense on the specimen surface in the interval when the surface is being examined with the SEM. Again, the same conclusion that was given for oxidation applies, i.e., that the contamination may affect the sputtering rate but not the form of the ridge structure.

Back-sputtered material

Material is sputtered from a crystalline surface in preferential directions, and where the surface is not flat, the material will be deposited accordingly. The influence of this back-sputtered material will depend on the surface etched and on the incident direction of the ion beam. No attempt was made to estimate its effect on the ridge structure because the data obtained in these experiments was insufficient. It should be noted, however, that it can have little effect on the early stages of ridge formation because the surface is initially flat. The vacuum conditions were such that the amount of material scattered onto the surface by collisions with gas atoms was negligible.

To gain conclusive information about ridge formation, experiments would have to be carried out at a vacuum level better than 10^{-7} mmHg, and the direction of the atoms ejected from the surface determined.

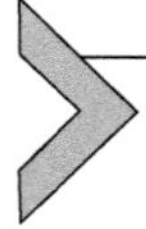

Ion etching of silicon

Introduction

Several silicon specimens (silicon slices) were etched with the beam of argon and oxygen ions from known crystallographic directions. In all cases, the silicon surfaces etched were (111) faces. None of the surfaces exhibited topographical structure except where the surface initially exhibited pin-holes or where shielding dust had been deliberately placed on the surface to measure the amount of silicon removed. The way the pin-holes etched was of interest.

Experiments were carried out to see whether crystallographic faults would be revealed by ion etching.

Several authors have investigated the preferential etching of crystallographic faults and their work is discussed in the Introduction to the section on the Ion etching of mechanically distorted aluminum crystals.

Experimental results

Four silicon slices were etched with 5 keV argon ions from various directions and for various lengths of time. One slice 350 μm thick was etched completely through and others were partially etched. In all cases, the surfaces, as observed in the SEM, remained smooth. This was true when an oxygen beam was used to etch the surface. The smoothness of the silicon surfaces should have made these silicon slices suitable specimens for investigating whether crystallographic faults, such as screw dislocations and stacking faults, would ion etch preferentially. To investigate whether this is true, the following experiment was carried out.

A silicon slice was etched to a depth of about 30 μm with a beam of 5 keV argon ions at a current density of about 1 mA/cm^2. The specimen was then removed from the specimen chamber and broken by pressing down on the silicon slice with the edge of a carefully cleaned glass slide while it rested on a pliant surface. It was hoped to make the slice break along a line passing through the center of the bombarded area, which was visible with the naked eye as a pit in the surface about 1 mm in diameter. This was achieved at the third attempt.

The broken slice of silicon was then placed in the SEM, and micrographs were obtained of the bombarded area near the edge along which the slice had broken. The specimen was mounted so that the fracture surface was also visible. The sample was then etched again with the beam of argon ions, and the surface re-examined. Both surfaces remained smooth, as seen in the SEM, even after a further 50 μm of silicon had been removed. It had been hoped that breaking the specimen would introduce crystallographic faults near the fracture and that the subsequent etching might reveal these faults. The surface was etched initially so that etch phenomena observed after fracturing the specimen and subsequent ion etching would be known to be caused by the crystal being broken and, therefore, likely to result from crystallographic faults.

On three of the specimens examined, pin-holes existed in the silicon surface. Fig. 3.17A shows several of these pin-holes after about 1.5 μm of silicon had been etched away with a beam of argon ions. Fig. 3.17E

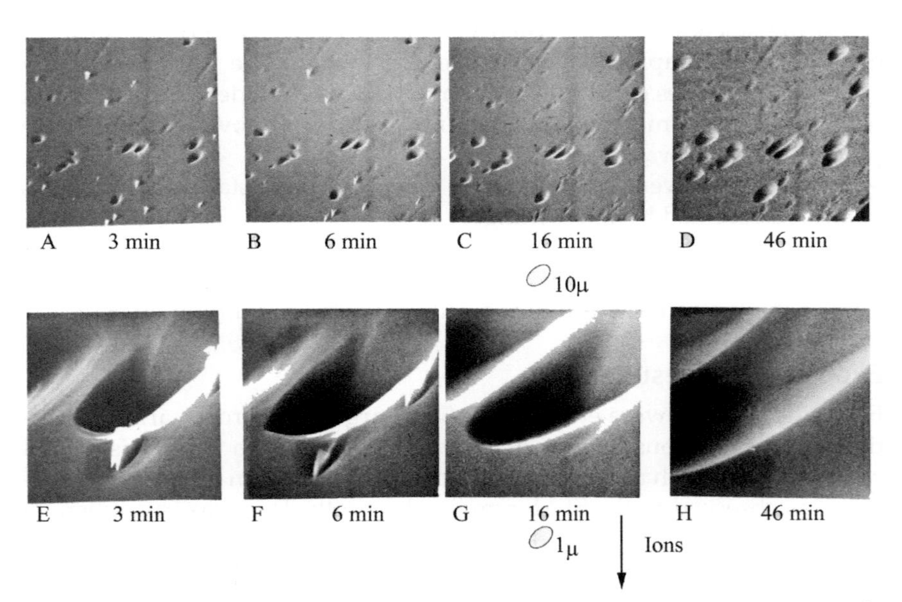

Fig. 3.17 Argon ion etching of a silicon crystal with "pin-holes" (5 keV, 0.5 mA/cm^2). *(Broers A., 1965. Selective Ion Beam Etching in the Scanning Electron Microscope, PhD Thesis. University of Cambridge).*

shows one of these pin-holes at higher magnification, and next to the hole, the shielded cones of the type used to measure how much material was removed. The cone near the edge of the micrograph still retained the shielding particle under which it formed and showed that the surface had been etched to a depth of about 1.5 μm. Fig. 3.17B–D and F–H show the same area of the surface as the etching continues. It is seen that the holes broadened out into shallow pits after the surface had been etched for 46 min.

Approximately 23 μm of silicon had been removed from the surface in this time, and the pits had continued to broaden as the etching continued. The pits are approximately 0.5 μm deep in Fig. 3.17H, and they still persist on the surface after it has bee n etched for an additional 55 min, corresponding to the removal of about 28 μm from the surface. The pits still exist after all traces of the shielded cones shown in Fig. 3.17E have disappeared (see Fig. 3.17H).

To emphasize that, as the etching continued, the pits on the surface of the silicon specimens persisted longer than the shielded cones, nickel was scattered on the surface of a specimen already etched and exhibiting pits, and the specimen etched again. Fig. 3.18A shows the shielded cones that

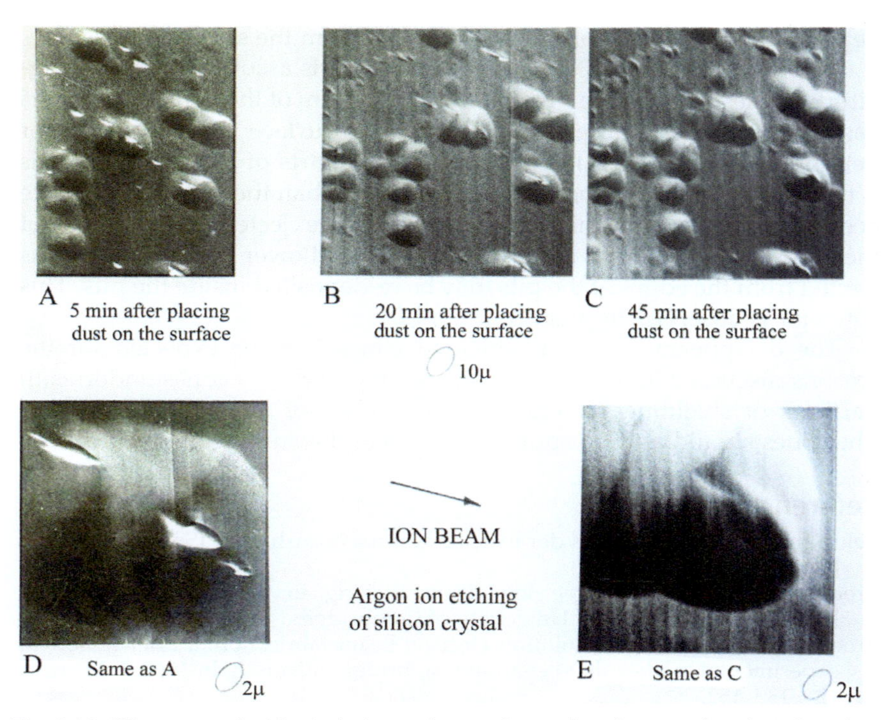

Fig. 3.18 Silicon crystal with pin-holes and cones formed under particles placed on the surface to allow measurement of the amount of silicon removed by ion bombardment. *(Broers A., 1965. Selective Ion Beam Etching in the Scanning Electron Microscope, PhD Thesis. University of Cambridge).*

were formed under the pieces of nickel dust on the pitted silicon surface, and Fig. 3.18B and C show the same area after further etching. It will be seen that not only do the shielded cones disappear, but they eventually end up as pits. Fig. 3.18D and E show the etching of the shielded cones that have been produced inside one of the pits.

Conclusions

(111) faces of silicon crystals were bombarded with argon and oxygen ions from different directions, and in all cases where the surfaces were initially smooth, they remained so after ion etching.

A fractured silicon specimen was ion etched to see whether crystallographic faults were revealed by ion etching, but the results of the experiments were inconclusive.

No selective etching of the surface was seen, and it was not known for certain, therefore, whether any faults were introduced when the sample was fractured. Pits in the silicon surface remained indefinitely as the ion etching continued. These pits occurred after ion etching, where pin-holes existed in the initial silicon surface. As the etching continued, the pits persisted long after shielded cones, whose heights were at least equal to the depth of the pits, were completely removed from the surface.

The way the pits etch can be explained if it is assumed that the magnitude of the angle through which the momentum of the bombarding ions has to be transferred to eject atoms from the surface is the major factor determining the etching rate of the different parts of the surface. If this is the case, the edges of the pits will etch faster than the rest of the surface because atoms from the inside of the edges can be ejected into the center of the pits. The pits should, therefore, become shallower because the atoms ejected from the edges of the pits may be re-deposited inside the pits. This behavior is observed in practice.

The disappearance of the shielded cones is to be expected for the reasons discussed in the section entitled "Formation of cones underneath particles of shielding material on the surface of ion-etched samples." The cones should and do sputter faster than the surrounding surface.

References

Behrisch, R., 1964. Ergeboises der Exacton Naturwissenshaften. (Sonderdruck aus band XXXV).

Broers, A.N., 1965. Selective Ion Beam Etching in the Scanning Electron Microscope. PhD Thesis, University of Cambridge.

Broers, A.N., 1985. High-resolution electron beam fabrication: a brief review of experimental studies that began in Cambridge university in 1962. J. Microsc. 139, 139–152.

Cunningham, R.H., 1960. J. Appl. Phys. 31, 31.

Cunningham, R.G.-Y., 1963. J. Appl. Phys. 34, 984.

Fluit, J.A., 1964. Phys. Ther. 30, 345.
Haymann, P., 1962a. Le Bombardment Ionic. (Numerous articles on regular surface structure. In *Editions du Centre National de la Recherche Scientifique.* Paris).
Haymann, P. (1962b). Selective Ion Beam Etching in the Scanning Electron Microscope, Thesis. Paris.
Kay, E., 1962. Adv. Electron. Phys. 17, 245.
Koedam, M. (1961). Electron Microscope Diagnostics of Thin Film Sputtering. PhD Thesis, State University of Utrecht, Netherlands.
Kubaschewski, O.A., 1962. Oxidation of Metals and Alloys. Academic Press Inc.
Stewart, A., 2004. Investigation of the topography of ion bombarded surfaces with a scanning electron microscope. Adv. Imaging Electron Phys. 133, 175–178.
Thomson, M., 1962. Brit. J. Appl. Phys. 13, 194.
Tsuge, H.A., 1981. J. Appl. Phys. 52, 4391–4395.
Wehner, G., 1955a. Adv. Electron. Phys. 7, 239.
Wehner, G., 1955b. J. Appl. Phys. 26, 1056.
Wehner, G., 1956. Phys. Rev. 102, 690.
Wehner, G., 1957. Appl. Sci. Res. Mague B5, 334.
Wehner, G., 1958. J. Appl. Phys. 29, 217.
Yurasova, V.E., 1957. Kristallografiya 2, 770.

CHAPTER FOUR

Microfabrication in an SEM

Contents

While studying the ridges that formed on the surface of metal specimens, I was always aware that the SEM electron beam was contaminating the surface between etching steps by cross-linking the molecules of the silicon oil used in the diffusion pumps. To reduce this effect, I would fill the liquid nitrogen cold trap that partially surrounded the sample. This reduced the pressure in the chamber by an order of magnitude and similarly reduced the contamination rate. I nonetheless noticed that the surface structure was perturbed slightly where the beam paused before scanning each line. This is mentioned in Chapter 3, where I noted that this seemed to trigger the formation of the ridges, and I carried out an experiment where I left the line scanning for a long time on a silver crystal to confirm this.

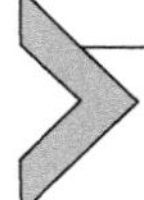

Micro-machining by ion etching through a mask of electron beam written contamination (Broers, 1964)

Because I could selectively "write" this contamination with the electron beam, I decided to try to use it in the way photo-resists were being used to make the fledgling planar integrated circuits that everyone was excited about at the time. I asked Les Peters, Professor Oatley's skilled technician, to make me some test samples by evaporating 3000 Å of gold onto a single silicon slice through an 80 bar/cm grid. The areas coated with gold were easily identified in the SEM because Au has a higher secondary

Advances in Imaging and Electron Physics, Volume 231
ISSN 1076-5670
https://doi.org/10.1016/B978-0-443-31462-9.00004-6

electron emission coefficient than silicon and appears brighter in the SEM. The bar pattern was useful in locating the tiny structures I was to make.

The test pattern shown in Fig. 4.1 was written by leaving the SEM beam scanning each of the eight 1.5 μm long lines in the two-dimensional array for 45 s. The beam current was about 10^{-12} A, so the dose delivered was very high at about 4.5×10^{-2} C/cm^2. The contamination is seen as dark

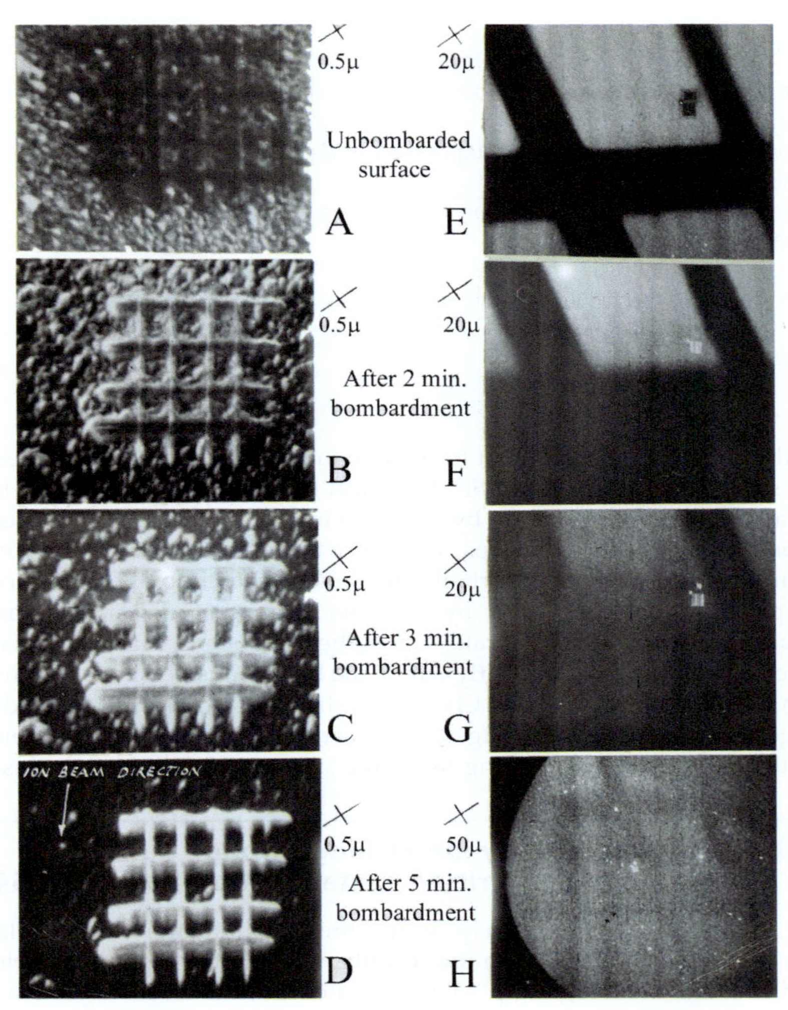

Fig. 4.1 Argon ion bombardment of a gold film on a silicon substrate. (A) shows a pattern of contamination lines written by the electron beam. (B–D) show how this pattern is transformed into a gold nanostructure as the unprotected gold is ion etched away. Reported at the First International Conference on Electron and Ion Beam Science and technology, 1964. © The Electrochemical Society. *(Reproduced by permission of IOP Publishing Ltd. All rights reserved).*

lines because it is largely carbon and carbon has a low secondary emission coefficient. The ion beam was positioned so that only the outer edge of its profile bombarded the area of interest. This slowed the etching process and allowed the mechanism to be studied in detail.

The high dose meant that the contamination remained in place until almost all of the unprotected gold had been removed leaving 75 nm wide and 300 nm thick gold wires despite the structure in the gold film. The ion beam was incident at 45° to the surface in the direction shown in Fig. 4.1 (D), which explains why the vertically oriented lines on the micrograph appear narrower; they are viewed end-on while the transverse lines are viewed from their sides. The roughness of the gold film was an early indication of the problems that would be encountered when dimensions below a tenth of a micron were needed. Nonetheless, these wires were a hundred times smaller than those used in the fledgling integrated circuits of the 1960s.

To test the resolution of the contamination-resist process, lines were also written on a 25 nm thick film of a AuPd which was known to produce smoother films than Au. These lines were written 30 times faster than the lines in Fig. 4.1, a dose of about 1.5×10^{-3} C/cm^2. The specimen was etched for 1 min at a current density of 0.3 mA/hcm^2. Fig. 4.2 shows the AuPd wires that were produced. They are about 40 ± 10 nm wide and

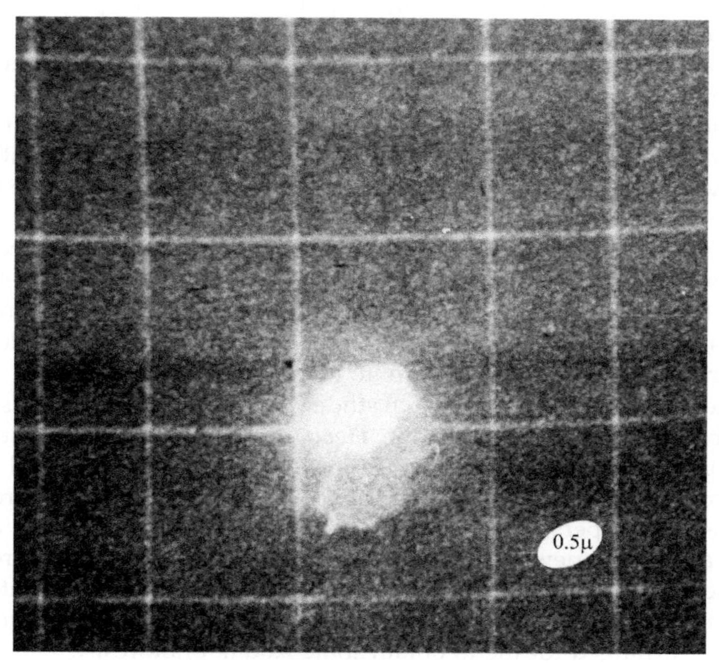

Fig. 4.2 Fine wires micro-machined from a 25 nm gold-platinum film.

it was assumed that their thickness was about 25 nm. This was the best resolution obtained with the Cambridge SEM.

The major disadvantage of the contamination fabrication process is the slow writing speed.

This was not limited by the electron beam current but by the arrival rate of the vapor on the surface. The 10 nm beam used to form the contamination in the above experiments delivered about 10^6 electrons/s to an area of 78.5 nm^2 on the specimen surface. The pressure in the specimen chamber was 10^{-5} mmHg if the liquid nitrogen baffle was not cooled. At this pressure, about 4×10^{15} atoms or molecules arrive on a surface per cm^2 per second, or $(4 \times 10^{15} \times 78.53)/10^{14} = 3142$ atoms or molecules on an area equal to that of the electron beam. So, there are more than 300 electrons for every particle arriving. If each electron was capable of completing one or more cross-polymerization acts between oil vapor molecules, then the rate of polymerization is determined by the number of particles, not the beam current.

This was found to be approximately true but no quantitative measurements of the thickness of the contamination were made although this might have been possible, for example by shadowing the samples with AuPd and then examining them in the SEM.

The most satisfactory way of increasing vapor pressure at the specimen would be to enclose the specimen in a separate, differentially pumped chamber into which a high vapor pressure of the polymerizable material was introduced. This would interfere with the collection of the secondary electrons to form SEM images so would need to be removed to examine the sample.

Other possibilities for increasing writing speed might have been to cool the sample or to find more suitable vapors but there was an alternative to using vapor resist and that was to use the photoresists that were being used in the fabrication of integrated circuits.

These experiments were carried out in 1963 and first reported at a Symposium on Electron Beam Techniques for Microelectronics, 6–8 July 1964, at the Royal Radar Establishment in Malvern Worcestershire, UK (Broers, 1965a,b) and at the First International Conference on Electron and Ion Beam Technology in Toronto in 1964 (Broers, 1964).

After my first year as a research student, Bill Nixon became my supervisor because Oatley became the Head of the Electrical Engineering Division of the Cambridge Engineering department and needed more time for the administration of the Division. Nixon, a Canadian, brought a trans-Atlantic perspective to Cambridge and made us all aware of the importance of the new integrated circuits. When I showed him the results of these experiments he suggested that I should use the SEM to write patterns in the photoresists that were being used to make these integrated circuits and arranged for me to obtain some Kodak photoresist.

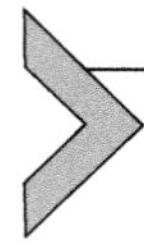

Micromachining by ion etching through a mask of electron beam exposed KPR photoresist (Broers, 1965a,b)

Kodak KPR photoresist was used because it was the most readily available resist that would have a faster writing speed than the contamination-resist used in the micromachining technique described above.

Photoresists are liquids that are spin-coated onto the surface of silicon wafers. A few ml of liquid photoresist are dispensed on the surface of the sample which is then spun at about 1000 rpm. Centrifugal force dispenses the resist into a uniform resist layer of the desired thickness. KPR is a negative resist: the parts exposed to light or an electron beam are rendered insoluble in the developer. After exposure, a solvent, called a developer is used to dissolve away the soluble regions to leave the patterned layer on the surface.

KPR is a polymer and on exposure becomes cross-polymerized. Reactive sites exist in the polymer chains which, when activated, form bonds between the polymer molecules. The cross-polymerization forms molecules of higher atomic weight that are insoluble in the developer.

When KPR is exposed with an electron beam, the minimum linewidth, assuming the beam is very small, is set by electron scattering in the resist layer and back-scattering of electrons from the underlying substrate. In the limit, it depends on exposure by secondary electrons and by the molecular size of the resist but in these early experiments it mainly depended on scattering in the resist layer and therefore on the resist layer thickness.

The thickness was varied by altering the spinning speed and by diluting the resist with KPR thinner. Measurements of the relation between spinning speed, resist concentration, and film thickness had been made by Corp. Westinghouse electric (1963). The KPR layer was baked after application to dry it thoroughly and baked again after development and washing to drive off any excess fluid and stabilize the resist pattern.

I carried out a wide range of experiments with KPR, first to understand how the dilution of the resist, the spin speed used in coating the samples, the subsequent baking cycles, etc. influenced the resist layers. I went on to explore the influence of electron beam energy, the writing speed, and the development process on the final patterned resist layer. Finally, I used these patterns to explore their effectiveness as a mask for ion etching and to explore the influence of the underlying substrate.

Fig. 4.3 shows a raster of gold-platinum wires on a silicon substrate. The resist pattern was written at a line speed of 750 μm/s, and after development and baking, the sample was etched with argon ions at a current density of about 0.2 mA/cm^2. It should be noted that the ion beam was incident at 45° to the surface, which explains why the wires transversely

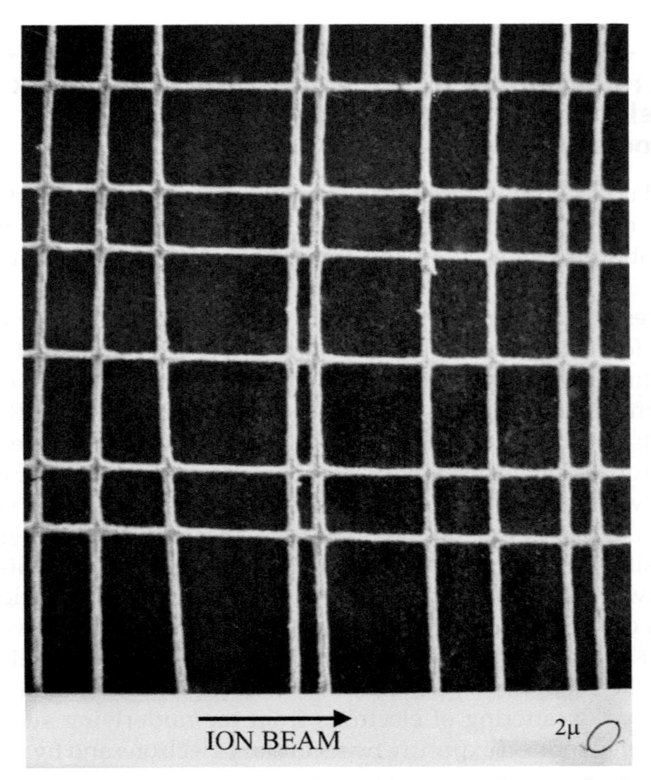

Fig. 4.3 Matrix of gold-platinum wires formed by ion etching with electron beam exposed KPR photo-resist mask. Reported at Symposium on Electron Beam Techniques for Microelectronics, Royal Radar Establishment, Malvern, Worcestershire, UK, 1964.

oriented on the micrograph appear thinner. They are viewed end-on, while those vertically oriented are viewed partly from their sides. The width of the narrowest lines is 300 nm.

Fig. 4.4 shows portions of the same raster shown in Fig. 4.3. (A) shows the wires after the initial 2 min of ion etching at 0.2 mA/cm^2 and (B) after a further 2 min. The thinning of the wires after the additional etching is clear, and the width of the transverse wires has been reduced from about 0.3 μm to about 0.13 μm in (B).

The following briefly describes my conclusions about the correct exposure dose for electron beam exposure of KPR photoresist. The electron beam diameter was 100 Å.

(1) Below an exposure dose of 5×10^{-6} C/cm^2, no resist remained after development. This may have been because the only completely exposed portion of the resist was near the surface of the resist film,

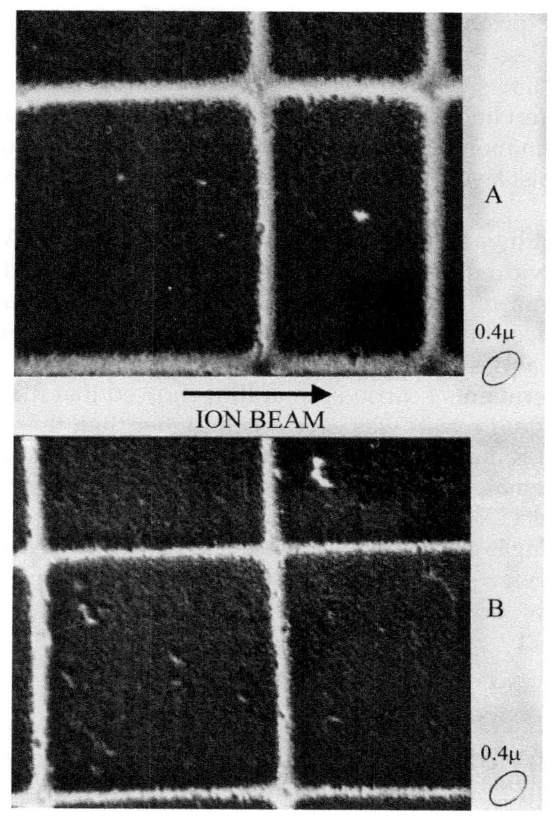

Fig. 4.4 (A) A portion of the pattern shown in Fig. 4.3 and (B) adjacent areas after further etching. The narrowest lines are now about 0.1um.

and this was washed away when the developer dissolved the insufficiently exposed resist underneath it. The exposure at the bottom of the resist film will have been slightly lower because of electron scattering.

(2) Between 5×10^{-6} and 3×10^{-5} C/cm^2 exposure was unreliable. When it was successful, the width of the developed stripe was 0.25 μm.

(3) The exposure between 3×10^{-5} and 8×10^{-5} C/cm^2 was correct, and a continuous stripe of developed resist 0.3 μm was obtained.

(4) Above 1×10^{-4} C/cm^2, the width of the exposed resist stripe became several microns. This gross spreading was due to the electrons backscattered from the substrate. I did not recognize it at the time, but it was to have serious consequences when writing complex patterns with varying densities. The exposure dose is varied today to compensate for this proximity effect.

The optimum exposure dose was between 3×10^{-5} and 8×10^{-5} C/cm^2 was about 30 times less than that of the contamination process. The line speed was about 750 times faster.

The minimum linewidth was about 0.3 μm compared to about 0.05 μm for the contamination-resist, but I felt that with the use of purified resists and thinner resist layers it should be possible to produce linewidths down to 0.1 μm.

The linewidth varied slightly with beam energy. At 15 kV on a silicon substrate, the width of the exposed resist line was 0.4 μm. At 25 kV it was 0.3 μm. I recognized that this was probably due to electron scattering in the resist layer but could not find any scattering data to confirm this. Looking back, this was almost certainly the case.

Future experiments I carried out at IBM showed that the ultimate resolution with liquid resists was very much higher than these early experiments with KPR indicated, especially when used on thin substrates that effectively eliminated electron back-scattering. This will be discussed in Chapters 9 and 12.

The following is a brief review of research on electron beam fabrication based on the review I included in my Ph.D. dissertation and on a comprehensive review of electron beam machining published in 1962 by Crawford (1962).

Summary of electron beam fabrication and recording techniques in 1964

There were two basic types of electron beam fabrication techniques: thermal processes and processes relying on electron beam-induced effects.

Thermal processes

Thermal machining processes could also be divided into two categories; those in which the material being machined was cut and those in which it was only shaped.

The narrowest slots cut through metal films by direct thermal machining with an electron probe are about 2 μm wide but this resolution was only obtained when machining very thin films, i.e., less than 1000 Å.

Higher resolution can be obtained in shaping films such as collodion using what I assume to be a thermal process. Mollenstedt and Speidel at Tubingen University in Germany used a 200 Å electron beam to write letters in a collodion film with comparable resolution. (Mollenstedt and Speidel, 1960a,b) The electron beam created a groove in the surface of the film, which they shadowed with platinum to make the grooves more visible in the electron microscope. They reported these and other experiments in which they used ions to produce a polymerized resist layer followed by chlorine etching to produce copper grids. They reported these experiments at the International Solid-State Circuits Conference in the USA in 1961 (Mollenstedt and Speidel, 1960a,b).

In 1964 Loeffler (1964) electron optically projected a reduced image of a grid onto a carbon film and produced a micro-grid with bar spacing of 1000 Å and bar width of about 500 Å. Again I assume that this was a thermal process.

Electron beams can be used to shape the surface of certain plastics in what are called thermo-plastic processes. The electron beam is used to lay down a charge pattern on the surface of a thermoplastic film. The film is then softened by heating and the electrostatic force between the charged area of the film and the substrate causes an indentation on the surface. The resolution of this technique is not high (2 μm Glenn, 1960) but the writing speed is fast and the process is reversible. By further heating of the thermo-plastic, the indentations in the surface can be smoothed out.

Electron beam-induced processes

The ability of electron beams to trigger chemical reactions on the surfaces was first observed by Carr (1930). He suggested that they might be caused by "A chemical combination or dissociation between the surface film of gas and the metal itself". No one followed up on this at the time but about 20 years later electron microscopists became interested because the build-up of carbonaceous material was spoiling the resolution of their microscopes. Hillier (1948), Poole (1953), Ennos (1953), Ennos (1954) and others studied the way the electron beam decomposed the monolayers of pump oil and vacuum grease that coat the insides of vacuum systems. Ennos and Poole noted that heating the sample to 250C stopped the monolayers forming and prevented the contamination. I discuss this in Chapter 8 where I report how I used a heating lamp to stop contamination when using scanning transmission microscopy where the problem is especially severe.

While I was using contamination as a resist, Fabian Pease (1963) in the laboratory next to me in Cambridge used the 50 Å beam in his SEM to grow whiskers of polymerized silicon oil, contamination on the bars of a 1500 bar/in. silver grid. Some of these whiskers were only 100 Å thick.

Others used the ability of electrons to decompose or polymerize specific materials. For example, Fisher decomposed silver nitrate, cupric oxide, and lead carbonate while observing the transformation by electron diffraction (Fisher, 1954). It was hoped that this might be a way to directly produce material for microelectronics.

The earliest proposal to use electron beam techniques to make electronic devices was, to my knowledge, made by Buck and Shoulders (1958). They made several proposals including the use of contamination as a resist, but did not put these ideas into practice. They did cite an experiment carried out by Crawford and Baker at MIT who used a flood beam of 1200 V electrons to create a patterned resist layer on a thin film of molybdenum. They did this by contact printing through a copper mesh and using tetraethoxy-silane vapor as a resist A chlorine etch was then used to remove the unprotected molybdenum. Magnified pictures obtained in an electron microscope showed that the resolution was "finer than one micron."

The last process that will be mentioned is the simple exposure of photographic emulsion with an electron beam. This technique has been used in an electron beam computer memory built by Thornley et al. (1964), in which information is recorded with a 1 μm beam in units 6 μm square. The writing speed on photographic emulsion is high and a resolution of 2 μm should be possible.

Comparison of techniques for micro-machining

To the best of my knowledge, the micro-machining techniques using contamination resist and KPR resist followed by ion etching described in this chapter had the highest resolution of all the methods described at that time for micro-machining metal films.

The contamination-resist process produced wire widths of less than 500 Å and edge-definition of better than 150 Å. It also allowed relatively thick films, i.e., about 0.5 μm, to be machined with a resolution of better than 1000 Å.

KPR resist produced metal line widths of 0.2 μm with the ability to reduce the linewidth to 0.1 μm by over-etching the metal.

Of the two techniques, the one using the polymerized silicone oil resist is more suited to one-off applications where high resolution is required and the speed of writing is of no consequence. It also has the advantage that it is only a two-stage process. In the SEM-ion beam system, the sample could remain in vacuum throughout the process. The KPR method is a multi-stage process and the sample had to be taken out of the vacuum to develop the resist. The KPR resist, however, has a much higher writing speed and demonstrated how relatively high resolution can be obtained with a standard commercial resist.

Ph.D. Oral Examination

I nominally had to finish my Ph.D. research in October 1964, as I began in October 1961. The time allocated for a Ph.D. in England at that time was three years. In the USA and Germany, the time could stretch to six years.

I finished my experiments by the end of July 1964 and thought that it would only take two or three months to write my thesis, but like most students, I hopelessly underestimated how long it would take. I also discovered that there were more experiments I should complete and asked if I could return to the lab to complete these, but Bill Nixon persuaded me that if I did this, I would probably never finish. In any case, he said it was not necessary to complete a definitive piece of research to be granted a Ph.D., only to demonstrate that you could produce a respectable volume of original research and describe it professionally in a thesis that you finished.

My friend Sir Eric Ash, an electrical engineer who became the Rector of Imperial College, used to say when persuading industry to hire people with Ph.D.s, "At least you will know that they are capable of actually finishing something."

I finally submitted my thesis just in time for my oral examination on March 13th, 1965, the day before I was to fly to New York to take up the job I had been offered by IBM three months earlier. By chance, I was interviewed at the airport by a woman from the Board of Trade with a clipboard who was looking into the number of UK graduates leaving the UK for the USA in what was called the 'Brain Drain'. She asked me a series of questions, and when she got the question, "When did you get your degree?" I answered, "I hope I got it 18 hours ago." She smiled and said, "Well, I landed one this time. You obtained a PhD at Cambridge yesterday, and you are leaving the country within 24 hours!"

Part of the delay in finishing my thesis was that Mary and I were married on December 28th, 1964, and went off for our honeymoon to Austria and Italy to ski for two weeks in our brand-new MGB. We had met two years earlier on a Thames barge moored off Cheyne Walk in London, where she lived with a friend. She worked for Inghams Travel, a company that pioneered package holidays in Europe, especially ski holidays. They also arranged travel for incoming tourists to London, and she managed this overseas business. She had lived in Spain and Italy, and I had visited her in Elba in Italy when she was looking after Ingham's holidaymakers there in 1964. We had a lot of common interests, but our professional talents were very different. She was an accomplished linguist and spoke good Spanish, Italian, and French. I spoke two languages, Australian and English, and neither country liked, or like, my accent. I worked with numbers, and she with words, but that has been a good combination.

References

Broers, A.N., 1965a. Combined Electron and Ion Beam Processes for Microelectronics. Microelectronics and Reliability. 4 Pergamon Press, England, pp. 103–104.

Broers, A.N., 1964. Micromaching by sputtering through a mask of contamination laid down by an electron beam. In: Bakish, R. (Ed.), 1st International Conf. on Electron and Ion Beam Sci. and Tecnol. John Wiley, NY, pp. 191–204.

Broers, A.N., 1965b. Selective Ion Beam Etching in the Scanning Electron Microscope. PhD Thesis, Univ. Cambridge.

Buck, D.A., Shoulders, K.R., 1958. An approach to microminiature printed systems. In: Proceeding Eastern Joint Computer Conference. Special Publication, pp. 55–59.

Carr, P.H., 1930. Rev. Sci. Instrum. 1, 711.

Corp. Westinghouse Electric, 1963. Third and Fourth Interim Engineering Reports No. AF 33(657).

Crawford, C.K., 1962. Chapter 11, Electron beam machining. In: Bakish, R. (Ed.), Introduction to Electron Beam Technology. John Wiley and Sons, Inc, New York.

Ennos, A.E., 1953. The source of electron-induced contamination in kinetic vacuum systems. Brit. J. Appl. Phys 4, 101.

Ennos, A.E., 1954. The origin of specimen contamination in the electron microscope. Brit. J. Appl. Phys. 5, 27.

Fisher, R.B.J., 1954. Appl. Phys. 25, 894.

Glenn, W.E.J., 1960. Thermoplastic recording. Soc. Motion Picture Television Engrs 69, 577.

Hillier, J., 1948. Investigation of specimen contamination in the electron microscope. J. Appl. Phys. 19, 226.

Loeffler K.H. Thesis, University of Berlin. 1964.

Mollenstedt, G., Speidel, R., 1960a. Electronenoptischer Mikroschreiber Arbeitskontrolle. Physikalische Blattere 16, 192.

Mollenstedt, G.W., Speidel, R., 1960b. Microminiaturization developments. Digest of technical papers. In: 1961 International Solid State Circuits Conference. Irving Auditorium. University of Pennsylvania, p. 54.

Pease, R.F.W., 1963. High Resolution Scanning Electron Microscopy. PhD Dissertation, Cambridge University.

Poole, K.M., 1953. Electrode contamination in electron optical systems. Proc. Phys. Soc. B66, 542.

Thornley, R.F.M., Brown, A.V., Speth, A.J., 1964. Electron beam recording of digital information. IEE Trans. Electron. Comput., 36–40. Vol. EC-13 No. 1.

CHAPTER FIVE

My move to IBM and research on long-life cathodes

Contents

It was not surprising that, after working for the latter part of my Ph.D. on the fabrication of microstructures, I decided to join a company that was applying similar techniques to the fabrication of integrated circuits. I looked first at jobs in British companies, but none offered me this opportunity. I thought about returning to Australia and restarting my audio equipment business, but that seemed a backward step.

What finally helped me decide what to do was attending the First International Conference on Electron and Ion Beam Science and Technology Conference held in Toronto in 1964. Getting to this conference was not straightforward. There was no money in my PhD grant for travel, so I had to approach Sir Charles Oatley, who, as Head of the Electrical Division, had some money for travel but not very much. I remember him grilling me on why I wanted to go and why it was going to take almost a week to do so. "Are you sure Broers," he said, "that you would not get more done working in the laboratory than travelling off to Canada and America?" I hesitated and mumbled that I thought I would learn a lot at the conference from the other attendees, and it would be useful to see whether anyone thought my ideas about applying what I had been

Advances in Imaging and Electron Physics, Volume 231
ISSN 1076-5670
https://doi.org/10.1016/B978-0-443-31462-9.00005-8

doing to making integrated circuits made any sense. "All right," he replied, "and good luck. You can tell us about it when you return."

It soon became obvious at the conference that there were several companies in the United States devoting large resources to this revolutionary technology, and after I presented my paper, I was approached by several of them with offers of employment. IBM was one of these, although interestingly, it was the least ambitious. They were only planning to integrate a handful of transistors on a single piece of silicon. This was in contrast, for example, to Westinghouse, whose engineers wanted to make a 1000 transistor chip. They thought that it would be impossible to have all of the 1000 transistors work and wanted someone to build an electron beam system that would scan the chip to find the good transistors and then write a custom metal pattern to interconnect them. It had already been shown that SEMs could be used to sense voltages and to distinguish between the differently doped semiconductor regions, so it should have been possible to identify the good transistors. I was skeptical about this and indeed, it never happened. To this project day, the vast majority of chips are scrapped if a single device, or interconnection wire, is faulty.

Of the other companies working on the new microelectronics, I talked to AT&T Bell Labs, and Hughes Aircraft, but decided that IBM, although conservative, was probably the most realistic and decided to accept one of their offers, which was to work in their new research laboratory in Yorktown Heights, New York, 50 miles north of Manhattan. This laboratory was housed in a magnificent new building designed by the Finnish architect Eero Saarinen, and the idea of being close to New York City with its rich culture appealed to Mary and me.

When I told Sir Charles what had happened, I think he regretted ever allowing me to go to the conference, but he did very kindly arrange for me to become a Fellow of Trinity College when I returned to Cambridge 20 years later.

I started at the IBM Research laboratory in March of 1965, a few years before semiconductor integrated circuits were used in IBM's computers.

Programmable electronic computers such as Colossus and ENIAC had been around for about 25 years. These, of course, used vacuum valves, or tubes, as they are called in the United States. The first computers that used transistors were built in the early 1950s at Manchester University, and Bell laboratories and others soon followed. The Bell Laboratories computer was made for military applications. IBM's first transistorized computer, the IBM 7070, was launched in 1958.

When I was an undergraduate in Melbourne, Australia, from 1954 to 1958, most applications of electronics, including the audio systems I built, used valves. The teaching of semiconductor science and transistors only began in earnest while I was at Melbourne University.

The application of electronic computers had been rapid for military, science, and academic purposes, but they were not yet being used in business and banking. In these applications, tabulating and punch card machines persisted, and when, in 1956, Thomas J Watson Junior, the CEO of IBM, decided to abandon mechanical tabulators and commit IBM to building electronic computers, it was regarded as a brave but risky decision. He had just taken over from his father and made this decision against the advice of his father, who had run the company for the previous 42 years. But history showed it to be an inspired decision. He hired hundreds of electronic engineers, and by 1971, when he stepped down as CEO, the company had grown 10 times and dominated the business and large computer market and continued to do so until the end of the 20th century.

Valves and then single discrete transistors were used for digital computation and fast memory from the beginning, but it was not until the late 1970s, when it became possible to integrate about a million transistors on a single chip, that transistors were used for large-scale data storage. A fascinating collection of alternative technologies was used. Data were stored in mercury delay lines, on photographic film, briefly in Williams Tubes, but mainly on magnetic tapes and disks. A Williams tube was essentially a cathode ray tube (CRT) where bits were written and read by an electron beam on the face of the tube. Write and read times were satisfactory but the tubes could only store about 2000 bits. When rapid access was needed, magnetic core memories were used in which the magnetization of tiny rings of magnetic material called cores was switched and detected with wires that passed through the cores. These fascinated me. The cores were mostly hand-threaded with hair-thin wires, although a sort of knitting machine was also invented. An array looked like a loosely woven cloth. About 30 kB could be stored in a cubic foot. It sounds ridiculous today, but magnetic core memories were the dominant form of random access memory for 20 years, from 1955 to 1975, when they were finally replaced by Random Access Memory transistor chips. Large-volume storage, on the other hand, remained on magnetic disks and optical CDs well into the 21st century when a variety of silicon chips populated with various forms of field effect transistors began to replace them. Most of us have solid-state storage in our PCs now, rather than vulnerable magnetic hard disks.

Digital computation settled down earlier. It started in the 17th century with mechanical devices, but these began to be replaced with electronic systems in the late 1930s. Valves were made as compact as possible, but before a real attempt was made to seriously miniaturize them and replicate them inside a single vacuum envelope, the transistor arrived. It was difficult to make the transistor linear enough to challenge valves in linear circuits because of the leakage currents inherent when there are positive and negative carriers, and this took a couple of decades, but they were ideal for digital circuits where linearity did not matter. As an electronic switch, the simple three-terminal transistors rapidly replaced valves.

It was then only a small intellectual step to come up with the idea of building several transistors into a piece of semiconductor. On May 7, 1952, Geoffrey Dummer, who worked at the Royal Radar Establishment in Malvern, UK, read a paper at the US Electronic Components Symposium and, at the end of the paper, made the statement: "With the advent of the transistor and the work on semi-conductors generally, it now seems possible to envisage electronic equipment in a solid block with no connecting wires. The block may consist of layers of insulating, conducting, rectifying, and amplifying materials, the electronic functions being connected directly by cutting out areas of the various layers." This was barely 5 years after Bardeen, Brattain, and Schockley invented the first transistor. Nonetheless, in the 60s, there was a great furor over whether the integrated circuit of Kilby of Texas Instruments or that of Noyce of Fairchild was the first, and it wasn't until 1961 that the US patent office awarded the first patent to Robert Noyce, while Kilby's application was still being analyzed. Today, both men are acknowledged as having independently conceived the idea.

In my opinion, Noyce's integrated circuit was superior largely because it integrated all of the components, including the wires that connected the components, whereas Kilby's had fly wires connecting the components. Noyce also used silicon, and silicon oxide turned out to be the ideal mask for modifying the underlying silicon. Kilby used germanium, which has a less useful oxide and is more difficult to dope. Silicon transistors can also operate at higher temperatures—but comparing silicon with germanium is a subject for a book rather than some brief remarks about the early history of integrated circuits.

In any case, practical integrated circuits became available by the mid-1960s, and although IBM stayed for a few years with what they called Solid Logic Technology, described below, the rest of the world got on with integrated circuits. Since then, everything proceeded at a remarkably steady pace for 60 years, as observed by Gordon Moore in the 1960s. The pace was primarily sustained, until about 2010, by miniaturization.

Progress was not easy. Innumerable problems arose as things became smaller, such as electro-migration in the metal wires that led to discontinuities, radiation-induced errors that appeared in memory devices at high altitudes, the isolation of adjacent devices became very difficult, chips overheating due to inadequate power dissipation, electron scattering in highly doped layers slowing devices, ultra-thin gate insulators being needed to sustain gain, and hot electrons collecting where they interfered with device switching. Creative engineers have found ingenious ways around these difficulties, but they are struggling today as dimensions reach a hundredth of a micron, and all of these problems remerge and are joined by unwanted quantum phenomena that interfere with device operation.

IBM Thomas J Watson research laboratory

I joined a small group of about six engineers at the IBM Thomas J Watson Research Laboratory in March 1965, almost a year after the conference in Toronto. The group was managed by Alan Brown. Michael Hatzakis, Richard Thornley, and Al Speth were also in the group. They were using electron beams to write on photographic film to store information. As I have just described, IBM was using hybrid electronic circuitry rather than fully integrated circuits and did not start exploring the use of electron beams to expose photoresists until 1965 (Thornley and Sun, 1965).

IBM's hybrid electronics used tiny (0.025 mm square) silicon chips with only one transistor or two diodes on each chip. Single logic circuits containing about four transistors were made by mounting these tiny chips on 12 mm square ceramic substrates. Wires and resistors were screen-printed on the substrates. The resistors were trimmed to the correct values by sand-blasting. The ceramic substrates were then connected to Solid Logic Technology (SLT) cards with 12 pins. IBM's SLT cards were used in the revolutionary 360 system computers that dominated the computer market by the 1970s. SLT was far too expensive to be used for storing data. Millions of bits of memory storage were needed, and several transistors were needed to store a single bit, and each transistor cost tens of cents. SLT was said to be a thousand times more reliable than vacuum tube circuits, but its transistors were 100 million times more expensive than they are today, where a chip can contain hundreds of billions of transistors.

For memory, IBM used magnetic core memories when rapid random access was needed and magnetic tapes or disks when high volumes were needed and relatively long random access times could be tolerated. System 360 computers had the largest core memory made at the time and had a maximum capacity of about 80 million bits.

Starting in the early 1960s, Alan Brown's group had been exploring whether it would be possible to increase the number of bits that could be written with an electron beam, perhaps by making the bits smaller (Thornley et al., 1964). The beam size in a typical CRT, or Williams tube, was several tens of microns, and it was thought impossible to produce a beam size smaller than a few microns using CRT technology and certainly impossible to go below a micron. Alan discovered that Oatley's group in the Cambridge University Engineering Department had already produced beams well below a 10th of a micron for use in what was then the new scanning electron microscope, so he hired Richard Thornley, who had recently completed his Ph.D. thesis working on the SEM, to explore the possibilities for applying these smaller electron beams for writing on photographic film.

SEM technology was de-mountable, allowing the target to be changed, unlike CRT technology, where the target could only be replaced by breaking the glass tube. Their idea was to explore the application of SEM electron optics for computer memory systems using photographic film. They were successful in doing this and built a disk recorder that was used in a Russian-to-English translation machine. The electron beam, which had a diameter of about a micron, wrote carefully defined 6-μm squares on a servo track on a 25 cm diameter photographic film disk. The disk was then developed and read using a photo-multiplier and a CRT. The photomultiplier detected light passing through the film from the CRT spot to read the bits on the spinning disk, while the servo track was used to keep the spot aligned with the row of bits. The CRT spot did not have to be as small as the electron beam that was used to write the bits.

The next step was to increase the capacity further than could be accomplished with a single disk. It might have been possible to design a mechanism for changing the disk rapidly and having multiple disks and multiple readers, but IBM decided instead to adopt the format that had been used in a system that IBM had developed in California for the CIA for storing large numbers of images on small photographic film cards. Instead of images, there would now be an array of squares filled with digital bits using the same square marks on a servo-track as the e-beam disk recorder. The film cards were 2.5 cm by 7 cm, and the data were stored in an array of 32 squares. Each card contained 6.5 million bits. 32 cards were stacked in small boxes called cells, and the cells were placed in racks from which they could be accessed pneumatically. The target was to increase the capacity of the memory to an incredible 1 trillion bits with an access time of about 1 s, and they succeeded.

Although I was keen to continue the work on device fabrication that I had started in Cambridge, I was given the task of finding a better electron source for the electron beam column that wrote the data on the photographic film. The electron source they were using was a tungsten filament of the type used in electron microscopes. These simple wire filaments were made by bending a 125-μm diameter tungsten wire over a sharp edge. They were run at about 3000 K at which temperature the tungsten sublimated at a rate that led to them burning out after about 20 h, so eight were mounted on a turret so they could be successively rotated into position and keep the system running for several days.

My task was to find a cathode that would last at least 200 h and retain the ability to operate in the relatively poor vacuum condition found in a de-mountable system using rubber O-rings and oil diffusion pumps. To succeed, I had to find a replacement in months rather than years, as they were already seeking customers for the Digital Cypress memory systems.

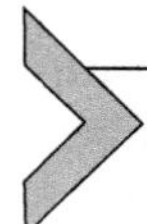

Long-life tungsten cathodes for use in electron beam memory systems

As discussed in Chapter 2, the major criterion for an electron gun used in an electron microscope or a high-resolution electron probe system is the brightness (A/cm/sr) of the electron beam it produces. Langmuir (Langmuir, 1937) considering the effect of the initial thermal velocities of the emitted electrons, found that the maximum brightness that could be obtained from any electron gun using a cathode at a given temperature was given by the expression

$$B = \frac{j_c eV}{\pi kT}$$

Where

B—brightness ($A/cm^2/sr$).

j_c—specific emission of the cathode (A/cm^2).

e—electronic charge (e.s.u.)

k—Boltzmann's constant.

T—absolute temperature of the cathode (K).

Up to cathode emission densities of about $2\,A/cm^2$ Haine and Einstein (Haine and Einstein, 1952) had shown that this brightness could be closely obtained in practice with simple tungsten wire hairpins in the standard triode electron gun used in electron microscopes. Above this figure, however, the efficiency of the gun deteriorates. For example, at $7.5\,A/cm^2$, a brightness equivalent to only 40% of Langmuir brightness is obtained (Haine and Einstein, 1952; Haine et al., 1958). At the higher emission densities, the efficiency with which electrons are focused into a cross-over is limited by space charge. If the source becomes larger, the space-charge limit occurs at lower emission densities.

The lifetime of the hairpin cathode is limited by three factors: evaporation of the tungsten, cathode sputtering, and gas erosion, but gas erosion and sputtering are only significant for pressures greater than 10^{-4} mmHg. In most systems, only evaporation is important. Evaporation is fastest at a hot spot that forms just below the apex of the hairpin because radiated heat is concentrated at this point, and when the wire here thins about 10%, the filament burns out. Bloomer (Bloomer, 1957) predicted that the lifetime at an emission current density of 2.7 A/cm^2 was about 40 h. The lifetime was not repeatable to better than 25%, so such a cathode could only be guaranteed to last for 30 h at this emission density.

Choice of cathode type

In seeking long lifetimes, I first considered the barium and strontium oxide cathodes used in vacuum tubes and cathode ray tubes. These can last tens of thousands of hours but require a vacuum level of 8×10^{-8} to 8×10^{-10} mmHg to operate satisfactorily. To obtain these vacuum levels, getter materials are vaporized by RF heating after the tube is evacuated. The glass tube is then sealed by melting the glass evacuation tube. The getter deposits on the inside of the glass envelope, where it reacts with and adsorbs residual gases. Such methods for obtaining and maintaining high vacuum are not practicable in de-mountable systems, so these efficient long-life cathodes cannot be used. The only materials capable of producing the emission densities required at such pressures are the refractory metals and Lanthanum hexaboride (LaB_6), and in 1965, LaB_6 had never been used in electron microscopes or electron probe instruments, so I decided initially to start with the refractory metals.

To obtain the electron brightness (10^5 $A/cm^2/sr$ at 12 kV) needed for Digital Cypress, the electrostatic field at the electron source needed to be as high as possible. This is why the tungsten wire cathodes were made from fine (125 μm) wire and bent into as sharp a hairpin as possible. This maximized the field concentration around the apex of the hairpin. Thick hairpins and flat cathodes are not suitable. For example, the maximum brightness that could be obtained at 12 kV accelerating voltage from a 375 μm tungsten wire hairpin in the electron gun used in these experiments, no matter how much the cathode temperature was increased, was 2.5×10^4 $A/cm^2/sr$, compared to a maximum of 15×10^4 $A/cm^2/sr$ for a 125-μm tungsten hairpin, and 25×10^4 $A/cm^2/sr$ for a pointed tungsten rod cathode. The three cathodes were operated under the same conditions with a cathode-to-anode spacing of 0.9 mm and an accelerating voltage of 12 kV.

Two refractory metal cathodes with potentially long lifetimes were investigated: the rod cathode and the cone cathode. In both cases, the cathode material was indirectly heated by electron bombardment. Indirect heating eliminates the hot-spot problem by making the heat supplied to the cathode almost independent of the geometry of the cathode as evaporation occurs.

The rod cathode was first developed by Bas (1962a,b), who called it a "Bolt" cathode. Here, the cathode is a metal rod heated by electron bombardment from a coil filament surrounding the rod. The end of the rod was ground to a sharp point to make the source small and the electrostatic field concentration at the tip high. I will refer to these cathodes as rod cathodes.

The second cathode was a hollow tungsten cone heated by bombarding the inside with an electron beam from a Pierce-type electron gun. The cone has the advantage that electron emission is drawn from the hottest part of the cathode, and it also has excellent mechanical stability.

Experimental arrangement for measuring brightness

The apparatus used to evaluate the cathodes measured the brightness they produced when used in a standard three-electrode electron gun. The temperature of the cathode was not measured directly so no direct check on the efficiency of the gun was available. Any error introduced in this manner would give a pessimistic value for the brightness and life of the cathode. The cross-over formed by the electron gun was defined by a small (50-μm) traversable aperture placed immediately after the anode. With the cross-over thus defined in size and position, it was possible to measure the brightness directly by measuring the current through a second aperture placed at a known distance from the cross-over defining aperture. The arrangement is shown in Fig. 5.1 for a rod cathode.

The brightness B is given by:

$$B = \frac{16IL^2}{\pi^2 d_1^2 d_2^2}$$

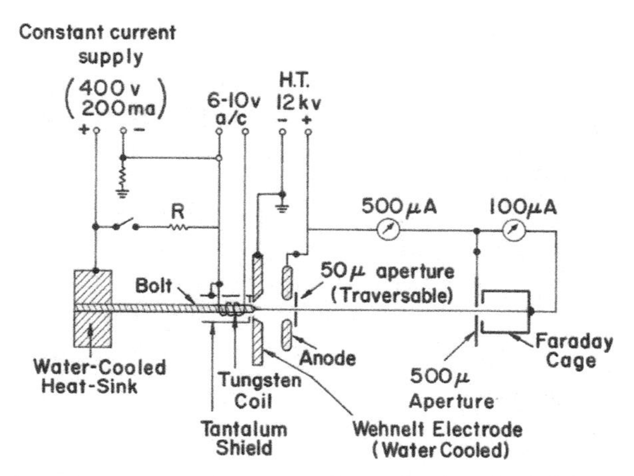

Fig. 5.1 Circuit and aperture arrangement for evaluating the brightness of long-life cathodes. The example shown is for a tungsten rod cathode. *(Reprinted Courtesy of IBM Corporation ©).*

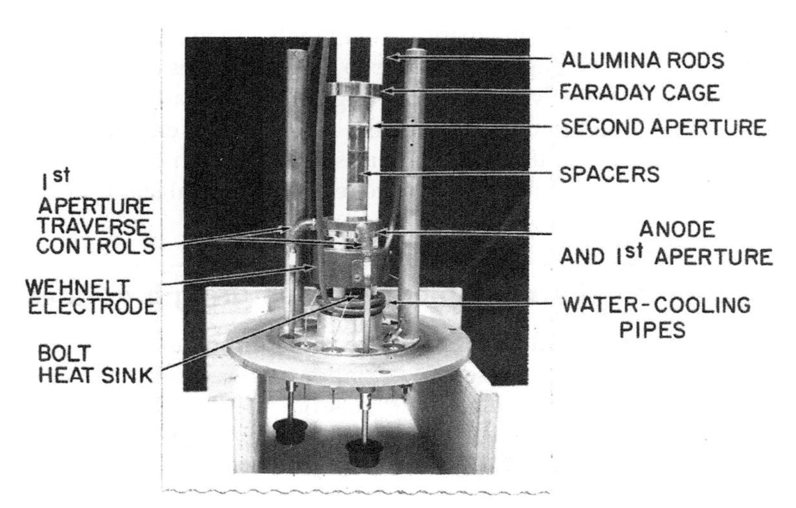

Fig. 5.2 Apparatus for measuring brightness and lifetime of long-life cathodes. *(Reprinted Courtesy of IBM Corporation ©).*

Where

I—current through the second aperture.

d_1—diameter of the cross-over defining aperture (50 μm).

d_2—diameter of second aperture (500 μm).

L—distance between the apertures (10.7 cm).

With these aperture sizes, the brightness was measured over a half-angle of 2.34×10^{-3} rad.

Brightness is current density $(4I/\pi d_1^2)$ into a given solid angle, which is a fraction of a steradian π. The solid angle is $(\alpha^2\pi)$ where $\alpha = d_2/2L$.

The disadvantage of this method of measuring brightness is that the cross-over is assumed to be at the location of the cross-over defining aperture after the anode, which may not be its optimum position. Any error introduced in this manner would again give pessimistic values of brightness and cathode life.

Fig. 5.2 is a photograph of the apparatus. The electrodes were mounted on three alumina rods. The nominal alignment was 25 μm, but in most cases, permanent magnets on the outside of the apparatus were found necessary to align the beam between the apertures.

Rod cathode

The lifetime of the rod cathode is set by the time it takes the rod to evaporate away at its hottest point. As would be expected, the hottest point is at the central bombarded region of the rod, and not at the tip. The temperature difference between the hottest part and the tip is generally between 100 °C and 200 °C, depending on the distance the tip extends outside the coil. This corresponds, for tungsten, to a factor of between 3 and 10 in the evaporation rate. It is possible from this ratio and the evaporation rate to predict the lifetime. The evaporation rate for tungsten was taken from (Reimann, 1938), and for tantalum from (Langmuir and Malter, 1939). A 1 mm diameter tungsten rod operating at 2800 K (emission density 3.5 A/cm^2) should have a lifetime of about 300 h, assuming that the rod can be 75% evaporated at its hottest point before its useful life is over. It should be possible to shape the rod so that it is thickest in the center of the bombarded region to extend its lifetime.

Another method used by Bas (1962a,b) to increase lifetime was to use a tungsten rod with a tantalum tip. Tantalum emits higher current density than tungsten for temperatures below 2800 K, and at these temperatures, the evaporation rate for tungsten is relatively low. Evaporation from the tip is not as important as it might at first appear because, in most cases, the general shape of the tip does not alter. The tip is the coolest part of the rod and evaporates slowest and the sharpness of the tip is sustained. A 1 mm tungsten rod with a tantalum tip should, based on a similar calculation to the one made above for the simple rod, last for about 1200 h at an emission density of 3.5 A/cm^2 (Fiske, 1942). The unknown factor, in this case, was the rate at which the residual gas in the vacuum attacked the tantalum tip.

The emitting area of the rod can be made smaller than the emitting area of a 125-µm tungsten hairpin. This is illustrated in Fig. 5.3, which compares a 1 mm rod cathode with a standard tungsten hairpin. This is an advantage because it allows brightness higher than 10^5 A/cm^2/sr to be obtained.

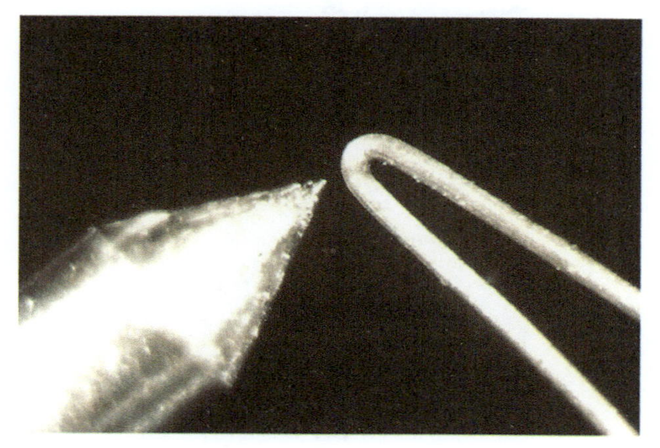

Fig. 5.3 The tip of a 1 mm tungsten rod cathode compared to a 125 µm tungsten hairpin cathode. *(Reprinted Courtesy of IBM Corporation ©).*

The lifetime of the heating coil is several times greater than that of the rod but is eventually limited because it collects material evaporated from the rod and becomes too thick to operate satisfactorily. This thickening has been discussed in detail by Bas et al. (1964).

Rod cathode experimental results

In all cases, the metal rods were 1 mm in diameter, and the end was ground to a point over about 1 mm. Several different rods were tested. The length of the rod was varied between 12 and 37 mm to determine the power required to heat rods of different lengths. The rods were mounted so that their tips extended about 2 mm outside the coils.

The heating coils (Fig. 5.4) consisted of 12–14 turns of 0.25 mm tungsten wire wound around a 1.5 mm former and flashed around a 2.15 mm diameter ceramic rod. The coil was flashed by rapidly heating it to about 2500 K. The heating current was reduced immediately this temperature was reached to prevent the ceramic rod from melting. The flashing relieved most of the stresses in the wire, and on subsequent heating to about 2500 K, the coil held its shape sufficiently to retain adequate clearance between itself and the rod. The coil was attached to insulated stand-offs mounted on the Wehnelt electrode of the electron gun.

Fig. 5.4 1 mm diameter tungsten rod cathode and its heater coil. *(Reprinted Courtesy of IBM Corporation ©).*

The rod was mounted separately on its heat sink. Both the Wehnelt electrode and the rod heat sink were water-cooled.

A constant current power supply was used to provide the accelerating potential between the coil and the rod. If the bombarding current was not stabilized in this way, it ran away. Heat was radiated back from the rod to the coil, and a positive feedback loop was closed. The most satisfactory way to control this circuit was to hold the electron current constant by varying the power to the heater coil. A constant voltage could then be maintained across the diode circuit, allowing the power to the rod to be controlled instead of just the current.

The following rods were tested:

(1) A 1 mm tungsten rod 3.8 mm long. This rod was run at a brightness of 3.6×10^4 A/cm^2/sr for 150 h in a vacuum of about 10^{-7} mmHg. The electron bombardment power required to heat the rod was about 36 W, and the power used to heat the coil was about 32 W. During the 150 h, the diameter of the rod at its hottest point was reduced by 50% by evaporation, and the diameter of the point immediately below the tip was reduced by 20%, indicating a ratio of 2.5:1 between the evaporation at the two points. Theoretically, for this ratio of evaporation rates, the rod should have taken 300 h to evaporate to this extent. The discrepancy between theory and practice was probably due to the inefficiency of the electron gun and inaccuracy in the brightness measurement; i.e., the gun was not producing Langmuir brightness for the particular temperature and emission density used, and the measurement of brightness was pessimistic for the reason outlined above. A life of 225 h would nevertheless have been obtained before the diameter of the rod in the central bombarded area was 75% evaporated. The top of the rod remained pointed during the 150 h.

(2) 1 mm diameter tungsten rod with a tantalum tip (see Fig. 5.5). This rod was made by butt-welding tungsten and tantalum rods together, cutting the tantalum so that about 2 mm remained, and grinding the tantalum to a point. The rod was run for 200 h at a brightness of 3.6 A/cm^2/sr in a vacuum of 10^{-5} mmHg. At this time, the diameter of the hottest part of the rod was reduced by 25% by evaporation, and the diameter of the point immediately below the tip was reduced by 18%. The total heat input to the system was 57 W, 27 W electron bombardment, and 30 W to the heater coil. During the 200 h, the tip of the rod became flattened and this reduced the maximum brightness that could be obtained from 2.5×10^5 to 4×10^4 A/cm^2/sr and also increased the total power required to obtain a brightness of 3.6×10^4 A/cm^2/sr to 63 W. This flattening of the tip may have been due to ion bombardment or a gas attack. In either case, a better

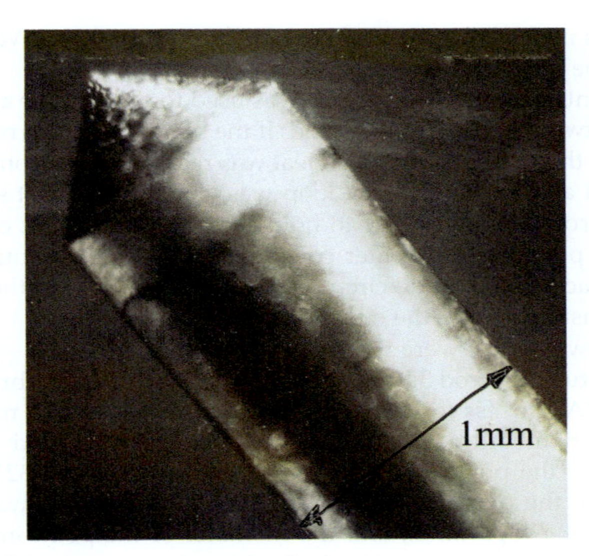

Fig. 5.5 Tantalum tip on tungsten rod cathode

vacuum would have improved the situation. If the evaporation rate of the rod had continued at the rate of the first 200h, a lifetime of 600h would have been obtained.

(3) 1mm diameter tungsten rod 1.25cm long. This cathode was only run to determine the power required to heat it. No brightness or lifetime measurements were made. The power required to heat is tip to 2800K was about 200W, a power that the coil bombardment system was well capable of delivering. The temperature of the cathode was measured by using s disappearing filament pyrometer. The mechanical stability of this rod was better than for (1) and (2) but the total power would be difficult to dissipate in a practical electron gun.

The cone cathode

Fig. 5.6 is a cross-sectional drawing of the cone cathode used in these experiments, and Fig. 5.7 is a photograph of the cone. The cones were made from solid tungsten by cylindrical grinding and spark machining. The thickness of the cone at the hottest point, the tip, was 200 μm, which should have taken about 600h to evaporate at 2800K. A lifetime of 450h should, therefore, be obtained if 75% evaporation was allowed. The cone could have been made thicker, but only at the expense of the power required to heat it.

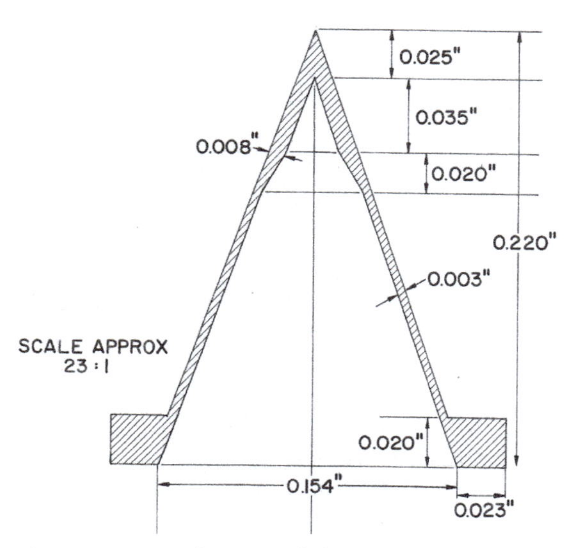

Fig. 5.6 Enlarged cross-section drawing of the tungsten cone cathode. *(Reprinted Courtesy of IBM Corporation ©).*

Fig. 5.7 Tungsten cone cathode. Scale is in millimeters. *(Reprinted Courtesy of IBM Corporation ©).*

Fig. 5.8 is a cross-sectional drawing of the Pierce gun (supplied by Microperv, Inc.) used to heat the cone. It had a spiral tungsten cathode with a life of over 1000 h at the normal operating temperature of 2400 K. The cone was mounted directly on the anode of the Pierce gun, as shown in Fig. 5.9 so that the minimum possible working distance was used and maximum power density was obtained in the electron beam.

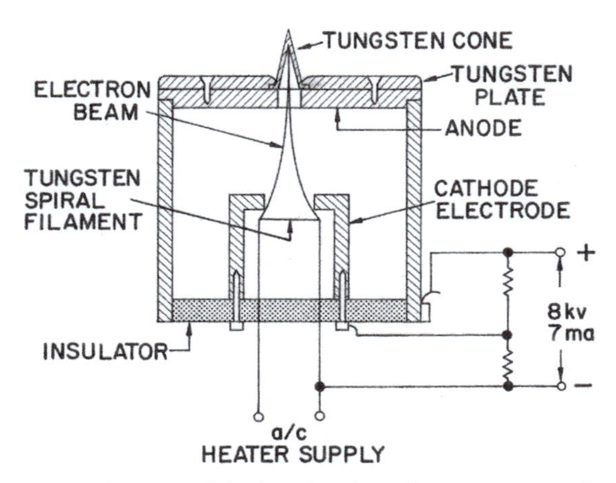

Fig. 5.8 Diagrammatic drawing of the bombarding electron gun used to heat the cone cathode and the cone cathode. *(Reprinted Courtesy of IBM Corporation ©).*

Fig. 5.9 Tungsten cone cathode mounted so that its apex is at the focal point of the Pierce electron gun used to heat it. *(Reprinted Courtesy of IBM Corporation ©).*

Cone cathode experimental results

As expected, the power needed to heat the cones was strongly dependent on their wall thickness. A cone with an 80 μm thick wall, as shown in Fig. 5.6, required 50 W to raise the tip to about 2800 K, a cone with a 100 μm wall required 70 W, and a cone with a 150 μm wall required 100 W. The maximum power from the Pierce gun was about 100 W.

This was obtained at 14 kV with the spiral tungsten cathode running at 2400 K. The beam diameter was of the order of 0.5 mm.

The gun provided 50 W at 9 kV with the cathode again at 2400 K. Water-cooling was used to carry heat away from a tungsten plate that held the cone against the anode of the gun. The particular geometry of the Pierce gun supplied by Microperv, Inc. was not suitable for long-life operation because the high-voltage insulator was too easily contaminated. This was only a function of the particular mechanical design, but it prevented the completion of any life tests on the cone cathodes. A more suitable gun was not designed and built because it was considered that the feasibility of the cone-type cathode was already sufficiently established. The major questions that originally had to be answered were: (1) whether a simple Pierce gun could supply sufficient power into a small enough area to heat the tip of a tungsten cone to 2800 K without any additional means of focusing and operating at a relatively low voltage (i.e., <10 kV), and (2) whether the tungsten cones could themselves be made. Both of these questions were adequately answered by the experiments carried out.

As cathodes, the cones behaved similarly to the rod cathodes. The maximum brightness that could be obtained, no matter how much the temperature was increased, was 2.5×10^5 A/cm^2/sr at 12 kV. A brightness of 3.6×10^5 A/cm^2/sr was obtained with the cone in Fig. 5.6 with input power from the bombarding electron gun of 55 W. Although no lifetime measurements were made, there is no reason to suppose that the cones would not last at least 50% of their lifetime as estimated from evaporation data, providing the pressure was maintained below 10^{-6} mmHg. This was the case for the rod cathodes. If the pressure was not maintained below 10^{-6} mmHg, cathode sputtering became significant and the life would probably have been limited to about 200 h.

Long-life tungsten cathodes: Conclusions

Both the rod and cone cathodes are capable of producing a brightness of about 4×10^5 A/cm^2/sr at 12 kV for at least 200 h at pressures of 10^{-5} mmHg and below. Longer lifetimes, up to 600 h for the rod cathode and up to 400 h for the cone cathode, should be obtained if the pressure is maintained below 10^{-6} mmHg. The advantage of the long life obtained with the rod cathode is somewhat offset by the extra complexity of the electronics needed to control the heater unit and also the relatively poor mechanical stability of the cathode compared to the cone cathode. The mechanical stability of the rod cathode, however, is still better than that of the tungsten hairpin. The cone cathode is not only mechanically very stable, but it can be positioned with great accuracy. The heating gun for the cone does not require any stabilizing circuitry other than the power supplies where the stability requirements are not high.

The total power input to both cathodes is much greater than that required to heat the conventional tungsten hairpin but could be readily dissipated by forced air cooling. Water cooling was used in the evaluation of the cathodes because it was more convenient, and the high accelerating voltage was only 10 kV.

As a final comparison, the rod cathode is more suitable where long life is the predominant need, but the cone cathode is better for mechanical stability, accurate alignment, and ease of operation.

Lanthanum hexaboride cathodes

I explored tungsten cathodes because I felt that they would certainly provide 200 h lifetime for a brightness 4×10^5 A/cm^2/sr at 12 kV, but I had also been searching through the scientific literature to explore alternative cathode materials. The most promising was lanthanum hexaboride. Lafferty (Lafferty, 1951) explored the possible use of rare earth borides as cathodes and showed that lanthanum hexaboride (LaB_6) had a more favorable ratio of electron emission density to evaporation rate than tungsten. The emission density was maintained at vacuum levels of 10^{-5} mmHg in de-mountable vacuum systems. However, it had proved difficult to use because it reacted with almost all substrate materials at the high temperatures required for electron emission. Favreau (Favreau, 1965) had cataphoretically coated rhenium hairpins with LaB_6 and used them in an ion gauge, but only at low emission densities of a few tens of mA/cm^2. I tried to use similarly coated hairpins at a current density of about 10 A/cm^2, but emission was only obtained for a few hours before the coating appeared to break up, and the cathode failed (see Fig. 5.10).

I also tried cataphoretically coating tungsten and rhenium rod cathodes with LaB_6, but these failed after a few hours. They created high voltage breakdowns, apparently because small quantities of LaB_6 flaked off the rod. However, I eventually obtained a 1 mm square LaB_6 rod that was 15 mm long and had been made by arc-melting LaB_6. Fig. 5.11 shows this cathode after I accidentally broke it after it had run for about 1100 h. I tested this cathode in the same brightness measuring system I had used for the long-life tungsten and tantalum cathodes. It was heated with 60 W to the coil and 20 W of bombardment power to give a temperature of about 1600 K and an emission density of about 5 A/cm^2 This corresponded to a brightness of 5×10^4 A/cm^2/sr at 12 kV, the same brightness I had used for the tungsten rod cathodes. The current from the cathode and through the measuring apertures was stable and remained stable for several weeks.

Fig. 5.10 1 mm diameter tungsten rod cataphoretically coated with lanthanum hexaboride showing cracks in the coating.

Fig. 5.11 1 mm square arc-melted lanthanum hexaboride cathode after 1100 h of operation—accidentally broken.

I did not know what to do as time went on. I did not want to jeopardize the experiment by turning the cathode off and taking the system apart so that I could examine the cathode rod or continue to build on the lifetime. In the end, after 1000 h, I turned it off and took the system apart. Unlike the metal rods that sublimated away at the hottest point on the rod, the LaB_6 rod remained unchanged.

After spending several months testing many metal rods and finding them sublimated away after 200 h, this was a remarkable result. To check that the result was repeatable, I reassembled and pumped down the system and turned the cathode on again to find it still emitted at the same level. To check that it had not been affected by exposure to atmospheric pressure, I ran it for another 100 h with no change. Finally, to be sure that there was nothing special about this particular LaB_6 rod, I replaced it with a second rod, this time made by sintering LaB_6 powder. It behaved the same way, and after 900 h, it too, was unaltered (Broers, 1967).

So, a replacement for the tungsten hairpin cathode used in the Digital Cypress memory systems, the IBM 1360 system, had been found. The lifetime target had been 200 h, and it was clear that a lanthanum hexaboride rod cathode could have a lifetime greater than 1000 h. Unfortunately, only five IBM 360s were built. Two were built for $2.1 million in the initial contract, one for the Livermore National Laboratory and one for the Lawrence Berkeley National Laboratory. Three more were sold later, two to the National Security Agency and one to the Los Alamos National Laboratory, but they were all identical to the original systems, and there was no opportunity to upgrade them, and the lanthanum hexaboride cathode was never tried.

Digital Cypress was the first trillion-bit computer store ever built with reasonable access time. It was vast and expensive, but it gave access in 1 s to 1 trillion bits. No other technology at the time could offer this performance. However, there was no continuing demand for the systems developed, and the last one was closed down in 1980.

This was disappointing but, in a way, fortunate for me because it left me free to pursue what I found most exciting about lanthanum hexaboride cathodes: their higher brightness. This made it possible to build an SEM/electron beam writer with better resolution than had been possible before. It was like the transition of my interests during my PhD research from trying to understand the physical processes involved in sputtering to pursuing ways to build a better scanning electron microscope and make smaller electronic devices.

I return to lanthanum hexaboride cathodes in Chapter 7 to discuss the directly heated LaB_6 cathodes that were invented by Vogel (1970) that were simpler and easier to use than the rod cathodes.

References

Bas, E.B., 1962a. Schweizer Archiv fur Angewandte Wissenschaft und Technik 28., p. 112.

Bas E.B. Transactions of the 8th Vacum Symposium and 2nd International Congress, 2. - [s.l.]: Pergamon Press, New York, 1962b. p. 112.

Bas E.B. et al. Proceedings of the First International Conference on Electron and Ion Beam Science and Technology. Ed. Bakish.-1964.

Bloomer, R.N., 1957. The lives of electron microscope filaments. Brit. J. Appl. Phys. 8, 83.

Broers, A.N., 1967. Electron gun using long-life lanthanum hexaboride cathode. J. Appl. Phys. 38, 1991–1992.
Favreau, L.J., 1965. Cataphoretic coating lanthanum boride on rhenium filaments. Rev. Sci. Instrum. 36, 856.
Fiske, M.D., 1942. The temperature scale, thermionics, and thermatomics of tantalum. Phys. Rev. 61, 513.
Haine, M.E., Einstein, P.A., 1952. Characteristics of the hot cathode electron microscope gun. Brit. J. Appl. Phys. 3, 40.
Haine, M.E., et al., 1958. Resistance bias characteristics of the electron microscope gun. Brit. J. Appl. Phys. 9, 482.
Lafferty, J.M., 1951. Boride cathodes. J. Appl. Phys. 22, 299.
Langmuir, D.B., 1937. Theoretical limitations of cathode-ray tubes. Proc. Inst. Radio Engrs. 25, 977–991.
Langmuir, D.B., Malter, L., 1939. The rate of evaporation of tantalum. Phys. Rev. 55, 748.
Reimann, A.L., 1938. The evaporation of atoms, ions, and electrons from tungsten. Phil. Mag. 25, 834.
Thornley, R.F.M., Sun, T., 1965. Electron beam exposure of photoresists. J. Electronchem. Soc. 112, 1151.
Thornley, R.F.M., Brown, A.V., Speth, A.J., 1964. Electron beam recording of digital information. IEE Trans. Electron. Comput. EC-13 (1), 36–40.
Vogel, S.F., 1970. Pyrolytic graphite in the design of a compact inert heater of a lanthanum hexaboride cathode. Rev. Sci. Instrum. 41, 585.

Further reading

Broers, A.N., 1966. Long-life Tungsten Cathodes [Report]. IBM Research Report RC1543.

CHAPTER SIX

LaB_6 cathode high-resolution electron probe and its application to microfabrication and surface microscopy

Contents

Our electron beam group at the IBM Thomas J Watson Research Center was in what was called Applied Research, inferring that we were working on problems related to IBM's products, but I soon realized that we could work on virtually anything provided it was original, even if it was advancing science rather than technology. Nonetheless, I thought it was too much to ask the company for the money to build an entire high-resolution electron probe system just because there was the possibility that it would have higher resolution than had previously been obtained. So, I quietly

Advances in Imaging and Electron Physics, Volume 231
ISSN 1076-5670
https://doi.org/10.1016/B978-0-443-31462-9.00006-X

designed the components for the new electron column, had them built in the machine shop, which had superb craftsmen, and stored them in a cupboard in my office. Workshop time was provided without the need for cost estimates and approval for expenditures in advance. These components included the LaB_6 electron gun, the final lens, which required the most precision machining, the lens spacers for the column and the sample chamber and sample stage, and finally, the secondary electron detector. I also had designs ready for the manifolds for the ion-pumped vacuum system and a heavy aluminum table for the microscope column. This was to be suspended on four steel springs to isolate the system from mechanical vibration.

I decided to use commercially available lenses for the first two de-magnifying lenses, and most importantly, obtain the scan and display electronics, and the beam deflection coil and stigmator unit from the Cambridge Scientific Instrument Company, all at a reasonable cost. I also found a suitable flying spot scanner that could be used to control the electron beam and generate the patterns needed when the probe was being used for microfabrication. Within 6 months I was ready to ask for the money to buy all of this and the high stability lens current and high voltage power supplies. Management granted my request, reaffirming what many of us had been saying about the IBM Thomas J Watson Research Laboratory; it was the best playhouse in the world for research engineers and scientists exploring the limits of technology!

Theoretical electron optical performance

Because this was to be a completely new system there were no physical constraints on the final lens as there had been with the SEM in Cambridge. This meant that I could use a shorter minimum working distance than in the Cambridge SEM where the focal point had to coincide with the axis of the ion beam. This reduced the aberrations of the lens and with the high brightness of the LaB_6 cathode Smith's formulae (Smith, 1956) predicted a minimum beam diameter of 27 Å, about half of that previously obtained for a surface SEM (Pease, 1963). Pease had used a tungsten hairpin cathode. The parameters were as follows: $C_S = 1.8\,cm$, $C_C = 1.1\,cm$, beam brightness $\beta = 1.07 \times 10^6\,A/cm^2/sr$ at 23 kV, beam current $= 10^{-12}\,A$ and $\alpha_{OPT} = 4.75 \times 10^{-3}\,rad$. The final lens had no detectable astigmatism and the energy spread δV was assumed to 0.25 eV.

LaB_6 electron gun

In the brightness measuring system used in the early long-life experiments, described in Chapter 5, the cathode was at ground potential and the anode at 12 kV. This made it easy to use water cooling to dissipate the heat delivered to the cathodes. However, in electron microscopes and electron probe microanalyzers, a negative high voltage is applied to the cathode so that the anode and the rest of the microscope are at ground potential. This meant that an electron gun had to be designed in which the abnormally high power required to heat the cathode was carried away from the high-voltage cathode region of the gun. This was accomplished with several copper rods that passed from the cathode region through the vacuum wall into an oil bath which formed an integral part of the gun. The high voltage cable containing the three wires to the heater coil and the rod terminated in this oil bath. The heat was transferred by convection currents in the oil to the walls of the oil container where it was readily dissipated into the air via aluminum fins attached to the can. The temperature developed at the base of the cathode rod was 200 °C for the maximum required input power of 80 W. The cathode assembly was designed to tolerate 500 °C, so the cooling system is more than adequate.

A cross-section of the new LaB_6 gun is shown in Fig. 6.1, and the gun itself is in Fig. 6.2.

LaB_6 material quality

At the time of my first experiments with LaB_6, only powdered material was available, and sintering had to be used to obtain solid material. The first supplier of powdered LaB_6 also had arc-melting equipment and was willing to try to produce an ingot of the material large enough to allow a 1 mm square rod to be cut from it. This was how I obtained the first LaB_6 rod cathode, but the supplier said that this was a one-off sample, and he could not repeat it. So, I had to live with sintered material. This was satisfactory for relatively low brightness, but it was difficult or impossible to grind or electro-polish the sintered material to a sharp enough point to produce the highest brightness. Unfortunately, the arc-melted rod that had a relatively sharp point broke when I removed it for examination.

The emitting properties of a sintered LaB_6 rod were subsequently measured, and the emitting surface was examined in the SEM, in a series of in situ experiments made in collaboration with Haroon Ahmed (Ahmed and Broers, 1972). Ahmed had previously examined dispenser and oxide cathodes in the SEM. Current-voltage characteristics were measured for the rod cathode at five different temperatures and a Richardson plot was made from the saturated emission density and temperature. These plots indicated a work function of 2.4 eV and Richardson constant of 40 A/cm^2 (K)^2. These are to be compared with the values published by Lafferty of 2.66 eV and 29 A/cm^2 (K)^2 (Lafferty, 1951). A current density

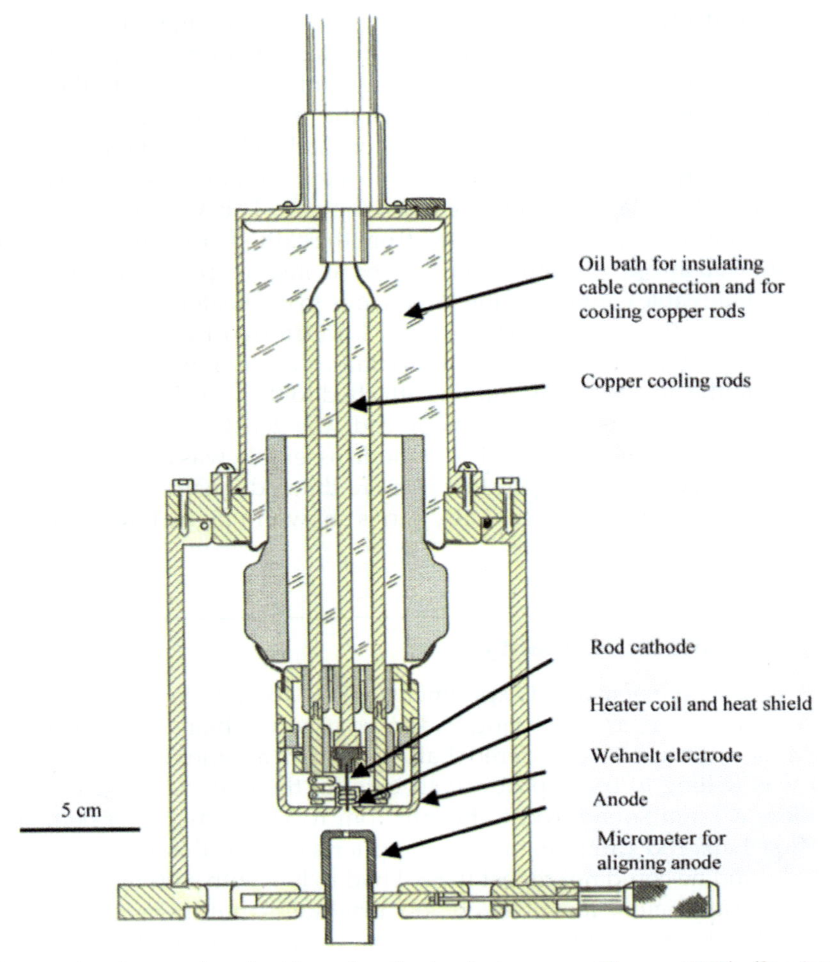

Fig. 6.1 Lanthanum hexaboride rod cathode electron gun (Broers, 1967). *(Reprinted from Broers (1967) with the permission of AIP Publishing).*

of $100\,A/cm^2$ was obtained from the rod at a temperature of 1680 °C, and lifetime tests showed a lifetime of greater than 200 h at $50\,A/cm^2$. Fig. 6.3 shows the configuration of the LaB6 cathode used in these experiments. Fig. 6.4 shows the emission characteristics at various temperatures, and Fig. 6.5 shows the Richardson Plot.

By 1980 LaB_6 cathodes were used in most high-performance electron probe systems and single crystals of LaB_6 of known orientation were available (Hohn et al., 1982). Several means had been found to directly heat these LaB_6 crystals by clamping them between graphite blocks (Schmidt et al., 1978; Vogel, 1970). These simpler configurations were easier to implement than the indirectly heated rod cathodes.

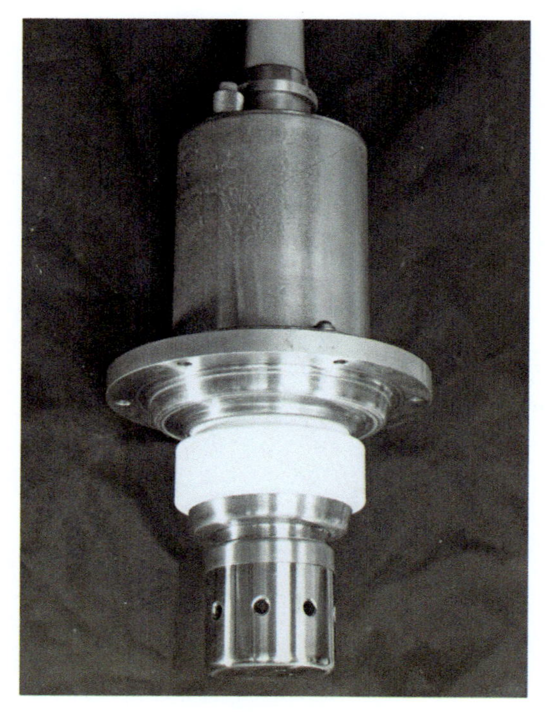

Fig. 6.2 Lanthanum hexaboride rod cathode electron gun.

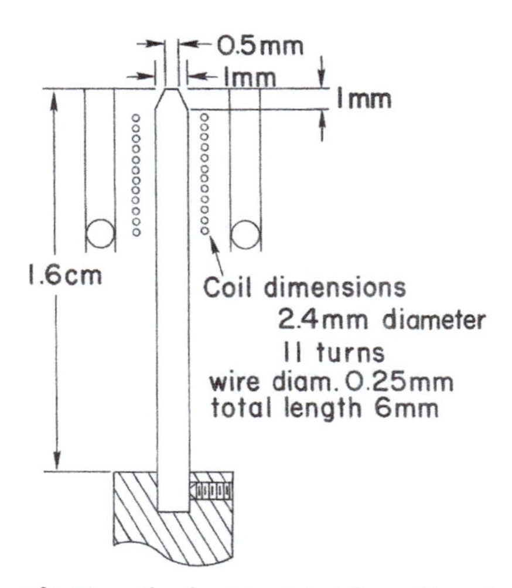

Fig. 6.3 Configuration of LaB_6 cathode. *(Reprinted from Ahmed and Broers (1972) with the permission of AIP Publishing).*

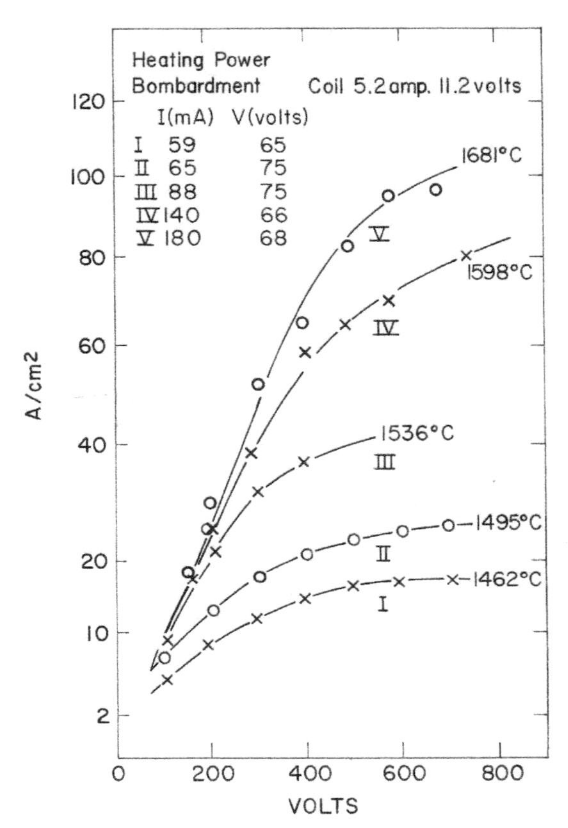

Fig. 6.4 Current vs voltage measurements. *(Reprinted from Ahmed and Broers (1972) with the permission of AIP Publishing).*

The brightness with LaB_6 cathodes, as with all thermal cathodes, depends on the sharpness of the emitting tip, because it determines the electric field at the emitting surface and the degree to which the beam diverges under the influence of space charge. In the limit, tips can be made so sharp that Schottky-enhanced emission occurs, but they are rapidly damaged by sputtering at pressures normally found in demountable systems. Tips sharp enough for enhancement by Schottky effect, however, can survive up to pressures of about 10^{-7} mmHg. The performance of thermal and field emission cathodes in probe systems is discussed in Chapter 7.

Electron beam column

Fig. 6.6 is a diagrammatic view and Fig. 6.7 is a photo of the new electron LaB_6 electron beam system. All the components of the column, including the electromagnetic scan coils, are pre-aligned to an accuracy of 0.0075 mm. The first two commercial lenses were re-machined to meet this accuracy.

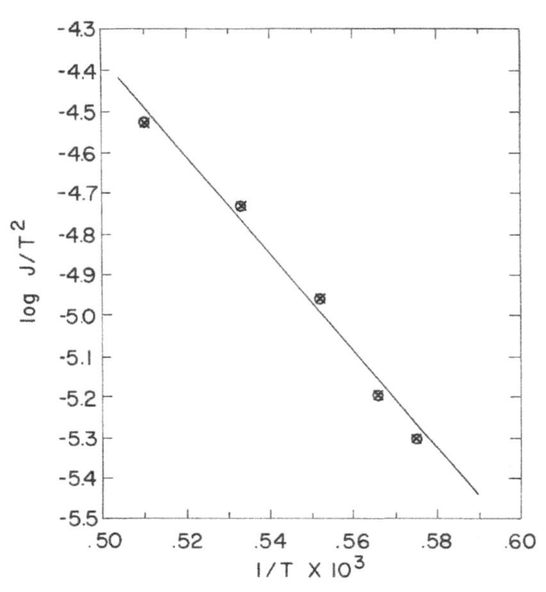

Fig. 6.5 Richardson-Dushman plot. *(Reprinted from Ahmed and Broers (1972) with the permission of AIP Publishing).*

They have pole-piece diameters of 2.8 mm and pole-piece gaps of 3 mm. The final lens has a first bore diameter of 5 cm, a final bore diameter of 1 cm, and a gap of 5 mm. The bores of the final lens were honed and lapped until they were round to better than 0.13 μm and aligned to better than 2.5 μm. As discussed in Chapter 3 in describing the design of the final lens for the Cambridge SEM, these tolerances were sufficient to ensure that the lens was anastigmatic as determined from Archard's (Archard, 1953) interpretation of Sturrock's (Sturrock, 1951) calculations.

The final lens protruded 12 mm into the specimen chamber, allowing the electron detector to be positioned so that its field "sees" the surface of the specimen, even for large flat specimens.

The lens spacers contained multiple Mu-metal tubes, ensuring that beam deflections due to external magnetic fields were negligible.

The first spacer included a set of deflection plates for blanking the beam. The only movable parts were the gun anode and the final lens aperture, both of which were controlled with micrometers from outside the vacuum.

The spacings in the column were as follows: gun anode to first lens gap 23 cm, first lens gap to second lens gap 20 cm, and second lens gap to final lens gap 33 cm. The minimum focal length of the first two lenses was 3 mm allowing a total de-magnification of 105,000 for a final working distance of 8 mm. The gun cross-over diameter was estimated to be about 20 μm, so this demagnification was more than sufficient for the smallest beam diameter.

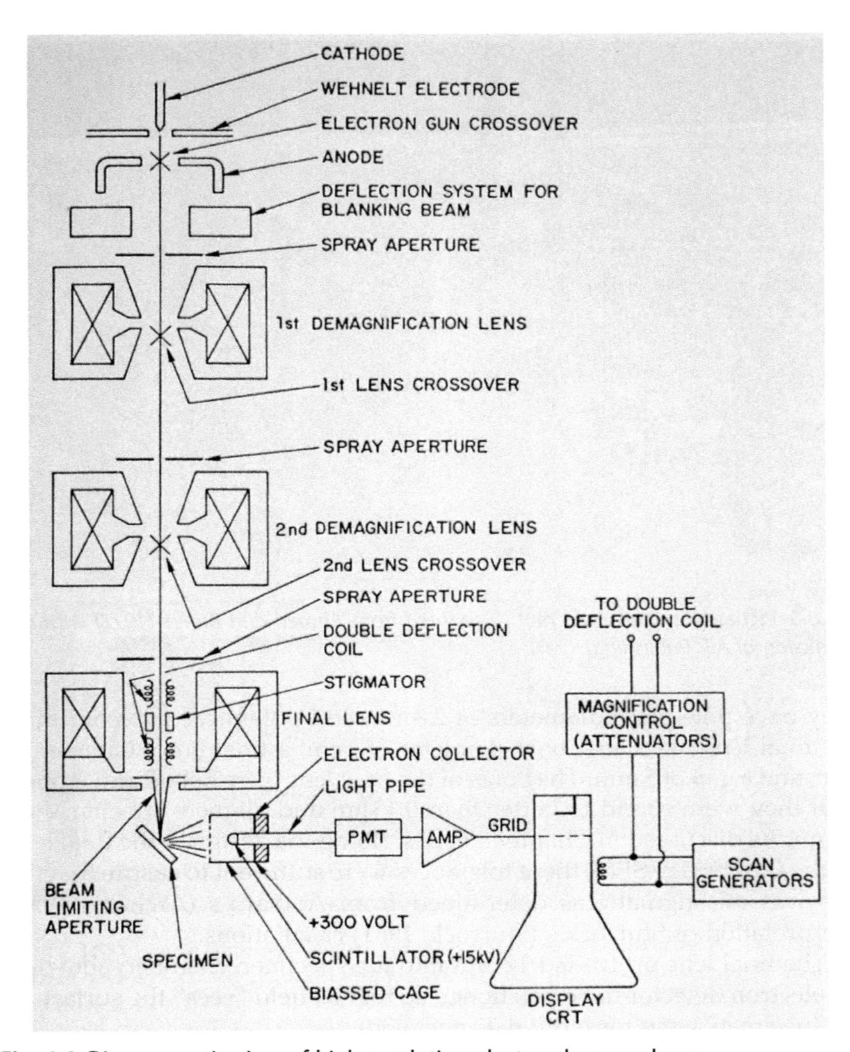

Fig. 6.6 Diagrammatic view of high resolution electron beam column.

Vacuum system

To reduce the effects of contamination as much as possible without the inconvenience of extensive cryogenic pumping or of a bakeable system and metal gaskets, a 270 L/s ion pump was used, and a combination of mechanical and absorption pumps for roughing. The mechanical pump

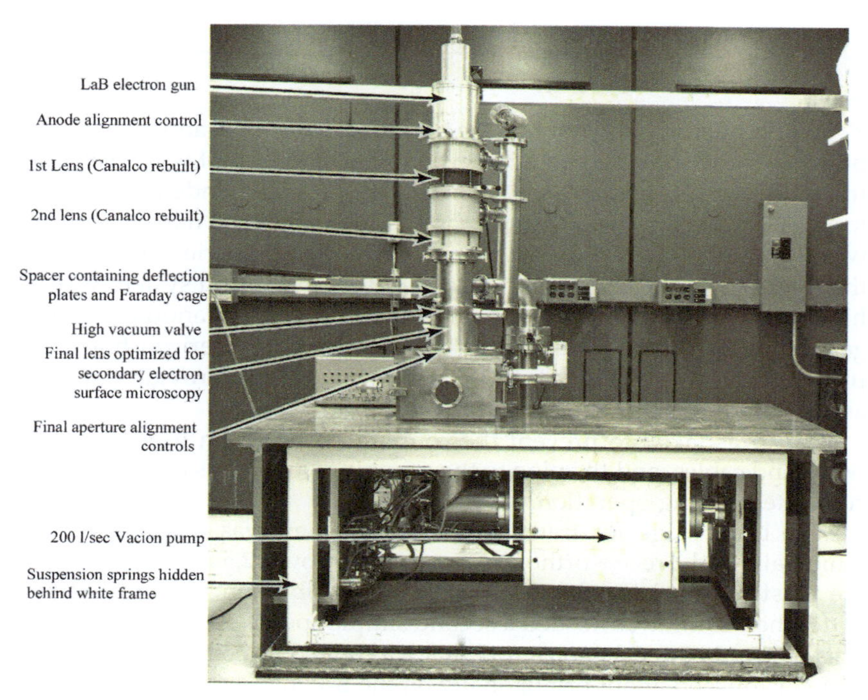

Fig. 6.7 High resolution lanthanum hexaboride cathode electron probe system. *(Reprinted from Broers (1969) with the permission of AIP Publishing).*

was used to reduce the pressure to about 10^{-2} mmHg, which took less than 1 min. It was then turned off, and the system opened to the absorption pump that reduced the pressure to $2 \times 10–4$ mmHg, at which point the system was opened to the ion pump. This minimized the exposure of the system to the oil in the mechanical pump and allowed the sorption pump to be used many times between bake-outs. Under normal conditions, the pressures in the gun and specimen chamber were 10^{-6} mmHg and 5×10^{-7} mmHg. Overall pump downtime was 8–10 min. Valves above the final lens and between the ion pump and the specimen chamber allowed the chamber to be accessed without turning off the electron gun. This was important with the first version of the LaB_6 gun, which took some time to stabilize.

Despite the relatively high specimen chamber pressure, no specimen contamination was observed unless the specimen itself was coated with something that could be polymerized by the electron beam, such as silicon wafers with remnants of a resist layer, or organic samples where the sample itself was the source of the contamination.

Vibration isolation

The microscope column, the specimen chamber, and the vacuum pump were all mounted on a heavy aluminum table that was suspended from a steel frame by four steel springs. The springs extended 20 cm under load, providing a resonant frequency of 1.2 Hz. The total mass suspended was about 0.9 metric tons. The springs are hidden from view in Fig. 6.7 by the steel frame, which is white. The Q of the suspension system was too high to leave the table un-damped without the risk of random impulses generating oscillation at the resonant frequency. Rubberized horse hair placed between the table and the frame was used to provide damping, and this reduced the attenuation from 25:1 to 20:1.

Care was taken to prevent the coupling of vibration to the table via wires and cables, and the pipe to the roughing pump was completely disconnected during operation.

The basic stage for translating the specimen was a ball-slide unit, which allowed precise orthogonal movement over 2.5×5 cm, but it was the most sensitive part of the microscope to vibration. When resolution better than 20 nm was needed a second simple specimen holder was used. This rested on three feet on a tray that was bolted directly to the final lens and was pushed about orthogonally on the tray by a square rod attached to the basic stage. The rod fitted into a square hole in the second holder that was slightly larger than the rod allowing the rod to be backed off after locating the specimen in the desired position. This eliminated all mechanical contact between the specimen and the bottom of the chamber. This stage, together with the spring suspension system, reduced interference due to vibration below a detectable level.

Electronics

The high-voltage power supply, made by Brandenburg Ltd., a British company, had a stability of better than 2 parts in 105 over 10 min. The circuitry for driving the LaB_6 gun was enclosed in a Plexiglass box filled with Freon. After warming up for about an hour, no focus drift was observed for periods of up to an hour.

Experimental performance

The measured electron optical performance was in close agreement with the theoretical predictions. The minimum beam diameter was measured to be 3.0 ± 0.7 nm from the edge sharpness of the scanning transmission images of portions of a silver calibration grid shown in Fig. 6.8. The accelerating potential was 23 kV, the beam current 10^{-12} A, the

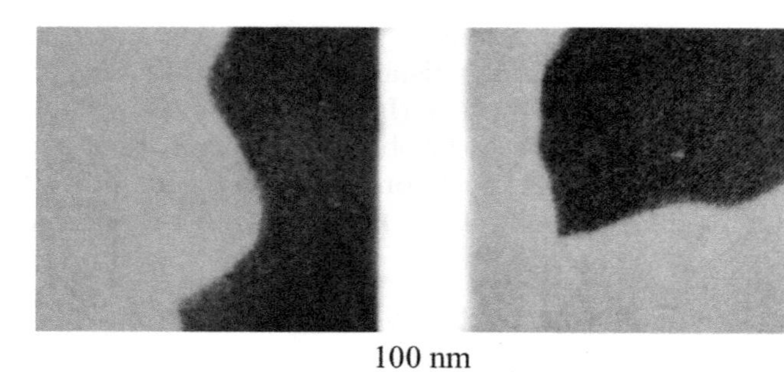

Fig. 6.8 Portion of a silver grid examined in transmission. Rise-time measurements across edges indicated a beam diameter of 3 nm. *(Reprinted from Broers (1969) with the permission of AIP Publishing).*

working distance 7 mm and the beam aperture 4×10^{-3} rad. No astigmatism correction was needed.

By operating in the transmission mode, the loss of resolution due to the range of the secondary electrons in the sample was avoided. Image contrast was increased with a stopping aperture placed after the specimen that prevented most of the electrons that penetrated through the edges of the grid from reaching the scintillator detector. The first two lenses were operated at a focal length of 10.8 mm for this measurement. The total demagnification produced by the three lenses was 8200, giving an estimated final beam diameter of 2.5 nm for a gun cross-over of 20 μm. The gun cross-over diameter was measured from the observed beam size with only the final lens operating.

When operating in the transmission mode, it is also possible to assess the electron optics of the SEM by observing the quality of the Fresnel fringes formed when examining a sharp edge in the out-of-focus condition. These fringes offer a means for accurately and critically testing instrument performance before it is used either for transmission or surface microscopy. The test is independent of the particular sample used and of sample preparation or contamination. It was used by Crewe et al. in their field emission cathode STEM but had never been used in the surface SEM, so I decided to try it in the new LaB_6 cathode SEM (Broers, 1972).

Observation and measurement of Fresnel fringes in TEMs is one of the best methods for evaluating resolution, objective astigmatism, and source coherence (Cowley, 1969; Heidenreich, 1974). The fringes form as a result of diffraction at a sharp specimen edge and are observed by focusing the objective slightly above or below the plane of the sample. The fringes are sharp and exhibit high contrast. The distance (x) from the sample edge to the nth fringe can be calculated from the wavelength of the electron (λ) and the amount the objective lens is defocused (Δf): thus,

$$x = [\lambda \Delta f\,(2n - 1)]^{1/2}$$

In practice, the over-focused fringe is generally used for measurement, and the fringe width is taken as the distance between the center of the dark line and the center of the bright line. (Heidenreich, 1964).

The number of fringes that can be observed depends on the coherence of the illumination, which is most conveniently expressed in terms of the half-angle of the illumination (α_C). For the nth bright fringe to be obtained,

$$\alpha_C < [\lambda/4\Delta f(2n - 1)]^{1/2}$$

In the SEM, the ray paths through the sample are reversed (Cowley, 1970). The final lens in the SEM corresponds to the objective lens in the TEM, and the appropriate degree of coherence is achieved with a small "contrast aperture" placed between the sample and the electron detector. In practice, α_C has to be much smaller than the optimum lens aperture for minimum beam size, α_L. As a result, only a small fraction, $(\alpha_C/\alpha_L)^2$, of the beam current at the specimen is available for the formation of Fresnel fringes, and they, therefore, provide a more stringent test of source brightness than other types of image formation. The unwanted current, which inevitably bombards the specimen under these conditions, and other conditions similar to them, such as those required for the imaging of lattice fringes, leads to greater specimen damage than encountered in a conventional TEM. This increased potential for specimen damage is a disadvantage of scanning as opposed to projection transmission microscopy.

Figs. 6.9 and 6.10 show fringes around holes in a Formvar film. The fringes were obtained with a contrast aperture, which subtended a half-angle of 5×10^{-4} rad at the sample. The final lens of the SEM was operated at a focal length of 8 mm, which is shorter than that needed for secondary electron surface microscopy because there was no need to collect secondary electrons when operating in transmission. The aperture in the final lens was set to yield an incident beam half-angle of 6×10^{-3} rad. The theoretical beam size for these experimental conditions (final lens focal length, 8 mm; final lens spherical aberration coefficient $C_S = 9$ mm, final

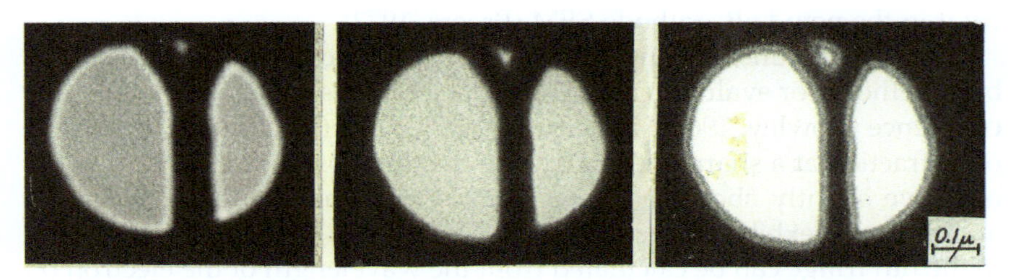

Fig. 6.9 Fresnel fringes around holes in a Formvar film. Central image is "in focus." *(Reprinted from Broers (1972) with the permission of AIP Publishing).*

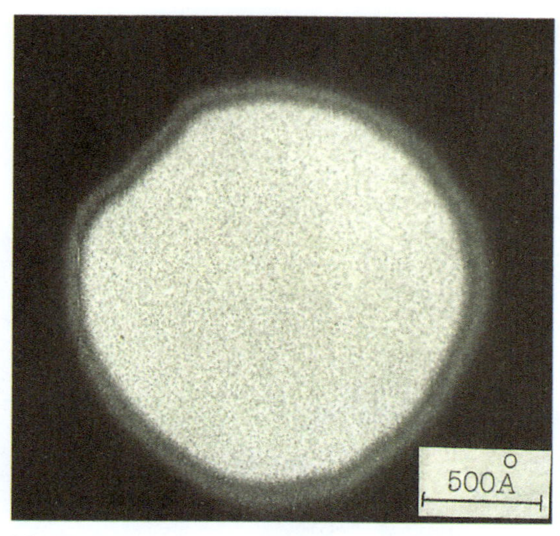

Fig. 6.10 Fresnel fringe width close to the theoretical beam size of 3.2nm. *(Reprinted from Broers (1972) with the permission of AIP Publishing).*

lens chromatic aberration coefficient $C_C = 6$mm; electron gun brightness $\beta = 1.1 \times 10^7$ A/cm^2/sr (lanthanum hexaboride cathode (Broers, 1968); beam energy spread $\Delta E = 2$eV; beam current $I = 4 \times 10^{-11}$ A) was 3.2nm. The actual beam diameter was estimated to be close to this value from the width (30Å) of the fringe shown in Fig. 6.8. This result also indicates that the actual brightness was close to the estimated value of 10^7 A/cm^2/sr (50kV).

The relationship between the "resolving power" of a microscope and the minimum measured fringe width is difficult to define because resolving power depends very much on the contrast obtained with each specimen. However, Haine (1961) has pointed out that, if a symmetrical fringe of width "y" can be observed, then it is also shown that all instrumental factors, including objective astigmatism, are adequate for a resolving power at least as good as "y."

Fig. 6.11 shows a catalase crystal stained with phosphotungstic acid. Catalase crystals have also been used to test the resolution of transmission electron microscopes. The clarity with which the half-lattice spacing of 88Å is imaged indicates a resolution of 20–30Å.

As expected, the point-to-point resolution on images of bulk samples using the secondary electron signal was not as good as that obtained in the transmission images, and having used the instrument in transmission mode to verify that it met theoretical expectations, I went on to explore the resolution limits when the instrument was used to obtain surface images of bulk samples.

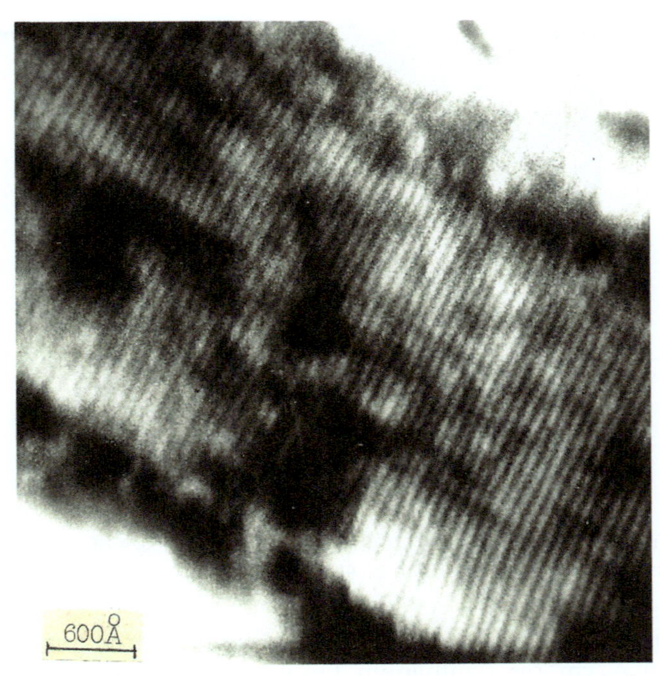

Fig. 6.11 Scanning transmission image of catalase crystal showing the 88 Å half-spacing of the crystal (Broers, 1974).

Measurement of the ultimate resolution of the surface scanning electron microscopy using the Everhart/Thornley secondary electron detector

All of the micrographs in the remainder of this chapter are surface micrographs obtained on bulk samples with the Everhart-Thornley secondary electron detector (Everhart and Thornley, 1960) located in the sample chamber at the bottom of the electron beam column. The Everhart-Thornley detector was the key advance made by Oatley's group, allowing the SEM to use the secondary electron signal to produce noise-free images at resolutions well beyond that of the optical microscope.

Secondary electrons are detected by accelerating them onto a scintillator, where each electron generates several photons. This photon signal is then amplified noise-free by a photomultiplier, the output of which is amplified to produce the video signal for the microscope.

The Secondary electrons are attracted to the detector by a relatively low voltage of about 300 V on a metal mesh that covers the entrance to a metal cage that contains a light pipe that is coated with a liquid scintillator. After

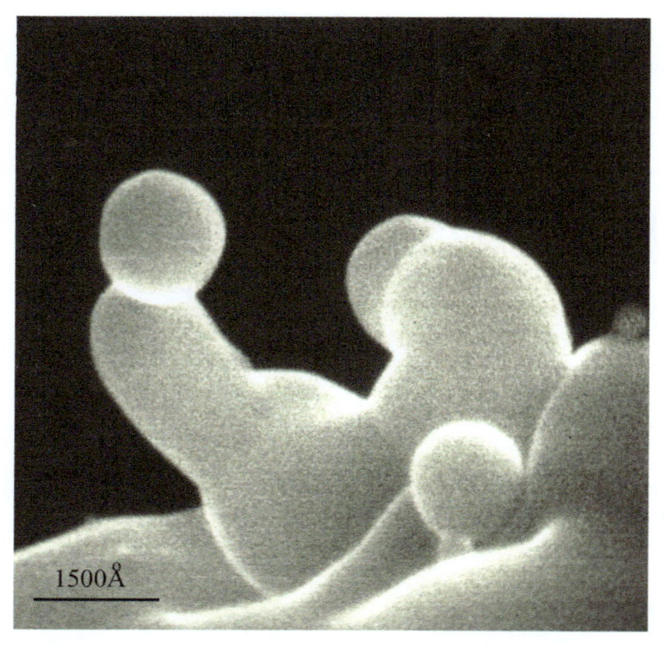

Fig. 6.12 SEM surface image of a tungsten particle on a lanthanum hexaboride cathode (Broers, 1970a).

passing through the metal mesh the electrons are accelerated onto the metal-coated scintillator with a voltage of about 15 kV. The mesh-covered cage prevents the high field from distorting the fine beam of the microscope.

Fig. 6.12 shows tungsten particles on the surface of a lanthanum hexaboride cathode. Rise-time measurement across the edge of the particles shown in Fig. 6.13 indicated a beam diameter below 3 nm. Fig. 6.14 shows that it was possible to resolve two converging edges until they were 2.5 nm apart.

After puzzling over this problem for some time I found myself talking to Leo Esaki about his research on semiconductor "superlattices." He was building layered semiconductor samples to see if they would reproduce the tunneling phenomena that he had predicted (Esaki and Tsu, 1970). These alternating layers of Ga As and GaAsP were grown using a vapor growth technique developed by Blakeslee (1971) and could have spacings between 10 and 100 nm. This was just the range of dimensions of interest for scanning electron microscopy, so we went ahead and made a four-band sample specifically for evaluating the resolution of the surface SEM. The sample was prepared by cleaving and then etching in a solution that attacks the GaAs more than the GaAsP, yielding surface ridges, see Fig. 6.15. The bands had nominal lattice spacings of 120, 60, 30, and

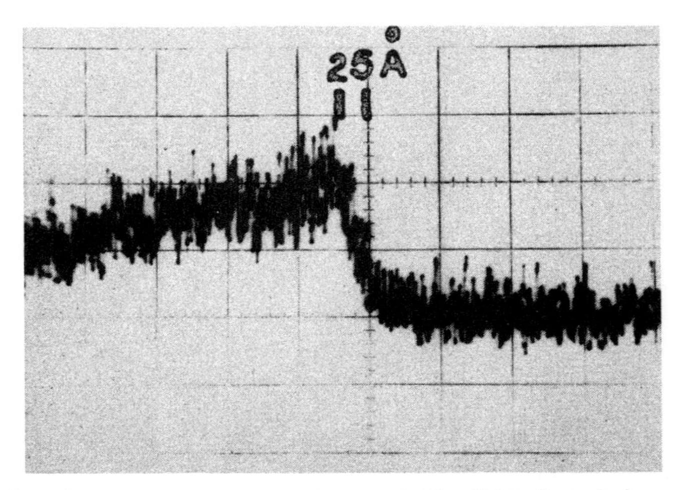

Fig. 6.13 Rise-time measurement across an edge in Fig. 6.12 indicated a beam diameter below 30 Å (Broers, 1970b).

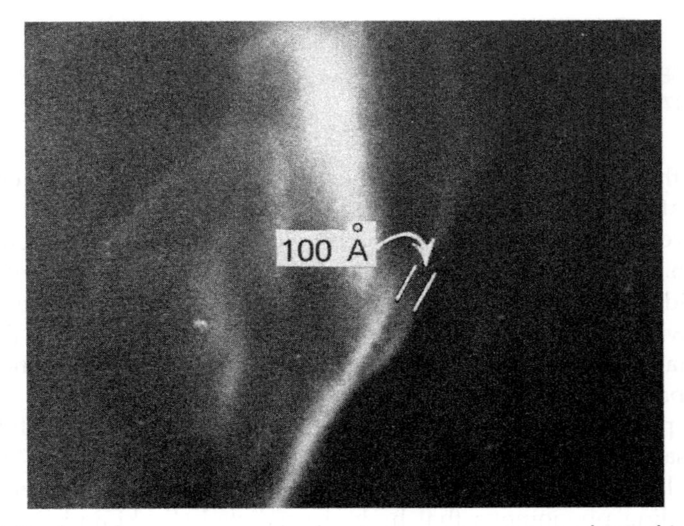

Fig. 6.14 Portion of a copper sample showing two converging edges which can be resolved until they are 25 Å apart (Broers, 1970b).

15 nm. The tolerance on the band dimensions is ±25% but the exact dimensions could be measured exactly by careful magnification calibration. Fig. 6.16 shows the 15 nm band at higher magnification. The 120 nm spacings were clearly visible.

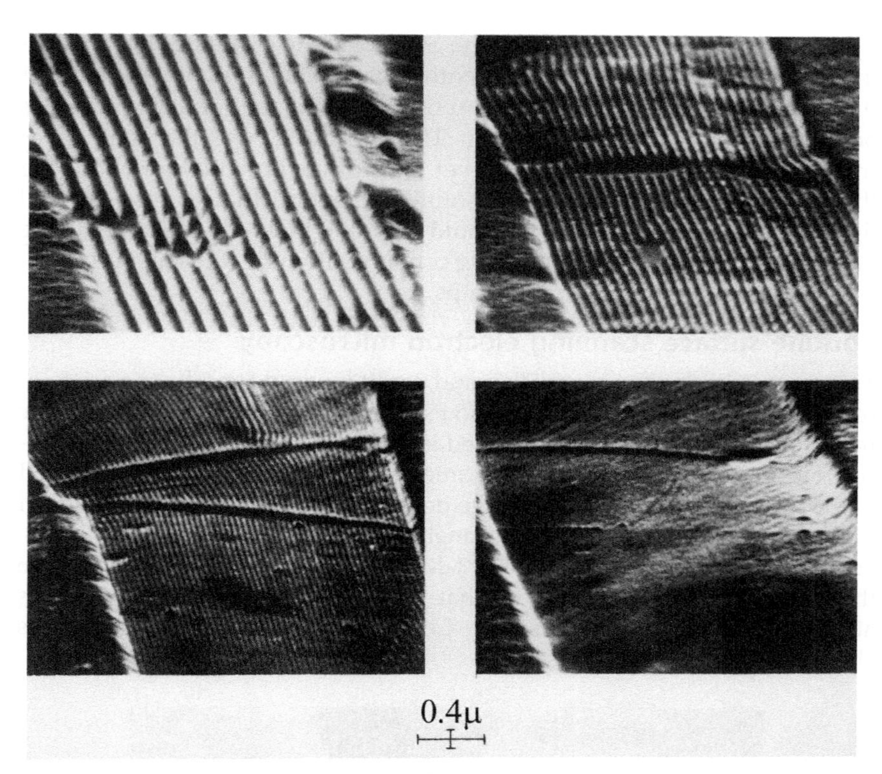

Fig. 6.15 GaAS–GaAsP superlattice resolution SEM test sample. Band spacings are 120, 60, 30 and 15 nm.

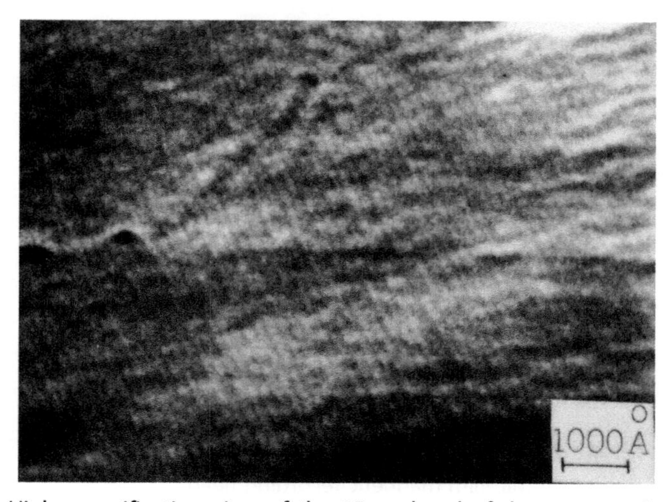

Fig. 6.16 High magnification view of the 15 nm band of the test sample shown in Fig. 6.13 (Broers, 1974).

We offered these samples to the SEM users to evaluate their instruments but many of them were not enthusiastic. They said that the smaller dimensions proved difficult or impossible to image, and the samples became contaminated very quickly. This, of course, was the point I was trying to make, that the SEM was yet to fulfill its potential for exploring the surface of bulk samples at resolutions beyond that of conventional light microscopes and that they could improve their resolution by using brighter electron guns and reducing contamination by replacing oil diffusion pumps with ion or turbo pumps.

Routine surface scanning electron microscopy

This microscope was now reliable and could be used for a large variety of routine tasks. One of the first was to examine the latest thin film magnetic recording heads that were being used in IBM's most advanced disk memories. Fig. 6.17 shows one of the experimental magnetic recording heads used in the IBM 3370 direct access storage device introduced in 1979 for the IBM 4331 and 4341 and System/38 midrange computers. It had 7 fixed 14″ disks, and each unit had a capacity of 571 MBytes. It was the first HDD to use thin-film technology; research on that technology started in the late 1960s at the Thomas J Watson Research Center. The gap in these heads was about 1 μm.

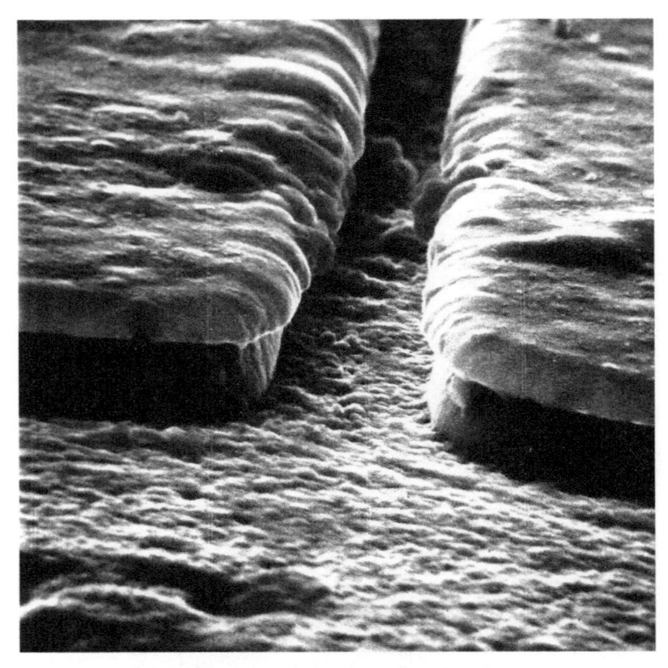

Fig. 6.17 Experimental thin-film magnetic recording head.

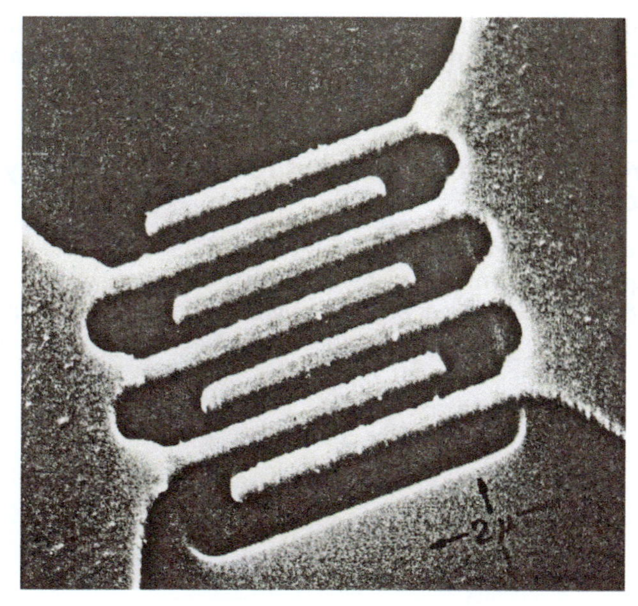

Fig. 6.18 0.5 μm finrt width transistor structure fabricated in 1969 to demonstrate the resolution of electron beam lithography.

Fig. 6.19 Experimental 1 μm integrated circuit.

Fig. 6.18 shows an experimental transistor structure that was made to demonstrate that electron beam lithography was capable of making devices with dimensions below 1 μm. These electron beam lithography methods were used to make the first operating 1 μm circuits at the IBM Thomas J Watson research laboratory in the late 1960s. Fig. 6.19 shows one of these circuits. At the time, commercial integrated circuit transistors were being fabricated using optical proximity printing. They had dimensions of 3–5 μm.

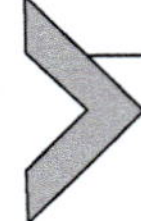

Microfabrication probe used for making prototype silicon circuits

A two-lens electron beam column was used to make the first micron and subsequently sub-micron transistors and circuits such as those shown in Figs. 6.18 and 6.19 (Fang et al., 1973; Magdo et al., 1971; Thornley et al., 1970). This system, which is shown in Fig. 6.20, used the same LaB_6

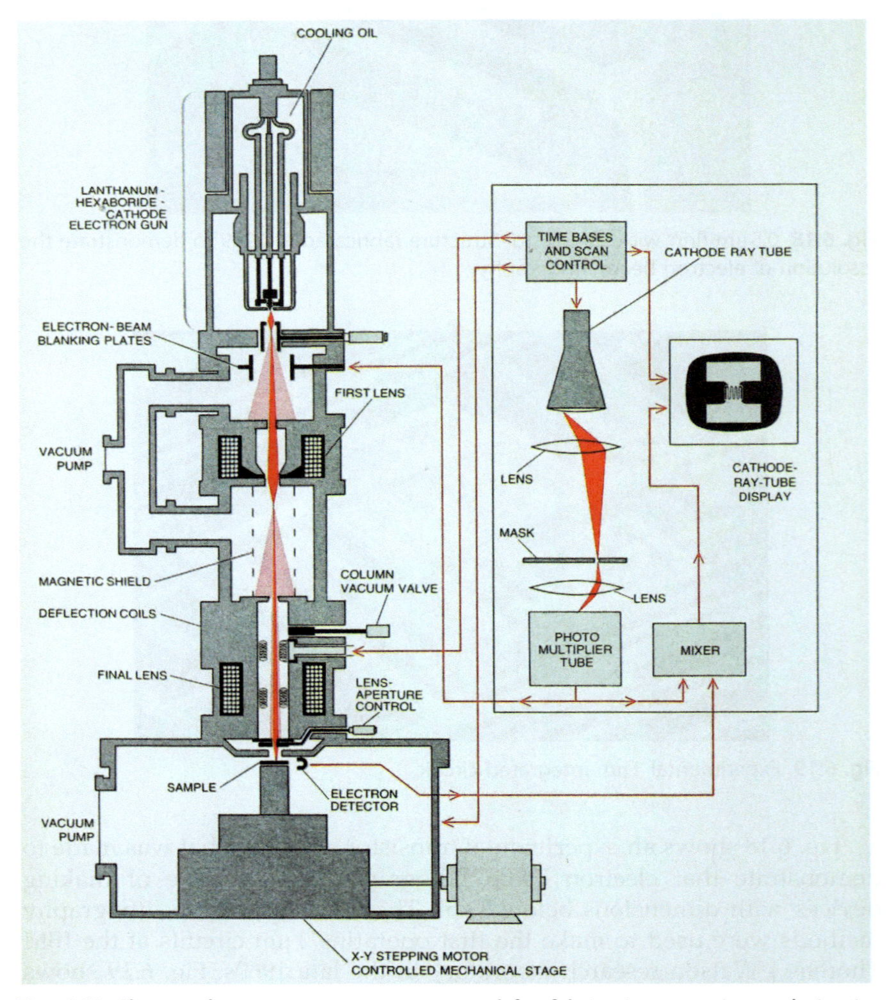

Fig. 6.20 Electron beam exposure system used for fabricating experimental circuits used in studying the scaling of FETs. *(Reproduced from Broers and Hatzakis (1972) with permission. Copyright © (1972) SCIENTIFIC AMERICAN Inc. All rights reserved).*

Fig. 6.21 Photograph of two lens electron beam column used in the fabrication system shown in Fig. 6.20.

electron gun and a similar final lens as the high-resolution SEM but only had two lenses as there was no need to reach beam sizes of about 20 nm (Fig. 6.21). The minimum device dimensions of the devices were about 0.25 μm. IBM was far-sighted in supporting this research as its solid-state devices at the time had dimensions of 5–10 μm. Michael Hatzakis, working with Richard Thornley, had already developed many of the processes needed to make sub-micron devices and worked together with Bob Dennard, the inventor of the Dynamic RAM, and his colleagues, who led the world in defining the way FETs could be scaled down. Figs. 6.19, 6.22, and 6.23 show an experimental memory cell (Dennard and Broers, 1973).

This system also used a flying spot scanner to generate patterns. In 1976, most of the work on semiconductor devices was transferred to the Vectorscan System built by Philip Chang and Alan Wilson (Chang et al., 1977). This had higher throughput and provided automatic alignment of the patterns with the wafer. Philip Chang had already built a high-performance electron beam fabrication system at the Cambridge Instrument Company in the United Kingdom

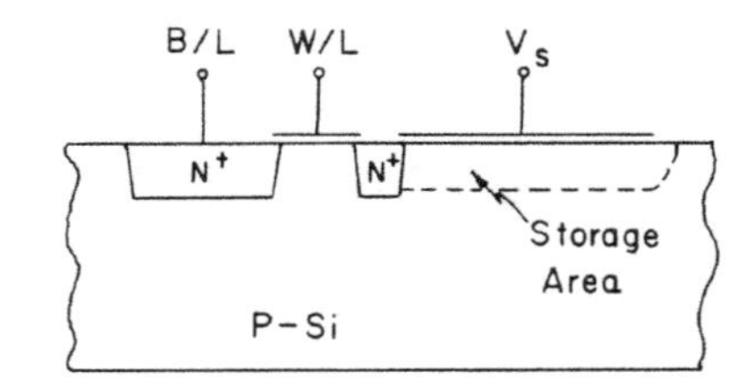

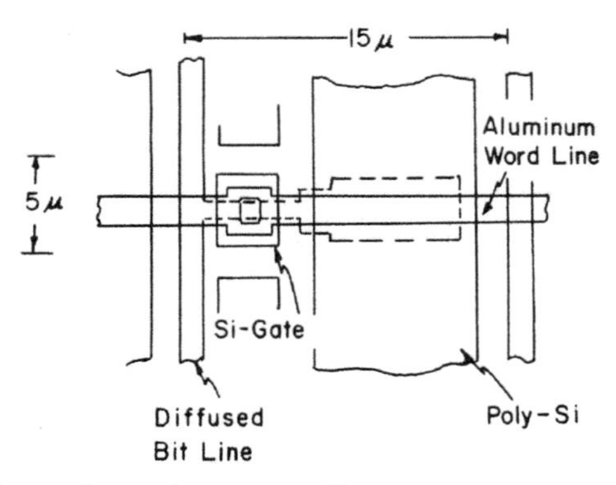

Fig. 6.22 Dennard experimental memory cell.

Microfabrication in the new SEM

While the two-lens column system was being used to fabricate semiconductor devices, I was using the three-lens system to work on a variety of other electronic devices and test structures that needed smaller dimensions. Fig. 6.24 shows a matrix of 0.14 μm wires. This was the first test structure fabricated with the SEM in 1968.

Acoustic surface wave transducers

The first devices fabricated with the high-resolution electron probe system were surface acoustic wave transducers. These interdigital transducers efficiently launched and detected acoustic waves on the surface of piezo-electric crystals and were used in signal processing, especially in radar systems. They were fabricated at the time with standard photoresist techniques, which limited the line widths to about 1 μm, which corresponded to an operating frequency of about 1 GHz. Working with Eric Lean and using the lift-off process with PMMA resist developed by Mike Hatzakis, we were able in 1969 to fabricate a transducer capable

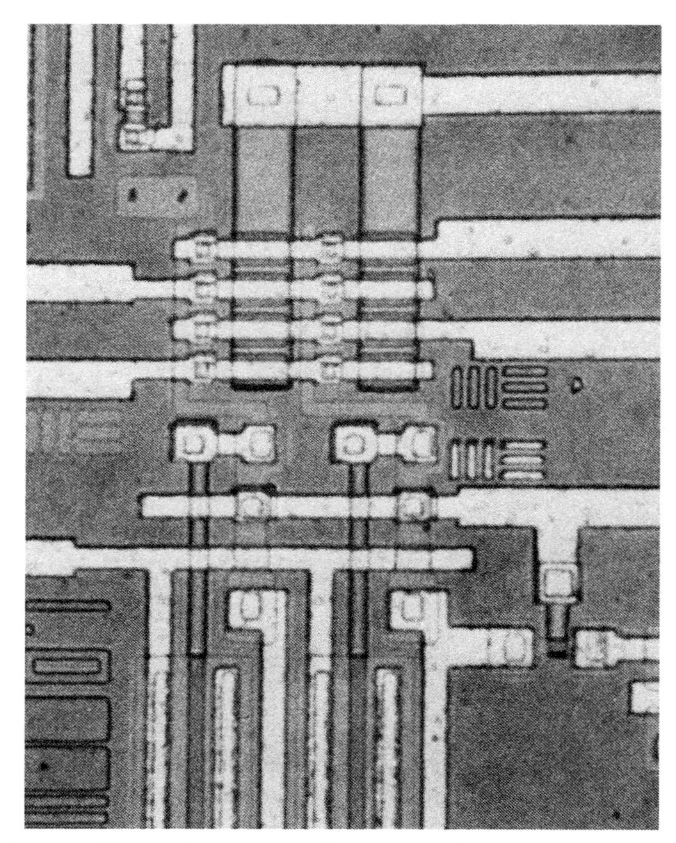

Fig. 6.23 Image of Dennard experimental memory cell.

of generating surface waves at a frequency of 1.75 GHz on a $LiNbO_3$ substrate (Broers et al., 1969), see Figs. 6.25 and 6.26. These transducers had 25 pairs of fingers which were 250 μm long, 0.3 μm wide, and 70 nm thick.

From the insertion loss measurements between two transducers, it was possible to estimate the attenuation of the surface acoustic waves on the y-cut z-oriented $LiNbO_3$ surface. It was measured to be 3 db/μs at 1.75 GHz. The conversion efficiency of the transducer, which is defined as the ratio of total acoustic input and output radio frequency energies, was measured to be −6 db.

In 1970, we went on to make and test 3.5 GHz transducers with a 0.15 μm finger width, as shown in Fig. 6.27 (Lean and Broers, 1970). They had comparable performance. We believed that the fingers of these surface transducers were the smallest features ever used in a working electronic device. The smallest experimental transistors had minimum

Fig. 6.24 Matrix of 0.14 μm gold wires frabricated with the lanthanum hexaboride cathode SEM.

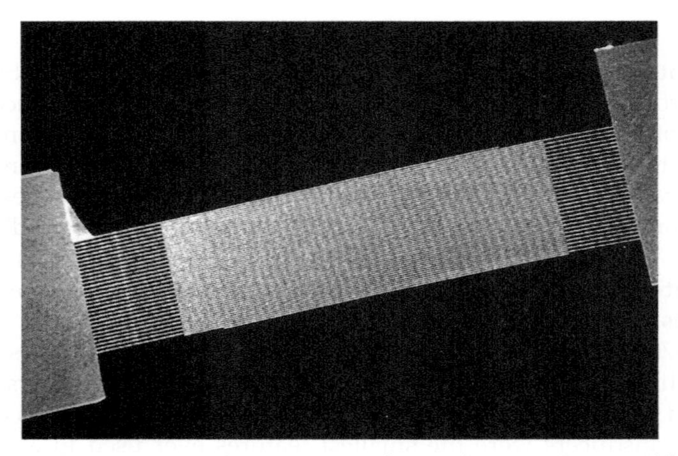

Fig. 6.25 1.75 GHz acoustic surface wave transducer. Aluminum finger width 0.3 μm.

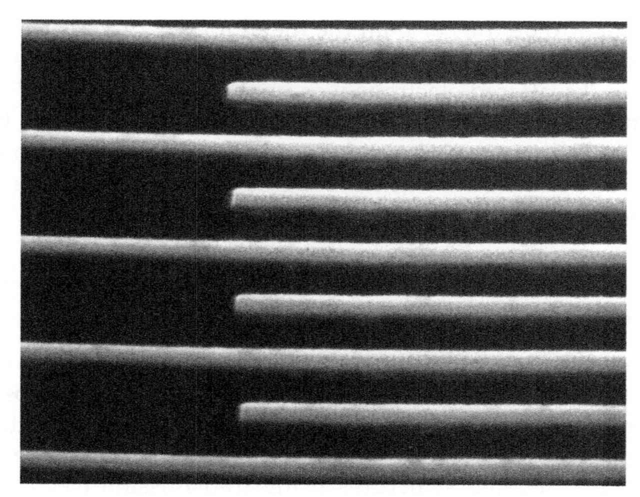

Fig. 6.26 High magnification view of the transducer shown in Fig. 6.25.

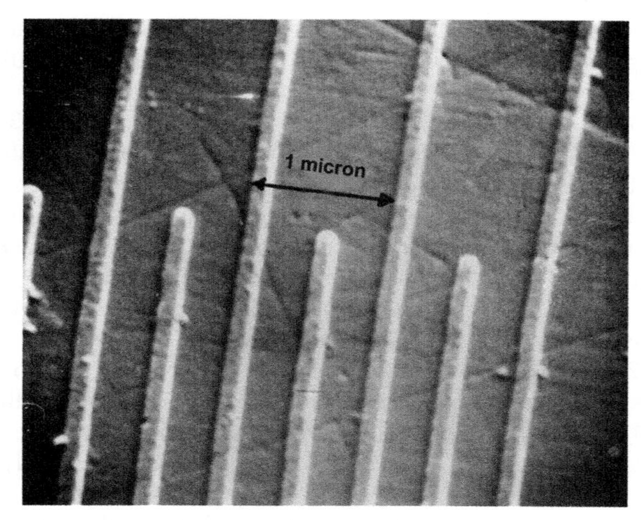

Fig. 6.27 3.5 GHz SAW transducer with finger width of 0.15 μm (Lean and Broers, 1970).

linewidths of 1 μm, and commercial integrated circuits had minimum dimensions of about 2 μm.

From the beginning, it was clear that electron scattering was important in electron beam lithography. By 1970, electron scattering in the imaging layer and the loss of contrast created by the electrons back-scattered from the substrate were limiting linewidth, especially for dense patterns such as those needed for surface wave transducers. With the lift-off process, the resist layer had to be thicker than the metal for the process to work, and with devices that had to be built into or on a bulk substrate, the exposure due to the backscattered electrons was unavoidable.

It was not possible to avoid these limitations for devices on bulk substrates, although by using subtractive etching rather than lift-off and thinner, higher-contrast resists that were highly resistant to etching, we were able to produce linewidths below 0.1 μm.

Thin window substrate

However, in 1972, in collaboration with Tom Sedgwick, a colleague in the IBM Research Center, we developed a thin membrane substrate that largely eliminated the loss of contrast produced by the back-scattered electrons (Broers and Sedgwick, 1976; Sedgwick et al., 1972). It also allowed us to inspect the fabricated structures using transmission electron microscopy and circumvent the resolution limitations of secondary electron surface scanning microscopy. Electrical contact could be made to nanometer-scale devices such as SQUIDs and nano-bridges on the thin membranes via metal pads that extended from the membrane onto the bulk silicon where wires could be spot-welded to them.

The window substrates were fabricated by etching through 0.13–0.25 mm thick silicon wafers. The polished (top) side of the wafer was coated with a 30–159 nm thick insulator film of vapor-deposited Si_3N_4 or sputtered or thermal SiO_2. The lapped (bottom) side was coated with 150 nm Si_3N_4 for use as a masking film. A pattern of square openings was etched in this masking film, with the sides of the squares aligned with the (110) directions in the silicon wafer. The silicon under the windows was then etched away with aqueous 33.5 wt% (w/o) NaOH. at 90 °C, or with ethylene diamine 31.5 w/o pyrotechtol 3.7 w/o, water 61.2 w/o at 110 °C. The NaOH did not etch the Si_3N_4 films noticeably during the removal of the Si. However, the NaOH etched SiO_2 too fast for it to be useful as the masking film. The ethylene-diamine-pyrotechtol solution etched both the SiO_2 and the Si_3N_4 at about 10 nm/h.

The Si etchants were highly preferential and exposed <110> planes as walls of the holes. The size of the openings in the masking films was calculated to give a window size at the bottom of the hole that varied in size from about 0.1 mm to 1 mm square. An opening 0.35 mm on a side would etch down through a 0.13 mm thick wafer to produce a window about 0.1 mm on the side.

Fig. 6.28 is a three-dimensional view of a thin-window substrate showing how electrical contact can be made to nano-structures by bonding wires to the pads located over the bulk substrate. Fig. 6.29 shows how back-scattering of electrons from the bulk substrate was eliminated.

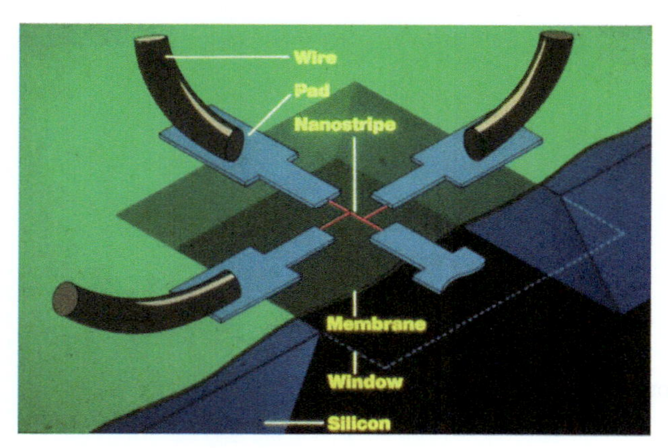

Fig. 6.28 Diagrammatic view of thin-window substrate Silicon nitride membrane is 150 nm thick. US Patent 3,971,860.

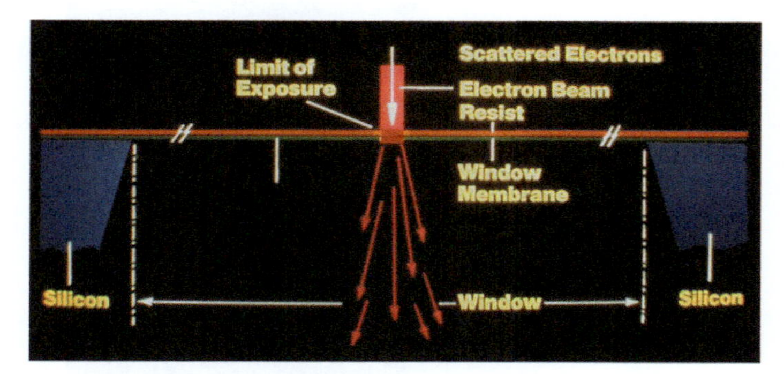

Fig. 6.29 Diagrammatic view of electron scattering with thin-window substrate.

Fig. 6.30 is an optical micrograph of a twin window substrate. Hundreds of these substrates could be made from a single silicon wafer, and it proved possible to evenly coat individual substrates with photoresist.

Fig. 6.31 shows 60 nm thick aluminum lines fabricated with a 10 nm diameter 25 kV electron beam using the PMMA lift-off process. The beam current was 1.5×10^{-10} A. The lines were written across the edge of a 150 nm thick Si_3N_4 window and their width can be seen to decrease in width from 100 nm over the bulk silicon to 60 nm over the window. It can also be seen that the secondary electron SEM signal is lower over the window where there are fewer secondary electrons generated by back-scattered electrons.

These window substrates came into their own with the scanning transmission version of the LaB6 SEM built in 1973, as is described in

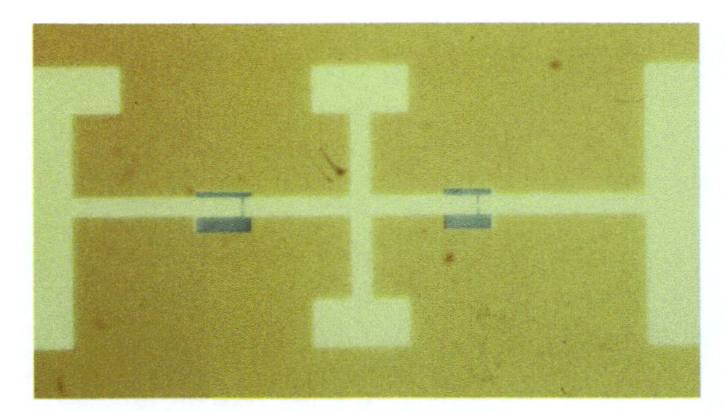

Fig. 6.30 Optical micrograph of thin window substrate.

Fig. 6.31 Aluminum wires written across the edge of a membrane window showing the effect of the back-scattered electrons. The wires over the window are narrower and less bright in this SEM image.

Chapters 7 and 8. The LaB_6 SEM, built in 1973, is described in Chapters 7 and 8. It made the fabrication and testing of SQUIDs and microbridges relatively straightforward, and the 0.5 nm beam diameter of the LaB_6 STEM, together with the PMMA resist's ultimate resolution limits, enabled the exploration of a variety of other patterning techniques.

Scanning electron microscopy of biological samples

In addition to using the microscope to monitor progress in the miniaturization of electronic and other technological devices, I decided that it would be interesting to examine biological specimens in the new SEM. The first commercial instruments had only been available for about 5 years, so this was relatively early in the use of the surface SEM for examining biological samples.

Red blood cells

I collaborated with J. A. Clarke of St Bartholomew's Hospital Medical College in London and A J Salisbury of Brompton Hospital (Clarke et al., 1971) also in London in an examination of the surface of normal human red blood cells.

Blood cells from a normal subject were fixed in suspension; some air air-dried and others were processed using the critical point drying method of Anderson (Anderson, 1951). After drying, they were coated with about 15 nm of palladium-gold using a multi-directional spinning substrate holder and examined in the SEM.

The blood cells showed a definitive surface pattern which was similar with both methods of preparation but was more pronounced with the air-drying method (Fig. 6.32) than with the critical-point method (Fig. 6.33). This was presumably due to the surface membrane collapsing more closely onto the elenin filaments that comprise the surface support structure. The dimensions of the elenin filaments bore a close relationship with those given by Dandliker (1950) (250 nm to 1 μm in length with the narrowest rods about 12.5 nm in width) and by Glaeser (1966) (20 nm in diameter) using red-cell-membrane preparations under the TEM. It, therefore, seemed reasonable to assume that, using high-resolution surface scanning electron microscopy, one was viewing a pattern of elenin filament lying just under the surface membrane. The finest structure visible was approximately 5 nm in size.

This preliminary report concluded that high-resolution SEM allowed the study of the overall surface pattern of a red blood cell in great detail and that the advantages of scanning the surface of erythrocytes at a resolution of up to 5 nm would allow a more detailed interpretation of red-cell defects than that obtained previously.

Human blood cells from patients suffering from sickle cell anemia were also examined in collaboration with Sidney Trubowitz, Hematology Research Laboratory, Veterans, Administration Hospital, New Jersey, USA, as shown in Figs. 6.34–6.36. The three-dimensional nature of these cells was visible in the scanning micrographs.

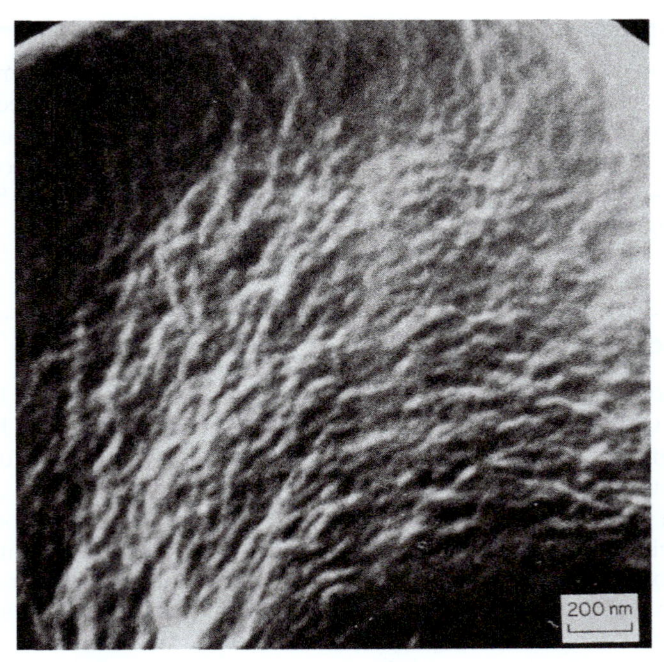

Fig. 6.32 Normal human red blood cell fixed in suspension and air dried.

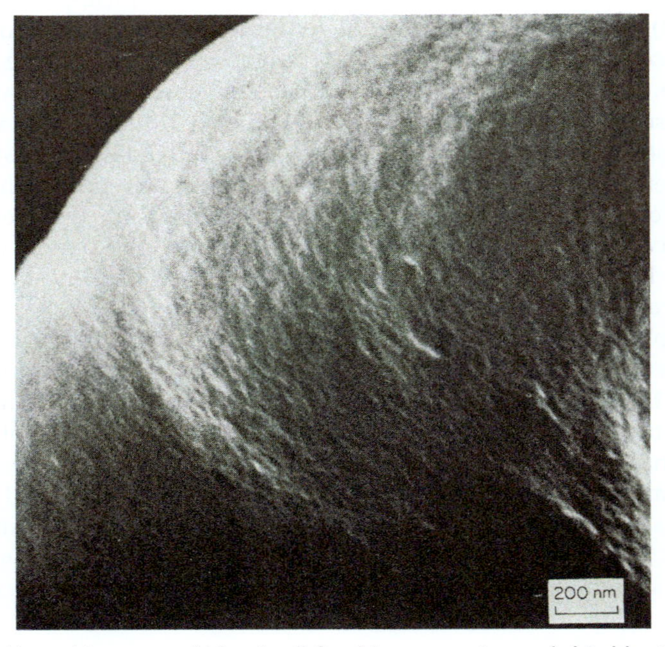

Fig. 6.33 Normal human red blood cell fixed in suspension and dried by the critical point method.

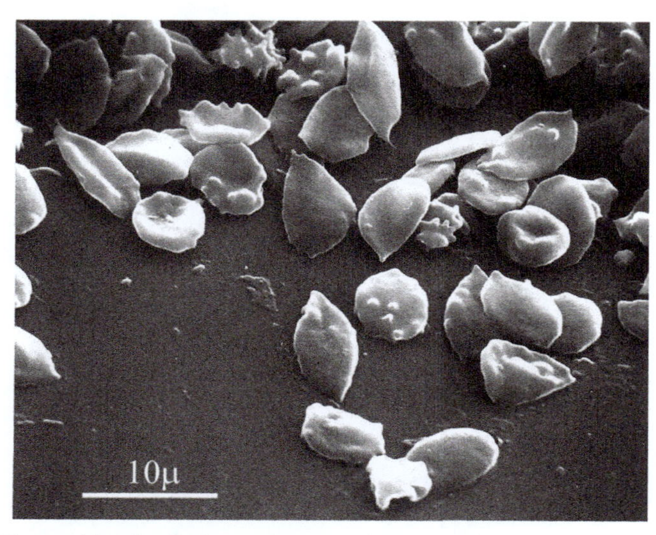

Fig. 6.34 Human blood cells from patients with sickle cell anemia.

Fig. 6.35 Human blood cells from patient with sickle cell anemia from the same sample shown in Fig. 6.34 but at higher magnification.

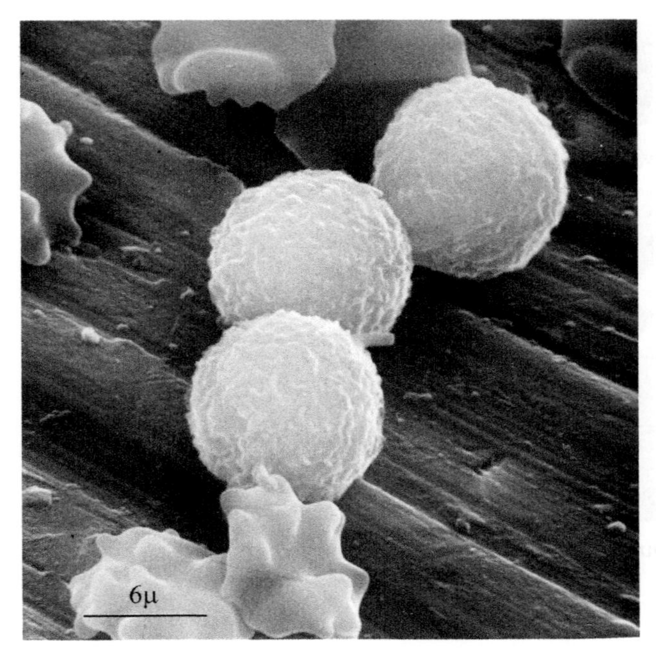

Fig. 6.36 Human white blood cells from patient with sickle cell anemia.

Human marrow

Human marrow tissue particles were also examined in collaboration with Sidney Trubowitz and Fabian Pease (Trubowitz et al., 1970). They were obtained from the sternum by the usual aspiration technique from hematologically normal individuals from a patient who had received radiation in a dosage of 3500 rad. to the mediastinum, and from a patient with chronic myelocytic leukemia. After fixing in 1% buffered osmic acid for 30 min, they were placed on aluminum specimen stubs and coated with a thin film of aluminum.

The scanning electron microscope produced a striking and dramatic three-dimensional image of intact cells and tissues.

Figs. 6.37 and 6.38 show the normal human marrow. It appeared as a loose aggregate of fat cells, about and between which the hemopoietic cells proliferate. The individual nucleated blood cells were not easily distinguished from each other, nor could the nucleated red cells be differentiated from the white blood cells.

In the irradiated marrow (Fig. 6.39), the fat cells varied in size but were, on average, larger than those of the normal marrow cellular content. It was hoped that the decreased cellular content of the irradiated marrow would

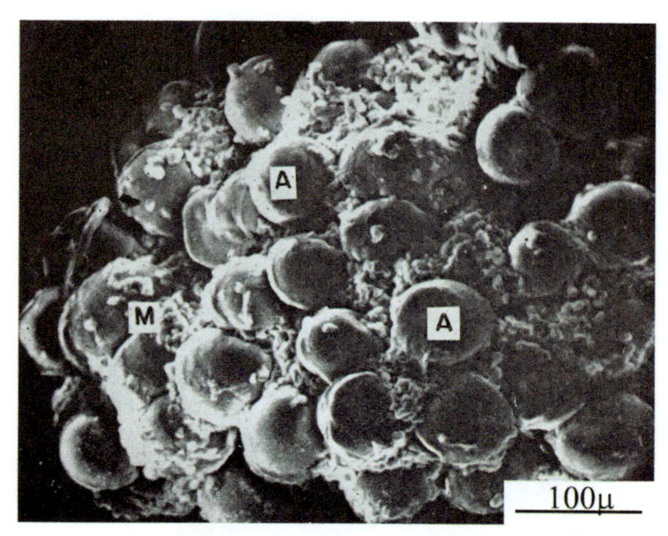

Fig. 6.37 Normal human marrow. A, Adipose cells; M, Marrow cells; V, Vascular structure.

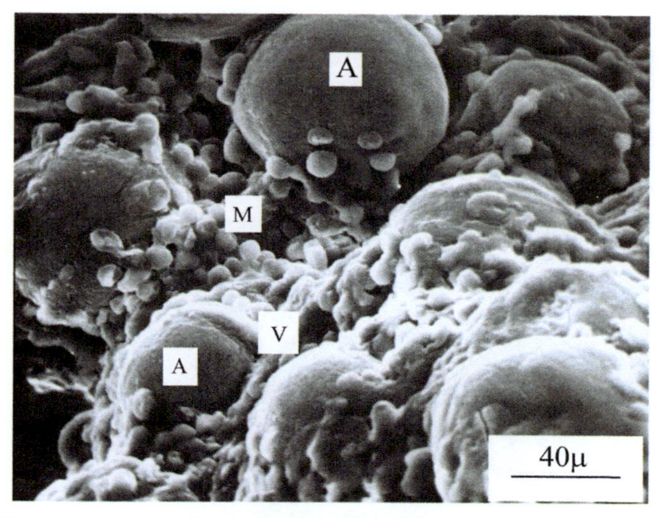

Fig. 6.38 The same sample shown in Fig. 6.37 at higher magnification.

unmask the vascular structure, but again, the nature of the cells in the described follicle could not be determined.

The marrow particles obtained from the patient with chronic myelocytic leukemia showed a marked decrease in the number of fat cells with concomitant extensive proliferation of the myelocytic cells in the form of mound-like excrescences (Fig. 6.40).

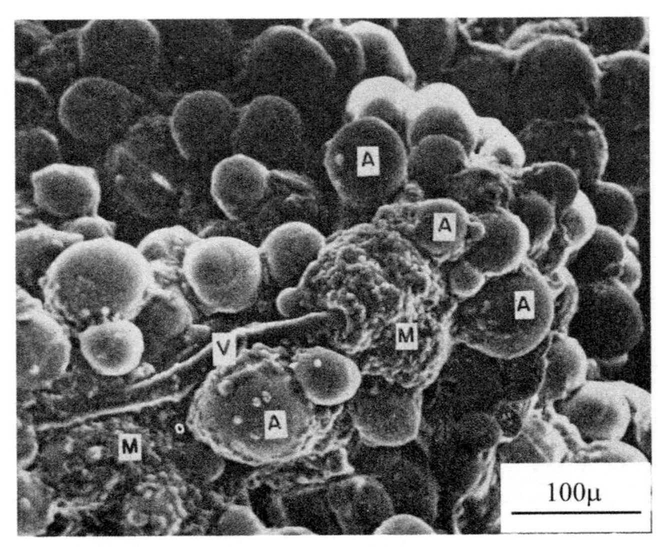

Fig. 6.39 Irradiated human marrow. Note vessel and surrounding follicular mass of cells. (A, Adipose cells; M, Marrow cells; V, Vascular structure).

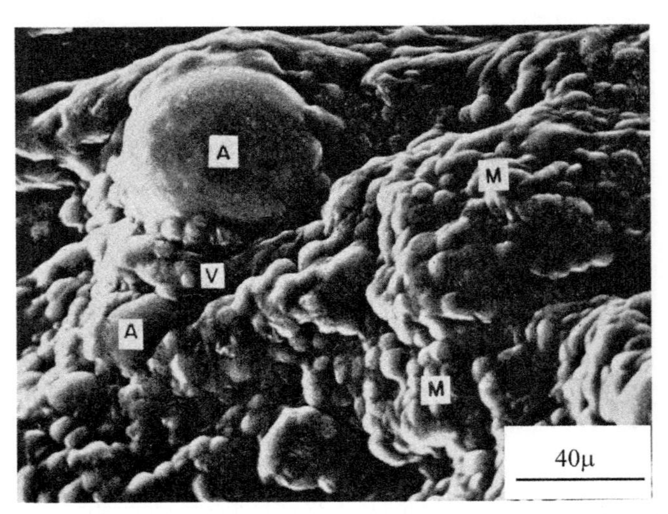

Fig. 6.40 Marrow from a patient with chronic granulocytic leukemia. (A, Adipose cells; M, Marrow cells; V, Vascular structure).

Despite obvious artifacts, particularly mild fracturing of the fat cells' surface, the marrow presented by the scanning electron microscope was apparently well preserved. We concluded that much work was needed to develop the information for cell recognition which was so necessary for future exploitation of this technique. The SEM would find useful application in the investigation of the morphology of the human marrow but at that stage, it was the three-dimensional nature of the images that was important. High resolution was not needed.

Mineralized tissue

I examined a set of fascinating specimens prepared by Boyde and Broers (1971) of University College London. Alan had pioneeered the use of the SEM to examine mineralized tissue working with Garry Stewart the early 1960s and prepared the samples showing the mineral matrix of animal bone and tooth enamel shown in Figs. 6.41–6.44.

Figs. 6.41 and 6.42 show an elephant molar tooth illustrating the great depth of focus available in the SEM. Fig. 6.43 the resolution we were able to achieve on such a biological sample. Some of the crystallites whose shapes are revealed are 50 nm by 10 nm. The sample was coated with 10–20 nm of palladium–gold in a spinning substrate holder.

We concluded that our micrographs demonstrated a resolution of about 5 nm, which was higher than had previously been obtained with biological samples and was as good as could be expected taking account

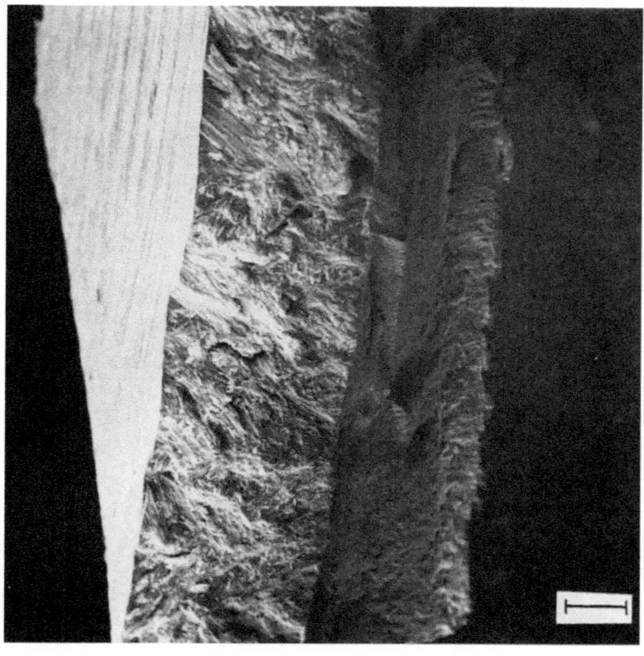

Fig. 6.41 Portion of an elephant molar tooth illustrating great depth of focus of the SEM. Scale dimension 100 μm.

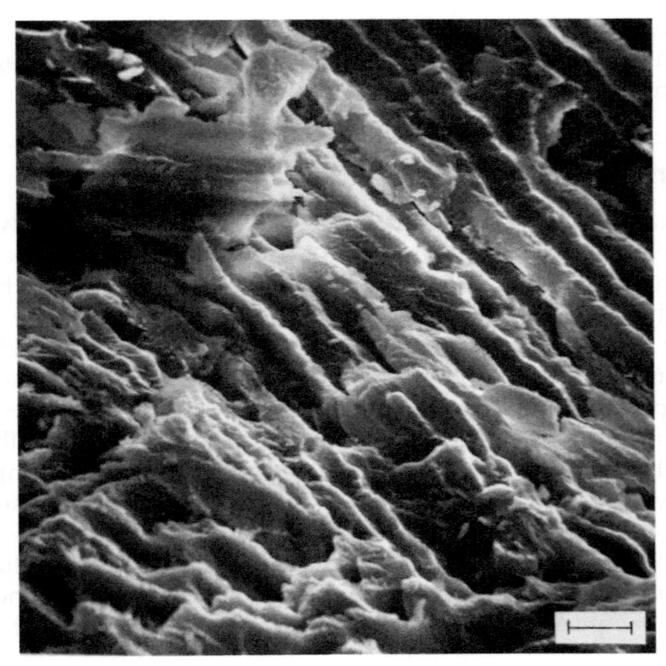

Fig. 6.42 Enlarged view of the elephant molar tooth shown in Fig. 6.26. Scale dimension 5 μm.

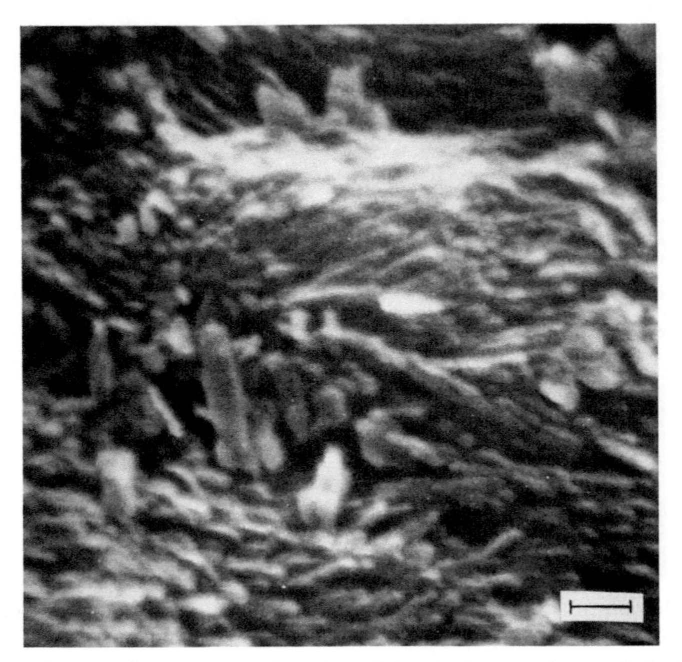

Fig. 6.43 High magnification view of portion of the elephant molar tooth. Scale dimension 100 nm.

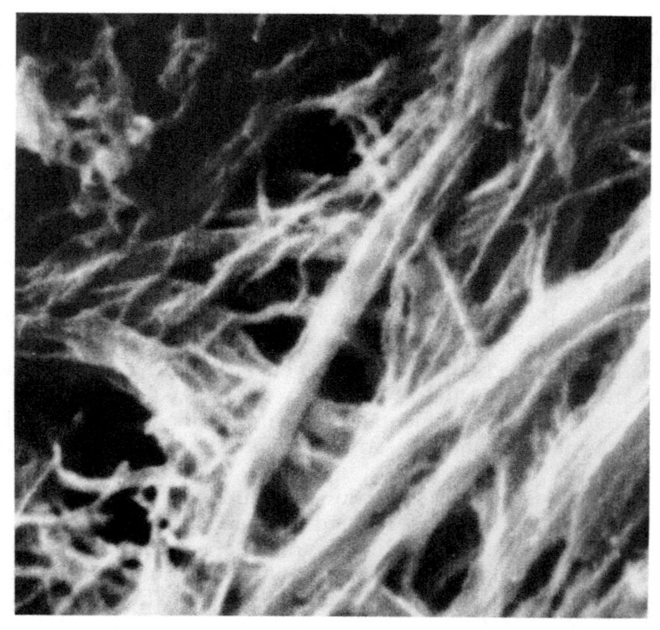

Fig. 6.44 Rat parietal bone, anorganic surface showing fine diameter "skeletons" of collagen fibrils. Scale, the same as Fig. 6.43.

of the nature of the specimen preparation methods. It was not far removed from the 2 nm resolution which was the best that could be obtained with the replica methods used for the transmission electron microscope on similar materials. The advantages of scanning microscopy were that it was easy the be sure of the exact location of the part with respect to the whole specimen, and that in most cases satisfactory replicas could only be obtained if the specimen was destroyed by total dissolution in suitable reagents. Thus, we felt that there was a good case for asking for a higher resolution on future SEMs built with biologists in mind.

Apart from merely demonstrating higher resolution several preliminary results were also obtained, for example, in the examination of rat parietal bone made anorganic by prolonged extraction with hot 1,2-ethane diamine in a soxhlet. In this specimen, shown in Fig. 6.44. The organic elements, particularly in collagen itself, had been entirely removed leaving the mineral material that was deposited within collagen fibrils. It was interesting to note that the diameter of the collagen fibrils at around 25 nm was considerably less than reported for dentine collagen, for example (Lester and Boyde, 1967). There were few values of the diameter of collagen fibrils in bone available in the literature and it seemed to be a valuable method for obtaining this information and thus, for example, studying changes in the dimensions of the fibrils with age and / or location.

References

Ahmed, H., Broers, A.N., 1972. Lanthanum hexaboride electron emitter. J. Appl. Phys. 43, 2185–2192.

Anderson, T.F., 1951. Techniques for the preservation of 3-dimensional structure in preparing specimens for electron microscopy. Trans. N.Y. Acad. Sci. 13, 130.

Archard, G.D., 1953. Magnetic electron lens aberrations due to mechanical defects. J. Sci. Instrum. xxx, 352.

Blakeslee, A.E., 1971. Method of making semiconductor superlattices free of misfit dislocations. J. Electrocem. Soc. 118, 1459–1463.

Boyde, A., Broers, A.N., 1971. High-resolution surface electron microscopy of mineralized tissue. J. Microsc. Pt. 3 (93), 253–257.

Broers, A.N., 1967. Electron gun using long-life lanthanum hexaboride cathode. J. Appl. Phys. 38, 1991–1992.

Broers, A.N., 1969. A new high resolution reflection scanning electron microscope. Rev. Sci. Instrum. 40, 1040–1045.

Broers, A.N., 1968. Some experimental and estimated characteristics of the lanthanum hexaboride rod cathode electron gun. J. Sci. Instrum. (J. Phys. E) Ser. 2. 2, 273–276.

Broers, A.N., 1970a. Factors affecting resolution in the SEM, scanning electron microscopy/1970. In: Proc. 3rd Annual Scanning Electron Microscope Symposium, IIT Research Institute. Keynote Paper, Chicago, Ill, pp. 1–8.

Broers, A.N., 1970b. The use of Schottky emission lanthanum hexaboride cathodes for high resolution scanning electron microscopy. In: Broers, A.N. (Ed.), Septieme Congres Internat. de Microscopie Electronique, Grenoble, Soc. Francaise de Microscopie Electronique, Paris, p. 18.

Broers, A.N., 1972. Observation of fresnel fringes in the conventional scanning electron microscope. Appl. Phys. Lett. 21, 499–501.

Broers, A.N., 1974. Recent advances in scanning electron microscopy with lanthanum hexaboride cathodes, scanning electron microscopy/1974. In: Proc. 7th Annual Scanning Electron Microscope Symposium. ITT Research Institute, Chicago, Ill, pp. 9–18.

Broers, A.N., Hatzakis, M., 1972. Microcircuits by electron beams. Sci. Am. 227, 34.

Broers A.N. and Sedgwick, T.O. Method for making device fo rjigh resolution electron beam fabrication - initiated 1971 [Patent]: 3,971,860. - USA, 1976.

Broers, A.N., Lean, E.G., Hatzakis, M., 1969. 1.75 GHz Acoustic Surface Wave Fabricated by an Electron Beam. Appl. Phys. Lett. 15, 98–101.

Chang, T.H.P., Hatzakis, M., Wilson, A.D., Broers, A.N., 1977. Electronics. McGraw Hill Inc., N.Y., May 12.

Clarke, J.A., Broers, A.N., Salisbury, A.J., 1971. High resolution scanning electron microscopy of the surface of red blood cells. J. Microsc. 93 (Pt. 3), 233–236.

Dandliker, W.M., 1950. The physical properties of elinen, a lipoprotein from human erythrocytes. J. Am. Chem. Soc. 72, 5587.

Dennard, R.H., Broers, A.N., 1973. Impact of Electron Beam Technology on Silicon Device Fabrication. Semiconductor Silicon 1973. ElectronChemical Soc, San Francisco, pp. 830–841.

Esaki, L., Tsu, R., 1970. IBM J. Res. Dev. 14, 61–65.

Everhart, T.E., Thornley, R.F.M., 1960. Wide-band detector for micro-microampere low energy electron currents. J. Sci. Instrum. 37, 246–248.

Fang, F., Hatzakis, M., Ting, T.H., 1973. J. Vac. Sci. Technol. A 10 (6), 1082–1085.

Glaeser, R.H., 1966. Membrane structure of OsO_4-fixed erythrocytes viewed 'face on' by electron microscope techniques. Expl. Cell Res. 42, 467.

Haine, M., 1961. Electron Microscopy. Interscience, New York.
Heidenreich, R., 1974. Fundamentals of Transmission Electron Microscopy. Interscience, New York.
Hohn, F.J., Chang, T.H.P., Broers, A.N., Frankel, G.S., Peters, E.T., Lee, D.W., 1982. Fabrication and testing of single crystal lanthanum hexaboride rod cathodes. J. Appl. Phys., 1278–1283.
Lafferty, J.M., 1951. Boride cathodes. J. Appl. Phys. 22, 299.
Lean, E.G., Broers, A.N., 1970. Microwave surface acoustic delay lines. Microw. J. March, 1–5.
Lester, K., Boyde, A., 1967. Electron microscopy of predentinal surfaces. Calcif. Tissue Res. 1, 44.
Magdo, S., Hatzakis, M., Ting, C.H., 1971. Electron beam fabrication of micron transistors. IBM J. Res. Dev. 15, 446.
Pease, R.F.W., 1963. High-Resolution Scanning Electron Microscopy. PhD Dissertation, Cambridge University.
Schmidt, P.H., Longinotti, L.D., Joy, D.C., Joy, D.C., Ferris, S.D., Leamy, H.J., Fisk, Z., 1978. Design and optimization of directly heated LaB6 cathode. J. Vac. Sci. Technol. A 15 (4), 1554–1560.
Sedgwick, T.O., Broers, A.N., Agule, B.J., 1972. A novel method of fabricating ultrafine metal lines by electron beams. J. Electrochem. Soc. 119, 1769–1771.
Smith, K.C.A., 1956. The scanning Electron Microscope and Its Fields of Application. Ph.D. dissertation, Cambridge Univ., England.
Sturrock, P.A., 1951. The aberrations of magnetic electron lenses due to asymmetries. Phil. Trans. Roy. Soc. A 243, 387.
Thornley, R.F.M., Hatzakis, M., Dhaka, A.K.A., 1970. IEEE Trans. Electron Devices ED-17, 961.
Trubowitz, S., Broers, A.N., Pease, R.F.W., 1970. Surface ultrastructure of the human marrow, a brief note. Blood 35, 112–115.
Vogel, S.F., 1970. Pyrolytic graphite in the design of a compact inert heater of a lanthanum hexaboride cathode. Rev. Sci. Instrum. 41, 585.

CHAPTER SEVEN

Illumination systems and cathodes for electron probes and the design of a short focal length final lens electron probe

Contents

The resolution of the SEM described in Chapter 6 was adequate to reach the resolution limits of conventional resist processes and of scanning images obtained with the secondary electron signal because, in both cases, this is set by the size of the volume excited by secondary electrons and not by the electron beam size. However, I wanted to explore these limits in more detail and determine functions equivalent to the modulation transfer function for fabrication processes. I also wanted to explore what Oliver Wells had called "low-loss electrons" to form scanning images. Low-loss electrons are beam electrons that have been scattered from the surface of a sample without losing much energy and could possibly produce higher-resolution images than could be obtained with secondary electrons.

Advances in Imaging and Electron Physics, Volume 231
ISSN 1076-5670
https://doi.org/10.1016/B978-0-443-31462-9.00007-1

There were also fabrication processes triggered by high-energy electrons that had the potential to avoid the limits imposed by the excitation range of the secondary electrons.

To achieve these aims I needed a final lens with the shortest possible focal length. This would ensure that the lens had the lowest spherical and chromatic aberration coefficients and could operate in the condenser-objective mode with the focal point at the center of the pole-piece gap. It would be similar to the objective lenses used in TEMs, but with the pole-piece gap at the end of the lens so that it would be easy to detect electrons emerging from the lens.

With the sample positioned in the middle of the lens field, I hoped that it would be possible efficiently to detect electrons elastically scattered from the surface of an inclined sample. These low-loss electrons would be focused back onto the lens axis by the second half of the lens field and easily detected with a scintillator detector placed on the axis below the lens. Oliver Wells had already demonstrated the feasibility of forming images using low-loss scattered electrons with a conventional SEM. I was optimistic that the new probe would produced resolution comparable to the beam diameter of less than a nanometer in examining the surface of bulk metallic samples.

I also decided to increase the maximum accelerating voltage, as this would further reduce the beam diameter and perhaps be more efficient in fabrication processes triggered by high-energy electrons rather than secondary electrons.

Finally, I made it possible to introduce apertures and meshes into the electron beam, so that the microscope column could be operated with "critical" or Kohler illumination, where a source with w delta-function distribution of current is de-magnified onto the specimen, rather than the gun cross-over that has a Gaussian distribution.

The illumination options and the choice of electron source are discussed in this chapter.

Illumination systems (Broers, 1979)

Thermal cathode SEMs, electron beam X-ray microanalyzers, and round beam electron beam microfabrication systems generally used the cross-over formed in the electron gun as the object that was imaged onto the sample. This is satisfactory, provided the electron gun produces a single round cross-over, although an object with a delta function distribution of current pr0vides higher equivalent brightness. When the gun produces a multiple cross-over, or when the shape of the cross-over is not round, the efficiency of a cross-over imaging system becomes low, and the equivalent brightness can be considerably below the peak brightness in the emitted electron beam. This case is often encountered with sintered LaB_6 cathodes, particularly when they are deeply immersed in the accelerating field and relatively large areas of the cathode surface are emitting. Deep immersion in the accelerating field maximizes the field at the cathode surface, minimizes space-charge divergence, and increases brightness.

Two critical illumination methods can be employed to provide a sharply defined source. In both cases, a physical aperture becomes the object for the electron optical system. In the first case, the aperture is placed in, or as close as possible to, the plane of the gun cross-over. In the second case (Fig. 7.1), the aperture is located in the plane of an enlarged image of the gun cross-over produced by a condenser lens.

The first critical illumination case can be used successfully with tungsten filament electron guns where the cross-over is relatively large (>30 μm), and apertures of 5–20 μm are small enough to be uniformly illuminated. The power absorbed by the aperture is large, however, and the aperture and its support can reach such high temperatures that stability and reliability become a problem. With LaB_6 cathodes and accelerating voltages of about 20 kV, we have encountered melting of even molybdenum

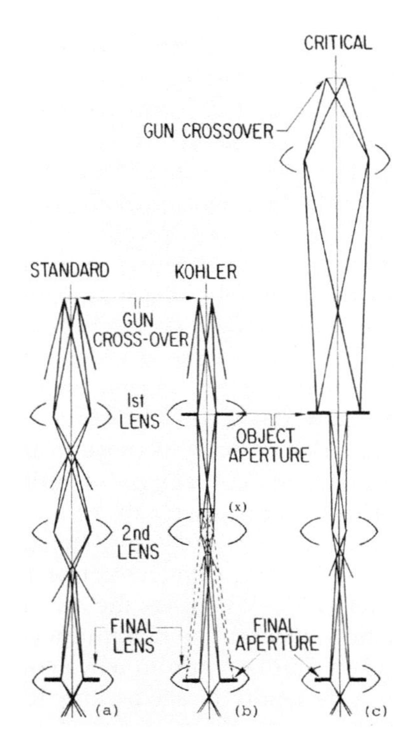

Fig. 7.1 Illumination systems for electron probes (Broers, 1979).

apertures, making this method intolerably unreliable. An additional difficulty with sintered LaB_6 cathodes is that fine structure exists in the cross-over, and the aperture had to be impractically small (<5 μm) for optimum performance. A final problem with any cross-over limiting aperture system is that the gun (Pearce-Percy et al., 1979) cross-over often falls between the grid and the anode and it is not possible to place the aperture in the same plane as the cross-over. This results in a loss of effective brightness.

When an additional lens is introduced to magnify the cross-over, many of these problems are resolved. The aperture can now be much larger, it can be placed exactly in the plane of the image, and the current density at the aperture plane is lower, thus removing the problem of aperture heating. The total power absorbed by the aperture may also be reduced to an acceptable value by using a suitable stopping aperture in the cross-over imaging lens. The disadvantages of this method are that the electron optical column becomes longer, and an additional lens may be needed to provide enough demagnification to reach the final beam size. A system of this type has been used by Pearce-Percy et al. (1979)

A Kohler illumination system (Fig. 7.1B) possesses all the advantages of the second type of critical illumination system but can be implemented with many existing three-lens systems without additional lenses or changes in lens spacing. It also allows the cross-over current distribution to be examined at high resolution (<0.5 μm) and the angular distribution of current from the cathode. A further advantage of Kohler illumination is that an image of an aperture placed in the plane of a cross-over image is projected into the pupil of the final lens, providing an ideal step-function distribution of current across the lens pupil. This is valuable in cases where it is not possible to place an aperture in the lens gap. Such cases are often found in electron beam fabrication systems where the electron beam is scanned considerably off the axis in the final lens.

Kohler illumination has been employed in electron optical systems designed to produce shaped electron beams for use in electron beam lithography systems. In these cases, the fabrication of square and rectangular apertures that are small enough to place in the gun cross-over is impractical, and critical illumination using a magnifying condenser lens is an inferior choice for the reasons given above.

The equivalent brightness of 3×10^5 A/cm^2/ster. (25 kV) obtained from a tungsten filament gun in the square beam electron beam fabrication system described by Pfeiffer (1971) illustrates the advantage to be obtained with a Kohler illumination system. The maximum equivalent brightness measured from a tungsten filament gun in a system in which the cross-over acts as the object is generally closer to 1.5×10^5 /cm^2/ster (25 kV). It is proposed that the higher brightness arises for two reasons: (1) the current density distribution of the source is a delta function; and (2) Kohler illumination makes it possible to operate the tungsten filament in a high total current, high peak brightness, mode without suffering a loss in equivalent brightness because the cross-over becomes larger and non-round under these conditions. As already mentioned, the cross-over becomes non-round because a relatively large and elliptical area of the filament is emitting.

In summary, although Kohler illumination has mainly been used in shaped beam and projection systems, I realized that it could also enhance the performance of round beam systems and be used in the new short focal length probe. Critical illumination also offers higher performance but only with a considerable risk of melting the object aperture, especially with LaB_6 cathodes.

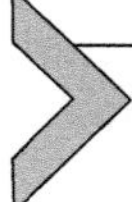

Comparison of thermal and field emission cathodes (Broers, 1972)

In this section, I consider field emission and LaB6 cathodes for the new short focal length probe and analyze the different options for each. In the next section, I describe the new column and how I optimized it for Kohler illumination. I had already designed and tested a cold field emission cathode electron gun with a tip that could be tilted mechanically to select the brightest emission lobe (Broers, 1968).

Pointed cathodes that operate at elevated temperatures in the thermally assisted and Schottky emission modes were to become widely used by the 1990s. They operate stably at higher pressures than cold field emitters and became practicable for electron microscopes and microfabrication systems as vacuum technology improved. As electron optical components, they are similar to cold field emission cathodes, and the optical systems created to use them are similar to those used with cold field emission in this chapter. As sources, they are 100 times smaller than thermal sources, 50Åversus 50 μm, so little or no demagnification is needed in many applications. On the other hand, the total current available is 100 times smaller, 1 μA vs 100 μA. If final beam currents of 1 μA or more are needed, gun optics become important, and the beam brightness falls. I discuss these cathodes in the final chapter when considering the ultimate limits of nanolithography and surface microscopy.

Cold field emission

Single-crystal tungsten cold field emitters are the highest brightness electron sources for beam currents below about 1 μA. Emission density up to $10^6\,A/cm^2$ is available, and the field gradient at the cathode surface is so high ($10^7\,V/cm$) that space charge divergence does not limit the brightness. The main difficulty with them is that stable emission requires a vacuum of about 10^{-10} mmHg. This is because emission depends strongly on the work function of the surface and the shape of the tip, both of which can be altered by surface contaminants and/or cathode sputtering. Emission densities of about $10^4\,A/cm^2$ can be obtained at pressures up to about 10^{-9} mmHg with sufficient stability for microscopy. The brightness is, therefore, between 50–100 times greater than with LaB6 cathodes, which have a maximum current density of about $50\,A/cm^2$.

Two anodes are generally used in field emission guns. The first anode controls emission and the second sets the overall accelerating potential. The effective source is very much smaller than with thermal cathodes, and it is important to reduce the aberrations in the gun so that the gun "sees" the source as close to its actual size. Several low aberrations guns have been designed, the first by Butler (1966).

The 20–60 Å in diameter allows very small probes to be obtained using the gun alone. The minimum beam size is set by the chromatic aberration of the gun and diffraction. Unlike the situation in an SEM using a thermal source and two or more lenses, the size of the Gaussian image of the source cannot be varied once the gun geometry is fixed, so the current in the beam is determined only by the operating aperture of the gun and the tip current. The following expressions for the diameter of the beam (d_G), the optimum aperture(α_{Gopt}), and current (I) describe the performance of this type of gun (See Fig. 7.2).

$$d_G^2 = A\alpha_G^6 + B\alpha_G^2 + \frac{C}{\alpha_G^2} + D$$

$$\alpha_{Gopt}^4 = \frac{\left[B^2 + 12AC\right]^{1/2} - B}{6A}$$

$$I = \pi\alpha_S^2 I_o = \pi M_G^2 V_r \alpha_G I_0$$

Where

$$A = \left(0.5C_{SG}M_G^4 V_r^{3/2}\right)^2$$

$$B = \left(C_{CG}\frac{\delta V}{V_1}M_G^2 V_r^{1/2}\right)^2$$

$$C = (0.9\lambda)^2$$

$$D = (M_G d_o)^2$$

C_{SG} = Spherical aberration coefficient of gun referred to the source
C_{CG} = Chromatic aberration coefficient of gun referred to the source
α_S = Beam angle at source (see Fig. 7.2)

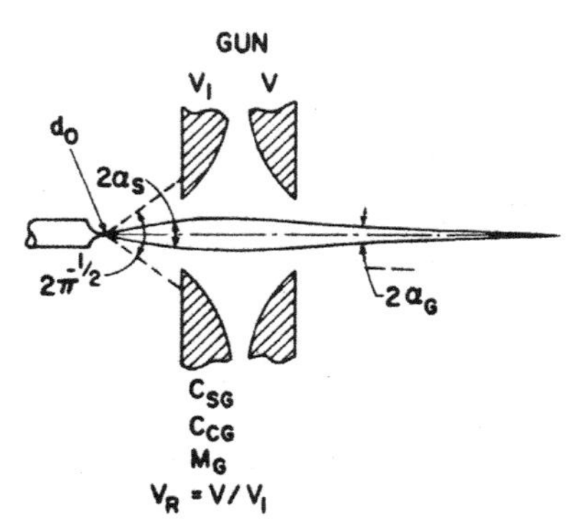

Fig. 7.2 Field emission gun probe (Broers, 1972). *(The Electrochemical Society. Reproduced by permission of IOP Publishing Ltd).*

α_G = Beam angle at image (see Fig. 7.2)
M_G = Magnification of gun
V_1 = First anode potential
V = Second anode potential
$V_r = V/V_1$
δ_V = Energy spread of emitted electrons
I_S = Total source current
I = Final beam current
d_O = Diameter of source (60 Å used in Figs. 7.3. and 7.10)

It is assumed that the current is emitted into a solid angle of 1 sr.

Fig. 7.3 shows the beam diameter and beam current obtained as the operating aperture is varied in such a field emission gun. Three operating conditions are given (Aberration data from (Crewe, 1970; Crewe and Wall, 1970)).

Curve A is for the typical field emission gun microscopes. The dotted curve shows the performance if the energy spread in the emitted beam increases from 0.2 eV, the theoretical spread set by tunneling, to 0.5 eV. Such an increase can occur when the emitter surface becomes contaminated.

Curve B is close to the optimum practical limit when the smallest beam diameter is needed. The limit is set in this case by the need to have enough space between the gun and the image for a deflection unit to scan the beam and a stigmator to correct for beam asymmetries. 2.5 cm is considered here to be the minimum practical working distance.

Curve C is close to the optimum practical operating limit if the maximum current is to be obtained at larger beam sizes. More current still can be obtained if V_r is increased further, but this leads to a smaller V_1 for a given accelerating potential (less than 1700 V for V_O = 25 kV). V_r = 15 is already optimistic compared with the normal values employed, which

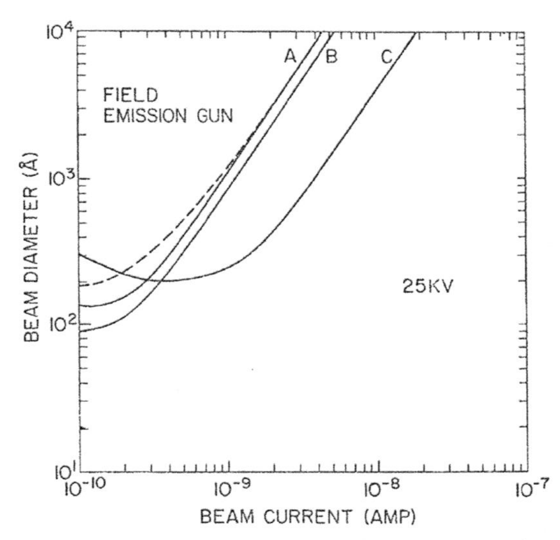

Fig. 7.3 Field emission gun beam diameter s beam current (Broers, 1972). *(The Electrochemical Society. Reproduced by permission of IOP Publishing Ltd).*

range between 3 and 6. Smaller V_1 means that smaller tip diameters are needed for a given emission density, and greater difficulty is encountered in maintaining stability.

The gun operating conditions for the three cases are as follows: Case A; Source distance 3 cm, Image distance 4 cm, $V_r=7.5$, $V_O=25\,kV$, $\delta_V=0.2$ V and 0.5 V, $C_C=0.5$ cm, $C_S=120\,cm$, and $M=3$. Case B; Source distance 1 cm, Image distance 4 cm, $V_r=15$, $V_O=25\,kV$, $\delta_V=0.2$, $C_C=1.3$ cm, $C_S=7.2$ cm, $M=1.85$. Case C; Source distance $=3$ cm, Image distance $=2.5$ cm, $V_r=8.6$, $V_O=25\,kV$, $\delta_V=0.2$, $C_C=5.3$ cm, $C_S=142\,cm$, $M=0.63$.

High brightness electron gun using a field emission gun

To gain familiarity with field emission cathodes and their application to high-resolution probes, I designed an experimental gun in which I would be able to explore the characteristics of cold field emission cathodes and measure their brightness. The gun was similar to that reported by Crewe (Crewe, 1970) except that the accelerating electrodes were plane rather than shaped, and the cathode was located by a gimbal mechanism, which allowed the cathode to be tilted over an arc of 70° in any direction and positioned laterally. The gun electrodes were precisely machined with the apertures round to within 0.25 μm and aligned with respect to each other to better than 10 μm. The second accelerating electrode was followed by scan plates, a test grid, and an electron detector, which

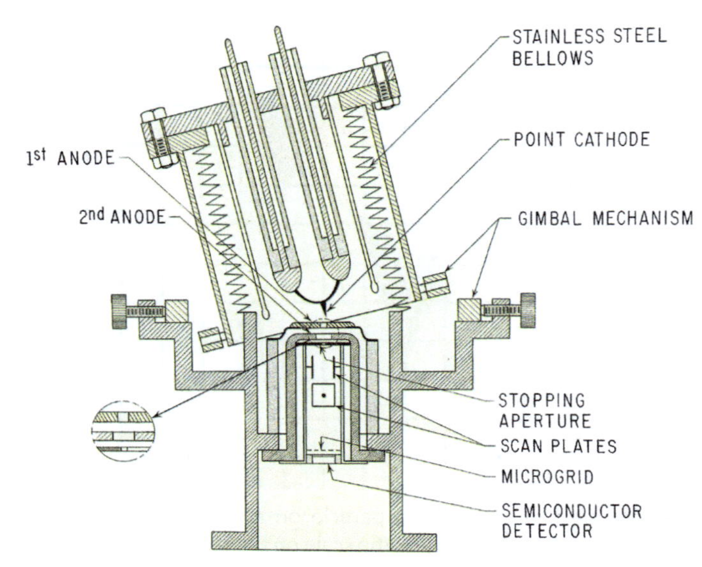

Fig. 7.4 Field emission electron gun (Broers, 1968).

together allowed the probe size to be measured in the usual scanning electron microscope mode. The arrangement is shown in Fig. 7.4.

The system was pumped with an ion pump and used metal gaskets throughout and could be baked to over 300°C. It would take several hours to pump down to the level needed for stable operation of the cathode, which had to be below 10^{-10} mmHg. The gun was mounted on a table that was isolated from mechanical vibrations by air-suspension units and coils surrounding the whole system largely canceled 60 Hz mains electromagnetic interference.

After aligning the system with the lateral positioning micrometers, the beam current to the semiconductor detector was maximized using the gimbal mechanism. These were the early days of field emitters and single-crystal wires of the optimum crystallographic orientation were not available. However, this was not a problem as it was easy with the gimbal to select one of the several "bright" lobes of current from the tip.

After selecting a bright lobe, the electrostatic deflection plates were used to produce a two-dimensional scanning image of the sample on the microgrid 7 cm below the second gun electrode. One of these images is shown in Fig. 7.5. It could be used to measure the beam diameter using the standard risetime method used in scanning electron microscopes. For a total beam current from the tip of about 1 μA, the beam diameter at the sample was between 50 and 100 nm, and the beam current was about 10^{-8}A. This yielded a brightness of about 2×10^7 A/cm^2.ster at 12 kV,

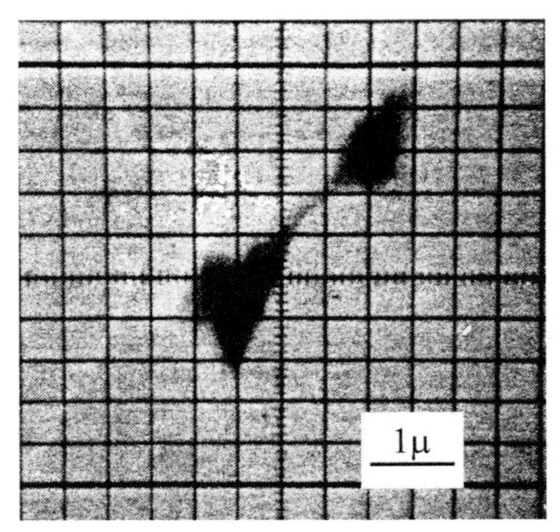

Fig. 7.5 Scanning electron image of a particle on a thin film supported on a TEM speciment grid. The grid on this image is the scale on the face of the oscilloscope screen used to display the image (Broers, 1968).

which was to be compared to a maximum at the time for a sintered LaB_6 of about 2×10^5 A/cm^2.ster. The brightness was roughly proportional to the tip current so was 10 times higher for a tip current of 10 μA, but the current stability was not as good.

The gun's aberrations were not calculated and would have been higher than those of Crewe and Butler's electron guns, which had shaped electrodes, but the brightness was still 100 times greater than LaB_6 and more than enough for any of the applications of electron probes we were planning in 1968.

Field emission gun and lens

To take full advantage of the potential of field emission cathodes for scanning electron microscopy, a demagnifying lens or lenses must be used in conjunction with the gun. With the additional lenses, the minimum beam diameter in effect becomes limited only by the aberrations of the final lens and diffraction. The aberrations of the gun are rendered insignificant by the demagnification of the lens. In practice for microscopy, neither the exact location of the lens nor the particular operating conditions of the gun are critical, provided they are correctly adjusted with respect to each other. That is, the lens must provide only enough demagnification to reduce the Gaussian image of the gun crossover, whether real or virtual, to a small but finite fraction of the overall beam diameter. This is the same with thermal cathodes, where beyond a certain point demagnification only reduces beam current without appreciably decreasing the beam diameter. The exact operating conditions and lens location become more

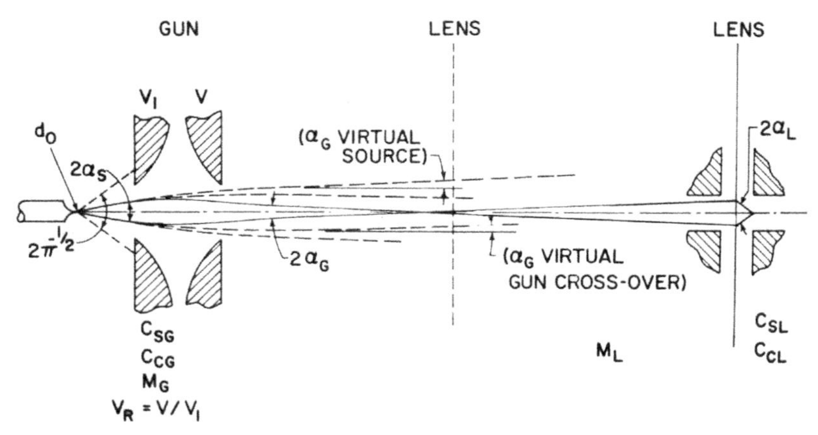

Fig. 7.6 Field emission gun plus lens (Broers, 1972). *(The Electrochemical Society. Reproduced by permission of IOP Publishing Ltd).*

important when larger beam currents and spot sizes are needed. In these cases, which are discussed later, care must be taken to transfer all the current available from the source into the final beam.

As with the simple field emission gun, the current available in the beam is set by the operating aperture subtended at the source and the total emitted current, according to the following expression (See Fig. 7.6).

$$I = \pi \alpha_S^2 I_O$$

This assumes again that the current is emitted into 1 sr. α_S can be calculated from α_L using the relation

$$\alpha_S = \frac{M_G}{M_L} \sqrt{v_r \alpha_L}$$

where M_G is the magnification of the gun, and M_L is the demagnification provided by the lens. α_L is chosen to give the best compromise between lens aberrations (predominantly spherical aberration) and diffraction. The minimum beam diameter considering lens aberrations alone

$$d_{\min} = 1.03 C_S^{1/4} \lambda^{3/4}$$

This beam diameter will only be obtained if the gun cross-over diameter is zero or if the lens provides infinite demagnification. In practice, this is approximately the case for beam currents below about 10^{-10} A when the demagnification can be kept relatively large. Figs. 7.7, 7.11, and 7.12 show the performance of gun/lens combinations.

In calculating the performance of all the gun-lens combinations shown in Figs. 7.5–7.7 except curve D, it is assumed that the optimum lens demagnification (M_{Lopt}) is used for each beam current. M_{Lopt} is calculated in a similar way to α_{opt} for thermal cathodes. That is by differentiating the expression for overall beam diameter with respect to M_L.

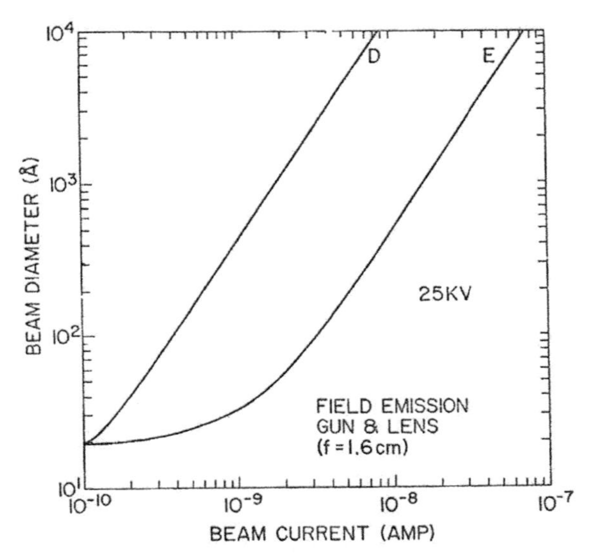

Fig. 7.7 Field emission gun and lens with f = 1.6 cm (Broers, 1972). *(© The Electrochemical Society. Reproduced by permission of IOP Publishing Ltd).*

The following expressions are used for the overall beam diameter (d), the beam current (I), and M_{Lopt}.

$$d^2 = AM_L^6 + BM_L^2 + \frac{C}{M_L^2}$$

$$M_{Lopt}^4 = \frac{\left(B^2 + 12AC\right)^{1/2} - B}{6A}$$

$$A = \left(0.5C_{SL}\frac{P^3}{M_G^3}\right)^2$$

$$B = \left(C_{CL}\frac{\delta VP}{VM_G}\right)^2$$

$$C = \left(M_G V_r^{3/2} 0.5 C_{SG} P^3\right)^2 + \left(M_G V_r^{1/2} C_{CG}\frac{\delta V}{V_1}P\right)^2 + \left(0.9\lambda\frac{M_G}{P}\right)^2 + (M_G d_0)^2$$

Where

$$P = \left(\frac{I}{V_r I_0 \pi}\right)^{1/2}$$

$$\alpha_L = M_L \alpha_G = \frac{M_L}{M_G V_r^{1/2}}\alpha_S$$

$$I = \pi\alpha_S^2 I = \pi\left(\frac{M_G}{M_L}V_r^{1/2}\alpha_L\right)^2 I_0$$

where the parameters are those given in Fig. 7.6.

In Cases D and E, the operating conditions for the field emission gun and the aberrations for the final lens are the same. They are; $C_{SL}=1.8$ cm, $C_{CL}=1$ cm, Source distance 0.82 cm, Image distance = 0.82 cm, $V_r=15$, $V_O=25$ kV, $\delta_V=0.2$ V, $C_{CG}=1$ cm, $C_{SG}=4.4$ cm, $M_G=3.5$. In Case D, M_L is fixed at 50. In Case E, M_L is optimized for each current I.

The gun operating conditions for D and E are close to optimum when maximum current is required at all beam diameters. Higher current is still available if the source distance is reduced and V_r increased. However, smaller source distances can lead to instability due to out-gassing from the anode. The difficulty in increasing V_r has already been discussed. D shows the performance when the lens demagnification is fixed at 30, and the beam current is altered by varying the beam aperture. E is the performance if the optimum demagnification is used for all beam diameters and currents.

If the gun and lens are fixed mechanically both in their own configuration and in relation to each other, the current in the beam can only be altered by altering I_O or α_L. In either case, the optimum operating conditions with regard to gun performance or demagnification are lost. This inflexibility can be overcome with a second demagnifying lens. The overall demagnification can then be altered without significantly interfering with the performance of the gin or the final lens. Fortunately, the "effective brightness" (it is very difficult to talk in terms of conventional brightness with field emission sources because of the strong variation of brightness with operating aperture) is so high that when only one lens is available, small losses of performance through excessive demagnification can be tolerated without reducing the beam current to an unacceptable level. Excessive demagnification is preferable to insufficient demagnification because, in the latter case, the spot diameter is increased.

When two lenses are used, care must be taken to keep the demagnification provided by the final lens high enough to render the aberrations of the first lens insignificant and to avoid increasing the spherical aberration of the final lens.

Fig. 7.8 shows how the beam diameter and current vary with lens demagnification when a single lens suitable for transmission microscopy is used. In this case a demagnification of about 100 is required. For of a relatively long focal length lens suitable for secondary electron surface microscopy a demagnification of about 35 is sufficient to reduce the beam diameter to within 10% of the limit set by the aberrations of the final lens.

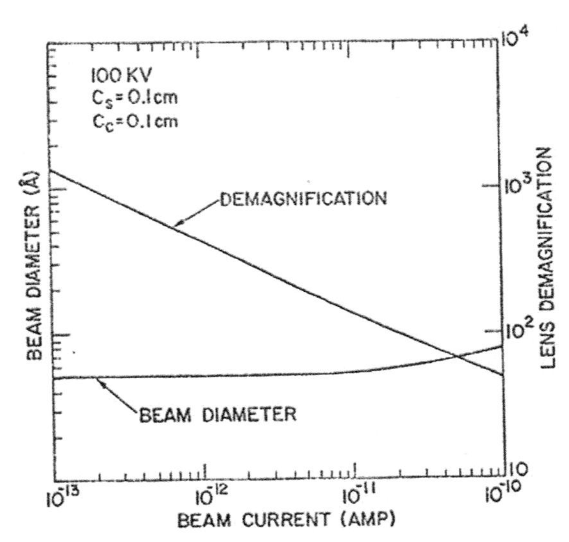

Fig. 7.8 Case for short focal length lens generally used for transmission microscopy (Broers, 1972). *(The Electrochemical Society. Reproduced by permission of IOP Publishing Ltd).*

Thermal electron sources

The formulae used to calculate the performance when using thermal cathodes is that used in Section "Theoretical estimation of beam diameter" in Chapter 2.

For a final lens with given aberrations, the electron optical performance is set by the brightness (current density per unit solid angle) of the electron gun. If space charge effects are negligible, brightness in thermal cathode guns, according to Langmuir is given by the relation.

$$\beta = 11{,}600\, J_C V/\pi T$$

In practical guns, however, space charge makes the electrons diverge as they leave the cathode surface, and the beam angle becomes greater than that taken by Langmuir who assumed that the beam divergence was set by the initial lateral thermal energy of the electrons alone. In order to obtain the highest brightness, therefore, the electric field at the cathode surface must be maximized so that the electrons are accelerated as quickly as possible and diverge as little as possible. The field at the cathode surface depends on the Wehnelt geometry, the cathode shape, the anode–cathode field, and the degree to which the cathode is immersed in the accelerating field, that is, the gun bias conditions. The sharper the cathode, and the more the cathode is immersed in the accelerating field, the higher the surface field. Plane or conical Wehnelt electrodes give higher fields than those with re-entrant apertures.

The degree to which the cathode is immersed in the accelerating field is limited by two factors; hollow beam formation and excessive energy spread due to the Boersch Effect (Boersch, 1954). As mentioned above, chromatic aberration is important in surface SEMs, and energy spread generation is, therefore, particularly detrimental to this type of high-resolution electron probe. Pfeiffer (1971) has shown that energy spread is proportional to $(\beta r)^{1/2}$ where r is the radius of the gun cross-over. Immersing the cathode further in the accelerating field increases the surface field but at the same time increases the emitting area and therefore the cross-over size. Measurements indicate that an energy spread of 4–5 eV can occur with tungsten hairpins at a brightness of 1.5×10^5 A/cm^2.steradian (20 kV) and 3–4 eV with LaB_6 cathodes (20 μ tip) at a brightness of 1×10^6 A/cm^2.steradian (Broers and Pfeiffer, 1971). Sharper cathodes allow higher brightness at lower energy spread because they do not have to be immersed so far into the accelerating field to produce a given field at the cathode surface.

In summary, high emission density and high surface accelerating fields are needed for high brightness. For these reasons, in 1970, LaB6 cathodes gave higher brightness than tungsten hairpins. The higher current density was available at a reasonable lifetime (10–200 A/cm^2 compared to 5–10 A/cm^2), and the LaB_6 cathodes could be formed into sharper points (1–20 μm compared with 125–500 μm).

As already mentioned, thermally assisted field and Schottky field emitters will be discussed later.

Fig. 7.9 shows the electron optical performance, beam diameter vs. beam current, for tungsten hairpin and LaB6 cathodes. The final lens was assumed to have CS = 1.8 cm, CC = 1 cm, and negligible astigmatism. The following brightnesses were used for the different cathodes. A planar grid electrode was assumed.

Tungsten hairpin
 Brightness: 1.5×10^5 A/cm^2.ster. (25 kV)
 Lifetime: 10–15 h
 Maximum acceptable pressure 10^{-4} mmHg

LaB_6 (20 μm diameter tip)
 Brightness: 1.1×10^6 A/cm^2.ster (25 kV)
 Lifetime: 100–200 h
 Maximum acceptable pressure 10^{-5} mmHg

LaB_6 (1 μm diameter tip)
 Brightness: 5×10^6 A/cm^2.ster (25 kV)
 Lifetime: 50–100 h
 Maximum acceptable pressure 10^{-6} mmHg

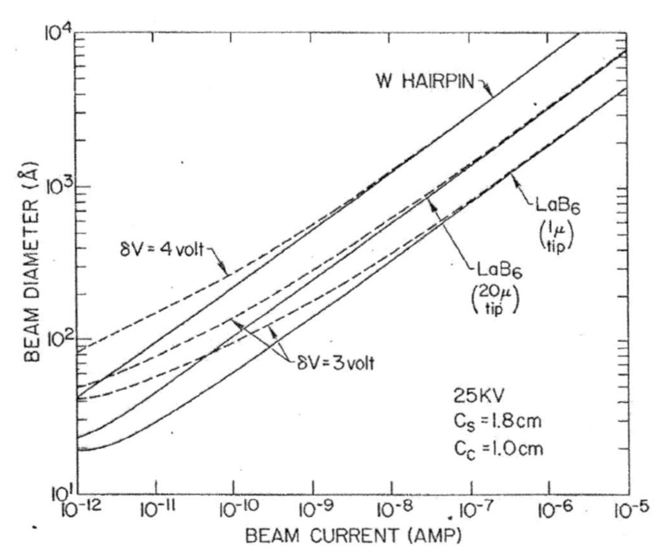

Fig. 7.9 Thermal cathode probes suitable for surface SEM (Broers, 1972). *(The Electrochemical Society. Reproduced by permission of IOP Publishing Ltd).*

The solid lines show the performance that will be obtained if the energy spread is due to the initial thermal energy alone (0.25 eV for tungsten and 0.2 eV for LaB_6). The dotted lines show the performance that will be obtained if the axial energy spread is increased by Boersch effect to 4 eV for the tungsten hairpin and 3 eV for the LaB_6 cathodes.

Figs. 7.10 and 7.11 show the performance of probes suitable for transmission SEMs at beam energies of 30 and 100 kV. The final lens is assumed to have a focal length of 0.4 mm (C_S = 0.4 mm, C_C = 0.4 mm) at 30 kV, and a focal length of 1 mm (C_S = 1 mm, C_C = 1 mm) at 100 kV. The difference between thermal and field emission cathodes is less for transmission microscopy than for surface microscopy particularly when higher accelerating potentials are used. This is because chromatic aberration becomes less important and because the "effective brightness" of field emission sources does not increase as much as the brightness of thermal cathodes at higher accelerating potentials.

In Fig. 7.12, the performance of thermal and field emission cathodes with a surface microscope lens are superimposed for comparison. For beam diameters up to about 30 nm the field emission gun plus lens combination provides superior performance. It is therefore the best choice for surface microscopy. For beam diameters between about 13 and 80 nm, the performance of the simple field emission gun exceeds that of the conventional column using a tungsten hairpin cathode. Above and below this

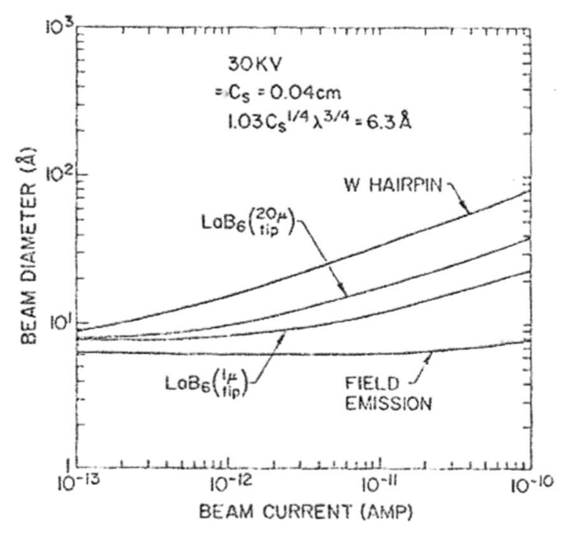

Fig. 7.10 30 kV electron probes formed with short focal length final lens (Broers, 1972). *(The Electrochemical Society. Reproduced by permission of IOP Publishing Ltd).*

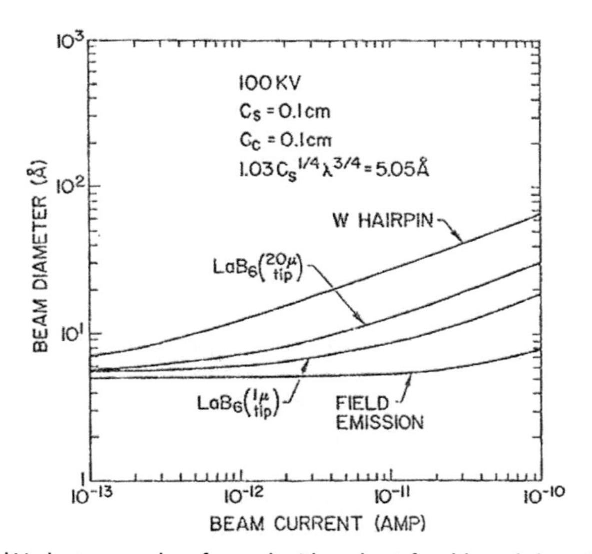

Fig. 7.11 100 kV electron probes formed with a short focal length lens (Broers, 1972). *(The Electrochemical Society. Reproduced by permission of IOP Publishing Ltd).*

range, a tungsten hairpin cathode is better. For beam diameters between about 13 and 20 nm the simple gun has approximately the same performance as a conventional column with a 20 μm LaB_6 cathode. A LaB_6 cathode with a 1 μm tip provides superior performance to the simple gun for all beam diameters.

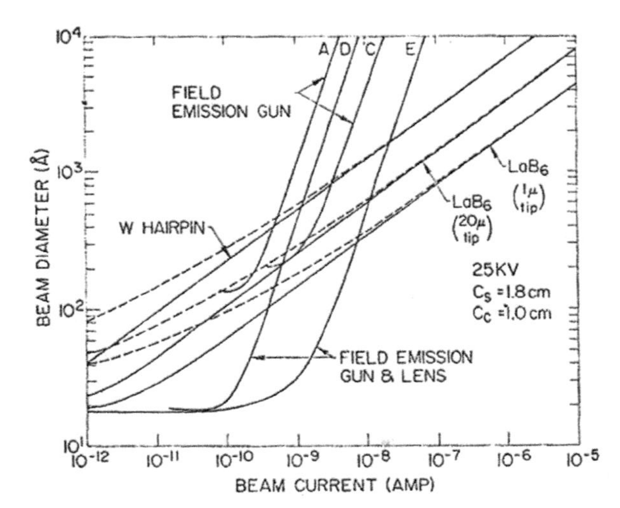

Fig. 7.12 25 kV Electron probes suitable for secondary electron surface microscopy (Broers, 1972). *(The Electrochemical Society. Reproduced by permission of IOP Publishing Ltd).*

Microfabrication

Electron beam systems suitable for microfabrication, especially for making 5X masks for semiconductor devices, must cover areas that are much larger in terms of pixels than imaging systems such as microscopes and micro-analyzers. Chip patterns typically have 100,000 times the complexity of microscope or microanalyzer images and larger fields have to be scanned if the mechanical. Stepping time between fields is not to dominate exposure time. Systems where the beam writes on a continuously moving table avoid this delay, but are difficult to implement for dimension below 0.1 μm. Where large-area scanning is needed, longer focal length lenses operating at smaller beam apertures and beam deflection angles can be used. Fig. 7.13 compares the performance to be expected when a field emission gun and a standard LaB_6 thermal cathode are used with such a lens. The performance of the field emission gun alone is also included. In Fig. 7.14A, the field emission gun is assumed to be operating at a 4 cm working distance. The focal length of the lens is about 5 cm. The dotted lines show the performance that will be obtained if the optimum beam aperture set for the on-axis aberrations of the final lens alone is used for each beam current. The aperture itself determines the current.

The solid lines show the performance that will be obtained if the beam aperture is held at 0.006 rad and not increased for larger beam sizes. 0.006 rad is a conservative limit for the beam half-angle if the beam is to

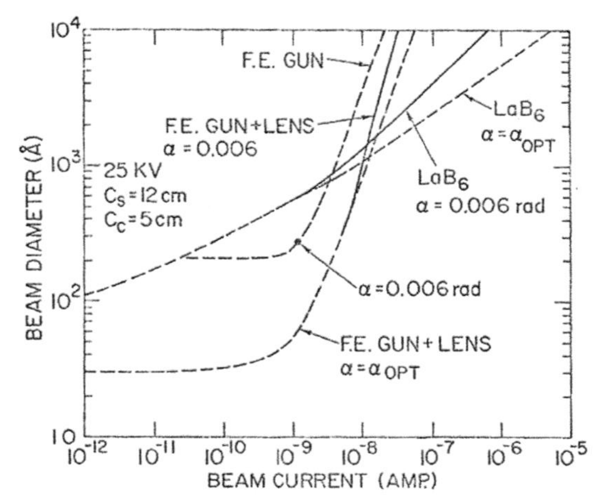

Fig. 7.13 20 kV electron probes suitable for microfabrication (Broers, 1972). *(The Electrochemical Society. Reproduced by permission of IOP Publishing Ltd).*

be scanned over fields 10,000 times larger than the ban diameter with only dynamic focus correction. Dynamic astigmatism correction will probably be necessary if larger field sizes are to be covered.

If a total deflection angle of 4×10^{-2} rad is used, the 5 cm focal length lens will allow a 2 mm diameter field. The field emission gun plus lens will give better performance than the simple gun or the LaB6 thermal cathode for beam sizes up to about 1500 Å. For beam sizes greater than 1500 Å the LaB_6 cathode is superior. The simple field emission gun, when operated at $\alpha = 0.006$ rad is superior to the LaB_6 cathode only up to beam diameters of about 600 Å.

The decision to use LaB_6 Schottky cathode in the new high-resolution electron probe

Field emission cathodes provide the highest brightness and are the most suitable for ultra-high resolution scanning transmission microscopy because they can provide beam currents up to 10^{-10}A into the minimum beam diameter as limited by spherical aberration and diffraction. This is about a hundred times more than that available from thermal cathodes. These higher currents are needed when high-coherence imaging is needed, and the collected current becomes a small fraction of the incident beam current. They require ultra-high vacuum, which usually means that the system uses metal gaskets and is bakeable, but this level of vacuum may be needed in any case to minimize specimen contamination. Many specimens, if they can tolerate it, may still need to be heated to completely prevent contamination. Differential pumping makes it possible to tolerate higher residual pressures elsewhere in the system, for example, to facilitate specimen changing.

The demands upon the electron optical system are less severe for surface scanning electron microscopes and microfabrication. The highest resolution in surface scanning microscopy using secondary electrons is about 5 nm, so the beam size becomes irrelevant if it is below 1 nm. With low-loss scattered electrons, the resolution with metal specimens can approach 1 nm, but with biological specimens and other specimens that have to be coated with a thin layer of metal, the resolution in surface images is above 2 nm.

With microfabrication processes capable of making operational devices, the smallest dimensions fabricated are still about 2.5 nm. Smaller features can be created for example by over-etching, shadowing onto edges, or using the vertical dimension of thin films, but these are limited in their application.

In conclusion, a beam size of about 0.5 nm is adequate to probe the limits of surface microscopy and microfabrication and this can be achieved with LaB_6 cathodes, especially when they are sharp enough (<1 μm radius) and the vacuum level is high enough (10^{-7} mmHg) to operate in the Schottky emission regime with stability.

As discussed above, thermal cathodes can also work over a larger range of currents than field emission sources. With the short focal length used in this case, a Schottky emission LaB_6 cathode provides more current than a field emission cathode for beam sizes above 1 nm.

Because I did not intend to pursue scanning transmission microscopy at the limit of resolution and because I wanted to avoid the inconvenience and cost of an ultra-high vacuum system for the electron gun, I decided to stay with LaB_6 cathodes. However, I did add a new ion pump attached directly to the gun to ensure that the vacuum level in the gun remained below 10^{-7} mmHg.

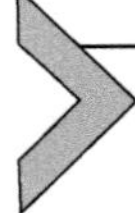

Short focal length LaB_6 cathode electron probe

Electron gun

The LaB6 gun was similar to the gun used in the long-focal length final lens probe but larger so that it could operate at voltages up to 100kv. See Fig. 7.14. The design was optimized for long-life operation with a minimum of thermal drift on warm-up, but any clean and mechanically stable conical LaB_6 emitting surface would have provided similar results. Both sintered and single-crystal cathodes were used with differences in emission distribution but with approximately equal peak brightness for a given operating temperature. The anode could be mechanically aligned from outside the vacuum system.

As with the earlier gun described in Chapter 6 the cathode was heated with radiation and electron bombardment from the surrounding tungsten coil (Broers, 1967). Electron bombardment powers up to 12 W were used and approximately 60 W was supplied to the tungsten coil. Auto-bias was provided by a series resistor connected between the Wehnelt electrode and the cathodes. Total beam currents of approximately 300 μA were used with an auto-bias resistance of about 10 MΩ leading to a bias of ~3000 V. It should be noted that detailed analyses of the rod cathode heating system had led to methods for reducing the power required to heat these rod cathodes (Ahmed et al., 1972; Ahmed and Broers, 1972). In particular, Verhoevan and Gibson (1977) showed that the total power required to reach typical tip temperatures can be realized with only 10 W.

Cathode tips were generally formed with precision mechanical grinding. This method produces tip diameters as small as 5 μm but leaves the cathode surface relatively rough. Single-crystal cathodes were prepared with essentially perfect conical tips (see Fig. 7.15) to confirm that the non-uniformity of emissions is largely due to this surface roughness. These tips, which were prepared using glass polishing techniques, produce smoothly varying angular emission patterns of the form shown in Fig. 7.17. The dark bands are assumed to correspond to crystallographic planes with relatively large work functions. Unfortunately, these cathodes were tested before the Kohler illumination mode was incorporated in the

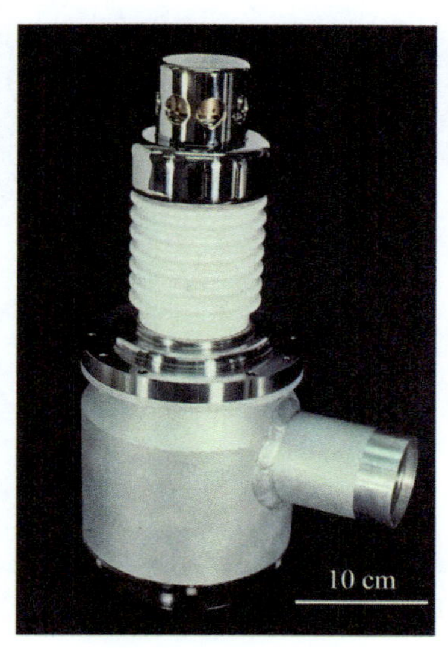

Fig. 7.14 100 kV LaB_6 cathode electron gun. *(Permission to publish granted by the Illinois Institute of Technology, SEM Symposia Office, Metals Research Division, Chicago, Illinois 60616, USA).*

Fig. 7.15 Tip of single crystal LaB_6 cathode (Broers, 1979).

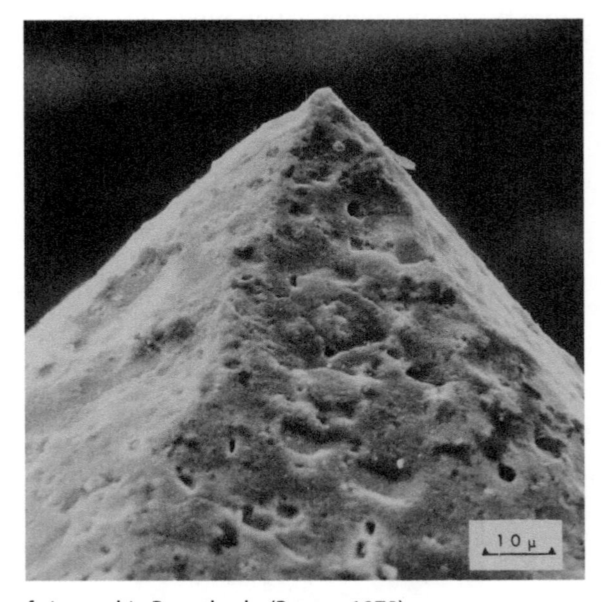

Fig. 7.16 Tip of sintered LaB_6 cathode (Broers, 1979).

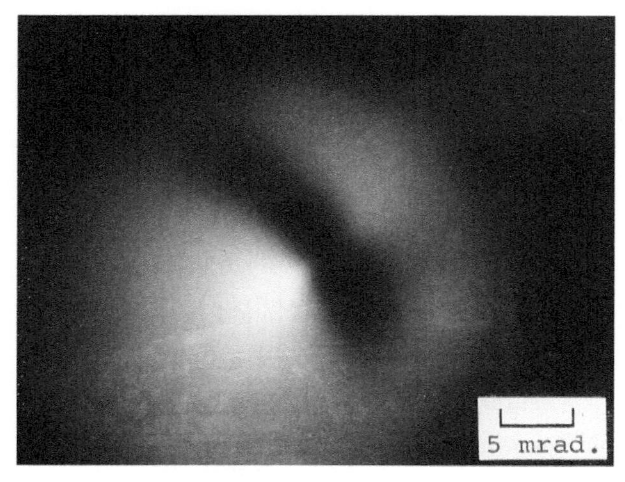

Fig. 7.17 Electron emission pattern for single crystal cathode (Broers, 1979).

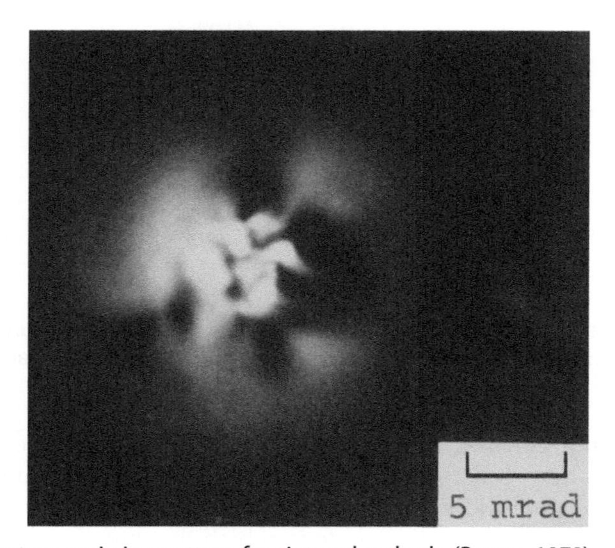

Fig. 7.18 Electron emission pattern for sintered cathode (Broers, 1979).

SEM, and no cross-over images were obtained. Fig. 7.16 shows a sintered cathode, and Fig. 7.18 shows the lobe-like angular emission pattern obtained from such cathode.

Fig. 7.20 shows the electron optical column of the short focal length lens electron probe diagrammatically, and Fig. 7.21 is a photograph of the system, including the anti-vibration table and the vacuum system.

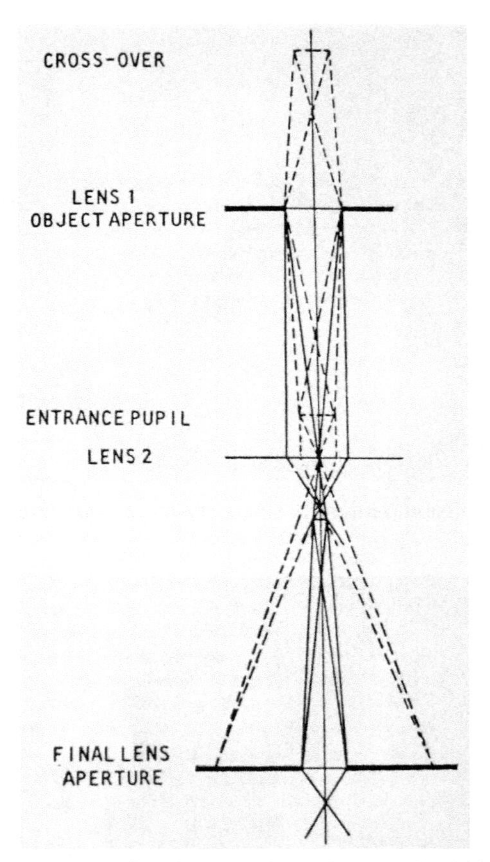

Fig. 7.19 Kohler Illumination. This diagram shows how, with Kohler illumination, the cross-over is imaged onto the final lens aperture, and the beam-shaping (object) aperture in the first lens is imaged onto the sample.

The column consists of the LaB_6 gun, three magnetic lenses, three sets of deflection coils, and an electrostatic blanking plate. Each of the lenses had pole piece bore diameters and gaps of 3 mm. Maximum demagnification is 1000 in the mode in which the gun cross-over is used as the object. Demagnification of 2.5×10^4 is available in the Kohler illumination mode. The object aperture in the Kohler mode is placed in the gap of the first lens (Fig. 7.19). For final beam diameters up to 30 Å, this aperture has a diameter of 5 μm. A 50 μm aperture is used for larger beam diameters.

The new short focal length lens, which I designed for beam energies up to 100 kV, was made by John Sokolowski with such perfection that it had no residual astigmatism. The vacuum system had valves that allowed the lens and chamber below the final lens to be isolated so that pump-down

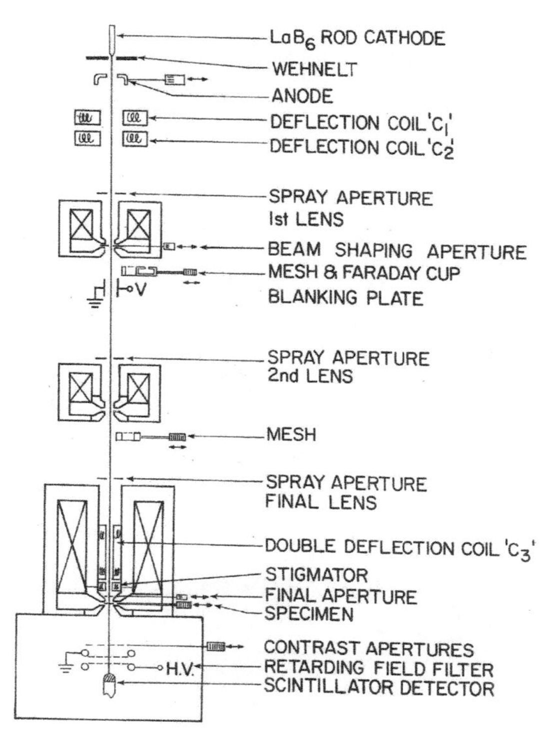

Fig. 7.20 Electron optical column of the short focal length final lens electron Probe (Broers, 1979).

after changing the sample was less than 5 min. The lens was much larger than the lens that had been used for secondary electron surface imaging and weighed 150 Kg, making it impossible to handle without hydraulic assistance. John Sokolowski designed a system that allowed the lens to be lifted hydraulically on two pinions that bolted to the outer diameter of the lens. The lens could then be rotated around a horizontal axis so that the pole pieces were uppermost and could be removed together with the specimen manipulation controls for maintenance and cleaning.

The final beam was scanned with a double deflection coil (C_3) located above the final lens pole-pieces and isolated from the vacuum Additional single deflection coils (C_1 and C_2) were placed between the gun and the first lens to provide means for: (i) examining the angular distribution of current from the cathode (C_1); (ii) examining current density in the cross-over (C_1); and (iii) positioning the brightest portion of the cross-over onto the entrance pupil of the lower half of the demagnification system in the Kohler mode (C_2). C_1 and C_2 contain coils for deflecting the

Fig. 7.21 Short focal length final lens, LaB_6 cathode, 100 kV electron probe. The high voltage cable attached to the front of the electron detector chamber at the base of the column supplies the retarding voltages to the energy filter meshes.

beam in two dimensions. The instrumental scan system was connected to either C_1 or C_3. The same SEM display system was used in each instance. An electrostatic deflection plate was placed between the first and second lenses to blank the beam when the system was used for microfabrication studies. With Kohler illumination, blanking was accomplished with a negligible shift in the final beam position.

Fine meshes could be introduced into the beam at various distances below the first lens. A scanning image of these grids was obtained when the microscope scan system was connected to scan C_1. The magnification of this image was used to calculate the magnification of the images obtained in the cross-over imaging mode described below. The cross-over size was measured from these images. When lens 1 is adjusted to focus the image of the cross-over into the plane of the mesh, an indication of cross-over size could also be gained from the sharpness of the mesh image. Interpretation of this image, however, was difficult compared to the cross-over imaging method, particularly when the cross-over contained several peaks.

Standard mode—Cross-over as the object for electron optical column

In the STANDARD mode (Fig. 7.1), the gun cross-over is imaged onto the sample. Under typical operating conditions, the cross-over is 5–20 μm in diameter, and a demagnification of 5×10^4 nominally yields, in the absence of aberrations, a 1–4 Å diameter image at the sample.

When the instrument scan signal is directed to scan coil C_1, an image of the angular distribution of current from the gun of the type shown in Figs. 7.17, 7.18, and 7.22 appears on the SEM display. (Fig. 7.23 shows the typical angular distribution when a LaB_6 cathode is operated at a temperature well below saturation.). The excitations of lenses 1, 2, and 3 remain unchanged during this operation, and the electron signal detected at the bottom of the microscope is used to modulate the brightness of the image. The angular resolution of this image is given by α_f/M (radian), where

α_f = half-angle of the final beam.

M = demagnification between the cross-over and the final beam.

Typically, $\alpha_f = 10^{-2}$ rad and $M = 50{,}000$ yielding an angular resolution of 2×10^{-7} rad. Angular magnification (cm per radian on the image) of this image (M_A) is given by

$$M_A = \frac{M_C + l_7}{M_1} \text{(cm/radian)}$$

where

l_7 = distance from the gun cross-over to deflection coil C_1,

M_1 is the magnification provided by lens 1 when the cross-over image is focused onto the entrance pupil of lens 2, and M_C is the magnification of the cross-over image as defined in the section on Kohler illumination.

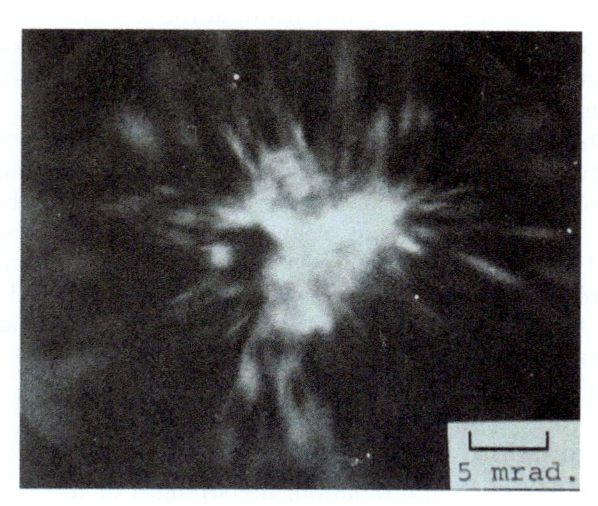

Fig. 7.22 Current distribution from unsaturated LaB_6 cathode (Broers, 1979).

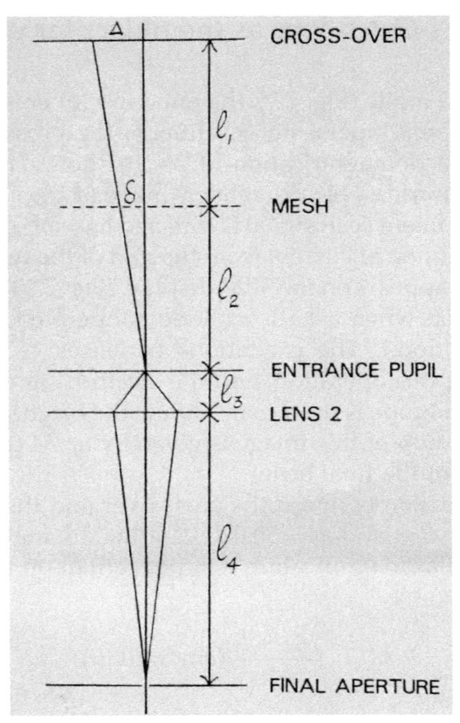

Fig. 7.23 Parameters used in estimating the magnification and resolution obtained in the Kohler illumination cross-over imaging mode (Broers, 1979).

The angular distribution image is used to select the portion of the emitted beam that produces the maximum current at the sample. Alignment coil C_2, or the anode translation mechanism, is used to bring this portion of the beam to the center of the image so that when the scan signal is returned to C_3, the maximum final beam current is obtained.

This method of alignment does not necessarily select the portion of the beam with the highest brightness because the angular distribution image does not contain information about the cross-over size or shape. It has been found, however, that with clean, fully activated cathodes, bright lobes located in the center of the current brightness distribution frequently originated from single, approximately round cross-overs, and the brightness is close to the maximum available. As the sintered cathodes age, for example, after more than 100 h of operation, the cross-over frequently becomes multiple and the method is not satisfactory.

Kohler illumination

To operate with Kohler illumination, the strength of the first lens was reduced until the image of the cross-over was in the plane of the entrance pupil of the demagnification system. A further enlarged image of the cross-over formed in the plane of the final aperture. The entrance pupil of the demagnification system is just above the gap of the second lens. The diameter of the entrance pupil is given by $d_f f_2/(l_4 - f_2)$, where f_2 is the focal length of lens 2, l_4 is the distance from the center of lens 2 to the final aperture, and d_f is the diameter of the final aperture. The adjustment to lens 1 is made while observing the image formed when the scan signal was sent to scan coil C_1. When the signal amplitude was maximized and a sharp image of the cross-over obtained, the cross-over was accurately focused onto the entrance pupil of the demagnifying system and, therefore, onto the final lens aperture. The correct focal setting was most accurately set by using a relatively large (>100 μm) aperture in the gap of the lens 1. Although scan coil C_1 simultaneously changed the angular portion of the beam accepted by the microscope column, this change was negligible for the deflection amplitudes needed to display the cross-over. Nakagawa has used a related method for displaying the cross-over, and Mumaw and Munger have obtained similar images of LaB6 gun cross-overs to those shown here in a conventional TEM (Mumaw and Munger, 1977). Sewell and Ramachandran observed cross-overs with a single-lens scanning method (Sewell and Ramachandran, 1977).

Optimum alignment at the illumination system was obtained by bringing the brightest portion of the cross-over to the center of the image using deflection coil C_2. Unlike the adjustment made with the angular current distribution, this procedure did select the portion of the electron beam with the highest electron brightness, at least for a particular gun alignment (anode placement).

Gun anode alignment was initially set by observing the cross-over image while operating the cathode at a low temperature. In this unsaturated condition, the cross-over consisted of a ring and, in some instances, a central spot. When the anode was accurately centered, the ring was approximately symmetrical (see Fig. 7.24A). Fig. 7.24B shows the case when the anode was misaligned, and the ring became asymmetrical. After the cathode was returned to its full operating temperature, small additional adjustments of the anode position were needed to reach optimum brightness. The highest intensity in the cross-over image was sought in this final adjustment, which effectively selected the optimum emission angle.

To find the size of the gun cross-over, it was necessary to know the magnification (M_C) of the cross-over image obtained on the SEM display. To obtain this magnification the sensitivity and exact effective position of deflection coil C_1 had to be determined. A scanning image of a mesh of a known period placed below lens 1 was used for this purpose. Lens 1 was deactivated, and the microscope scan signal was directed to C_1 to obtain the mesh image. The electron signal arriving at the bottom of the electron

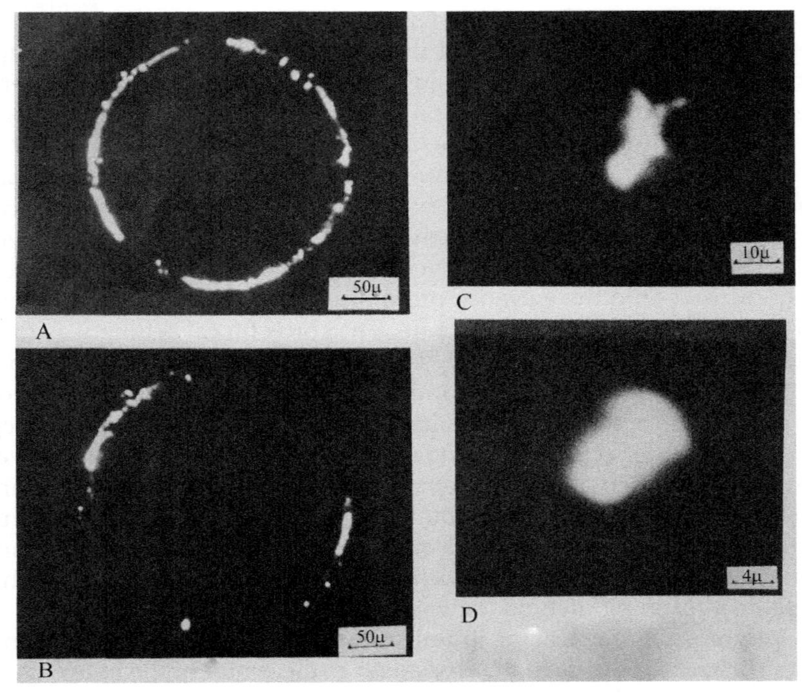

Fig. 7.24 (A) Cross-over image obtained at a low cathode temperature (1400°C) with the mode correctly aligned. (B) The same cathode but with the anode slightly misaligned. (C) and (D) are cross-over images from sintered cathodes at an operating temperature of ~1550°C (Broers, 1979).

optical column was used to modulate the brightness of the display. Deflection coil C_1 introduced an effective shift (Δ) of the cross-over away from the microscope axis, where

$$\Delta = \frac{\delta(l_1 + l_2)}{l_2}$$

and

δ = distance between the axis and the rays from the cross-over that reached the electron detector.

l_1 = distance between the cross-over to the mesh and l_2 = distance from the mesh to the entrance pupil of the demagnifying system, as shown in Fig. 7.24.

The angular portion of the beam reaching the electron detector changed during this deflection, but this is unimportant in determining the sensitivity and the effective position of deflection coil C_3.

If M_M is the magnification of the mesh image, then the magnification of the cross-over image M_C is given by

$$M_C = M_M \frac{\delta}{\Delta} x M_1$$

where M_1 is the magnification subsequently provided by lens 1 when the cross-over was imaged onto the entrance pupil of the demagnification system.

In these experiments

$$M_C = 0.4 M_M$$

Resolution R of the cross-over image was given by $R = d_f l_3 / l_4$. Under typical conditions $l_3 = 0.5$ cm, $l_4 = 31$ cm, $d_f = 20$ μm and $R = 0.3$ μm.

Figure 7.24A and B are images of cross-overs taken under various conditions. Fig. 7.24A shows the cross-over shape obtained from a typical sintered LaB_6 cathode when the operating temperature was too low. Under these conditions, a large area of the cathode was emitting, and a hollow cross-over, occasionally with a central spot, was formed. This was similar to the unsaturated emission condition for a tungsten filament electron gun and corresponds to the case where electrons were drawn from the sides of the cathode cone. As the temperature was increased, current density increased, and for a given bias condition, the emitting area was reduced (Fig. 7.24C). At a temperature of about 1450°C, the cross-over generally became singular for total beam currents below about 250 μA (see, for example, Fig. 7.24D). At higher beam currents, the side walls of the conical cathode again emitted, and the singular shape of the cross-over was lost. With Kohler illumination, this did not reduce effective brightness, although increased beam currents might have led to greater energy spread as a result of the Boersch effect (Boersch, 1954).

A disadvantage of the Kohler operating mode was that the alignment of the focused electron beam onto the entrance pupil of the second lens was very critical. If the ross-over image became displaced by a distance equal to the diameter of the image, the final beam current fell to zero. With a conventional cross-over imaging system, such an effective shift of the cross-over only results in a displacement of the final beam by a single beam diameter. The final beam current is not affected. A beam alignment servo can, in principle, be used to maintain adequate alignment for the Kohler mode. With the instrument used, the drift of the cross-over image fell to below 0.1 μm/min (referred to the cross-over) after 2 h of operation.

Measurement of brightness

To measure the maximum brightness accurately, two important precautions had to be taken.

1. A small enough ($<10\,\mu m^2$) portion of the cross-over had to be sampled to ensure uniformity of current density at the maximum value.
2. A small solid angle ($<5\times10^{-4}$ rad) had to be sampled in order to realize the highest available current per unit solid angle.

Independent adjustment to achieve 1 and 2 was possible.

The brightness (β) of the LaB_6 gun was measured while operating in the Kohler illumination mode. The following relationship was used

$$\beta = I_f\left[\frac{4l_4l_5}{\pi l_3 d_f d_k}\right]^2 \text{amp/cm.}^2\text{ster.}$$

where

I_f = the final microscope beam current—measured with a pico-ammeter connected to a Faraday cage placed below the final lens.

d_f = the diameter of the final lens-stopping aperture as measured with an optical microscope.

l_5 = the distance from the aperture in lens 1 to the entrance pupil of lens 2.

l_1 = the distance from the entrance pupil of lens 2 to the center of the pole-piece gap in lens 2. The focal length of lens 2 was calibrated by adjusting lens 2 focus while observing an image of a mesh introduced at a known distance below lens 2. The signal was directed to scan coil C_1 during this observation, and the signal was detected at the bottom of the microscope column. Lens 1 focus was reduced to a few mm for this measurement and lens 2 formed a demagnified image of the gun cross-over in the plane of the mesh.

l_4 = the distance from the center of len2 to the final lens aperture.

d_k = the diameter of the aperture in the 1st lens—measured electron optically.

For the dimensions employed, a 3 μm diameter circular section of the cross-over was sampled from a solid angle of 1.75×10^{-5} rad. Fig. 7.25 shows the brightness obtained for three cathodes for different heating powers. The accuracy of the experiment was estimated to be ±15%. Cross-over diameters were between 7 and 10 μm. It should be noted that neither lifetimes nor cathode temperatures were measured for these cathodes. Experience with rod cathodes indicates that lifetimes over a 100h can be obtained up to bombardment powers of ~9 watts (Ahmed et al., 1972). The single crystal cathode had a cross-section approximately half that of the sintered cathode and, therefore, reached a higher temperature for a given heater power. The lifetime available with this cathode for the higher heater powers used in these measurements, therefore, would probably be impractically short. The maximum practical heater power was probably about 5 watts, corresponding to a brightness of 10^7 A/cm^2.ster. (44.5 kV). A similar brightness was measured from the sharper of the sintered cathodes.

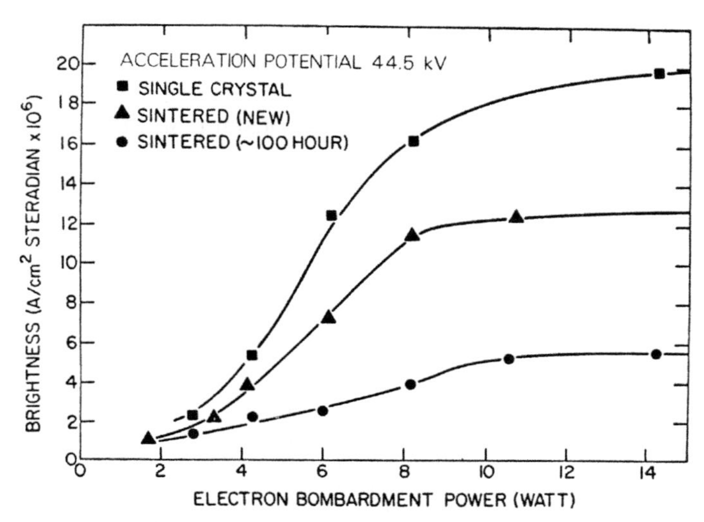

Fig. 7.25 Brightness measured for three LaB_6 cathodes. The single crystal cathode had a tip diameter of about 2 μm and approximately half the cross-section of the 1 mm square cathodes. It was round with a diameter of 0.8 mm. As a result, the temperature attained at a given heater power was higher. Unfortunately, it was not possible in these experiments to measure the cathode temperature. The "used" sintered cathode had been in routine use for about 100 h. The tip radius was approximately 10 μm compared to 4 μm for the new sinetered cathode. The coil heating power was about 60 W in all cases (Broers, 1979).

When the cross-over is multiple, as in Figure 7.24C, the ratio of the brightness measured from the brightest portion of the cross-over to the average brightness when 80% of the current in the cross-over image is accepted can be as much as 5:1. Even greater differences were measured between the lobes in the cross-over and the darker portion between the lobes.

Anode alignment was found to be critical and can affect the shape and intensity of the cross-over. Variations of a factor of 2–4 have been observed in measured brightness for an anode misalignment of 0.05 mm. This is believed to be due to a change in the accepted cone of radiation. In addition to anode alignment, brightness for a particular cathode varied with bias conditions, tip shape, cathode position in the Wehnelt, crystallographic orientation of the tip, and, of course, cathode temperature.

Despite the proliferation of variables that influence brightness, the portion of the beam with the highest brightness for a given cathode and Wehnelt geometry can be chosen when operating in the Kohler mode. To do this, the gun anode must first be aligned (in the SEM here, the anode was translated mechanically, but a deflection coil placed close to the anode could also be used) to obtain the portion of the cross-over with the highest intensity, and then the brightest portion of the cross-over is brought onto the axis of the microscope using alignment coil C_2 (Fig. 7.20). The intensity of the cross-over image provides a direct measure of brightness, provided the excitation of the second lens remains constant, and the first lens and final lens apertures are not changed,

References

Ahmed, H., Broers, A.N., 1972. Lanthanum hexaboride electron emitter. J. Appl. Phys. 43, 2185–2192.

Ahmed, H., Blair, W., Lane, R., 1972. Rev. Sci. Instrum. 43, 1048–1051.

Boersch, H., 1954. Z. Phys. 139, 115.

Broers, A.N., 1967. Electron gun using long-life lanthanum hexaboride cathode. J. Appl. Phys. 38, 1991–1992.

Broers, A.N., 1968. High brightness electron gun using a field emission cathode. In: Arceneaux, C.J. (Ed.), Proceedings 26th Annual Meeting Electron Microscopy Soc. of Am. Claitor's Publishing Division, Baton Rouge, LA, pp. 294–295.

Broers, A.N., 1972. Electron and ion probes. In: Proc. Internat. Electron Ion and Laser Beam Conf. Electrochemical Society, Houston, pp. 3–25.

Broers, A.N., 1979. Thermal cathode illumination systems for round beam electron probe systems. In: Proc. SEM/1979/1—Washington D.C. : AMF O'Hare, IL, pp. 1–10.

Broers, A.N., Pfeiffer, H.C., 1971. Minimum beam diameter obtainable in electron probe apparatus. In: Thornley, R.F.M. (Ed.), Record of 11th Symposium on Electron Ion & Laser Beam Technol. San Francisco Press, Inc, pp. 205–207.

Butler, J.W., 1966. Proc. 6th Annual Cong. Electron Microscopy, Rome., p. 191.

Crewe, A.V., 1970. The current state of scanning electron microscopy. Q. Rev. Biophys. 3, 137–175.

Crewe, A.V., Wall, J., 1970. A scanning electron microscope with 5A resolutions. J. Mol. Biol. 48, 375–393.

Mumaw, V.W., Munger, B.L., 1977. In: Bailey, G.W. (Ed.), Proc. 35th Annual EMSA Meeting. Claitors, Baton Rouge, Louisiana, pp. 64–65.

Pearce-Percy, H.T., Spicer, D.F., Abbot, M., Winborn, C., Varnell, G., 1979. High speed electron optics for direct slice writing. J. Vac. Sci. Technol. 16, 1794–1799.

Pfeiffer, H.C., 1971. In: Thornley, R.F.M. (Ed.), Record if the 11th Symp. on electron and Laser Beam Technol. Boulder, p. 239.

Sewell, P.B., Ramachandran, K.N., 1977. SEM1977/I. IIT Research Institute Ilinois, pp. 17–24.

Verhoevan, J.D., Gibson, E.D., 1977. In: Johari, O. (Ed.), SEM/1977/I. IIT Research Institute, Illinois, Chicago, pp. 9–16.

CHAPTER EIGHT

Performance of the high resolution short focal length final lens electron probe and low-loss surface scanning electron microscopy

Contents

Probe performance

The electron optical performance of the short focal length probe (Broers, 1973b) was first measured using scanning transmission micrographs. Figs. 8.1 and 8.2 are STEM images of gold particles near a 0.25 μm diameter latex sphere. The edge definition on the Au particles and details in the phase contrast image of the supporting carbon film indicate that the beam diameter was close to the theoretical minimum of about 5 Å set by spherical aberration and diffraction ($C_S^{1/4}\lambda^{3/4}$ where $C_S = 0.44$ mm, $C_C = 0.6$ mm) at 50 kV. Fig. 8.3 shows a catalase crystal negatively stained with phosphotungstate with lattice spacings of 8.8 nm and 6.9 nm (Broers, 1973a, b).

Advances in Imaging and Electron Physics, Volume 231
ISSN 1076-5670
https://doi.org/10.1016/B978-0-443-31462-9.00008-3

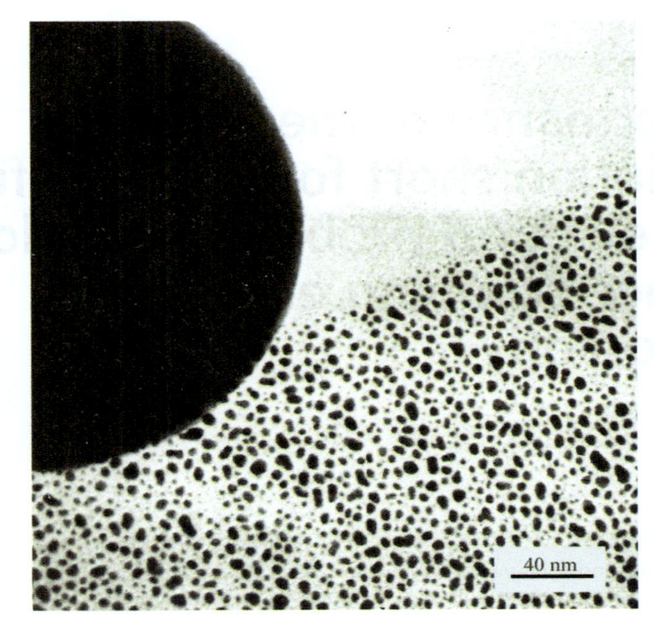

Fig. 8.1 Gold particles near 0.25 µm latex sphere.

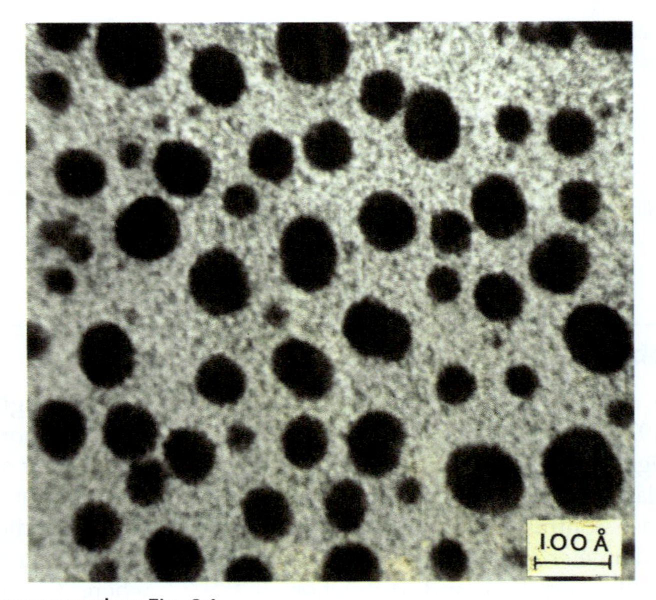

Fig. 8.2 Same sample a Fig. 8.1.

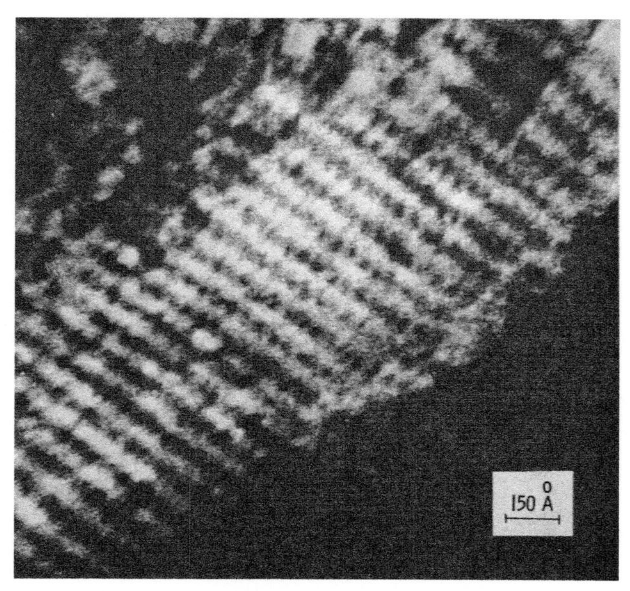

Fig. 8.3 Catalase Crystal. Used with permission of AIP. *(From Broers A.N. 1973a. High resolution scanning electron microscopy of surfaces. Microprobe Analysis. Anderson C.A. New York: John Wiley and Sons; Broers A.N. 1973b. High-resolution thermionic cathode scanning electron microscope. Appl. Phys. Lett. 22, 610–612 permission conveyed through Copyright Clearance Center, Inc.).*

The theoretical beam diameter versus beam current for a beam voltage of 50 kV using the formulae of Smith (1956) is shown in Fig. 8.6. The beam current for Figs. 8.1–8.3 was 10^{-12}A, the convergent angle of the beam was 2×10^{-2} rad, and the collector half angle was $\sim6\times10^{-3}$ rad. Fig. 8.4 is a dark-field image of the same sample as in Figs. 8.1 and 8.2 with a beam current of 5×10^{-11}A.

Fig. 8.5 shows images obtained with beam currents of (A) 10^{-9}A, (B) 10^{-8}A, (C) 10^{-7}A, and (D) 10^{-6}A. The beam diameters in each case were estimated to be (A) 30 ± 8 Å, (B) 80 ± 20 Å, (C) 200 ± 40 Å, and (D) 400 ± 100 Å. These experimental points are superimposed on Fig. 8.6, which shows the beam diameter vs beam current for the LaB_6 cathode probe for a brightness of 10^7 A/cm^2.ster. at 50 kV using the formulae of Ken Smith used in Chapter 2. The energy spread in the beam was assumed to be 2 eV.

The measurement of brightness is discussed in Chapter 7.

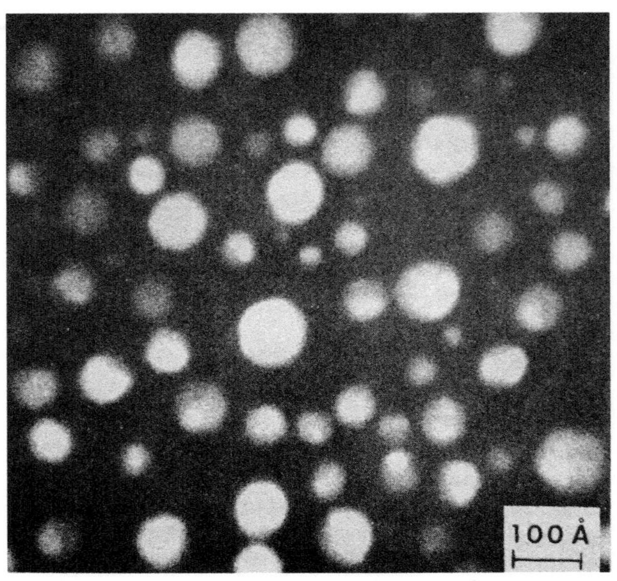

Fig. 8.4 Dark-field image. Beam current 5×10^{-11}A. *(Reprinted from Broers A.N. 1979. Kohler illumination and brightness measurement with lanthanum hexaboride cathodes. J. Vac. Sci. Technol. A, 16. 1692–1698, with the permission of AIP Publishing).*

Resolution limit and contamination

The fivefold decrease in beam size, compared to that obtained with the longer focal length lens needed for secondary electron imaging, was expected but gratifying. It was also pleasing that there was no detectable astigmatism. What was not so pleasing was that the rate of contamination with most samples was very high. This had been observed by others, especially by Albert Crewe with his field emission STEM. This was not expected in his case because the vacuum level in his microscope was 10^{-10} mmHg. It seemed that the electric field gradients that surround these very intense sub-nanometer electron beams attract surface contaminants to the beam, where the electrons disturb the stationary state by decomposing the contaminants into fragments. They are then cross-linked (polymerized) on the sample surface, and a carbon-rich film starts to grow (Hillier, 1948).

It had been reported that heating the sample reduced contamination, see for example (Ennos, 1953). Ennos (1954) so I decided to do this in situ by shining an intense light beam onto the sample. The simple apparatus I used is shown in Fig. 8.7. It consisted of a tungsten filament from a 50 W quartz lightbulb and a spherical gold-plated mirror. The apparatus was swung into a position between the final lens and the electron detector where the mirror projected the image of the filament onto the sample. The result of

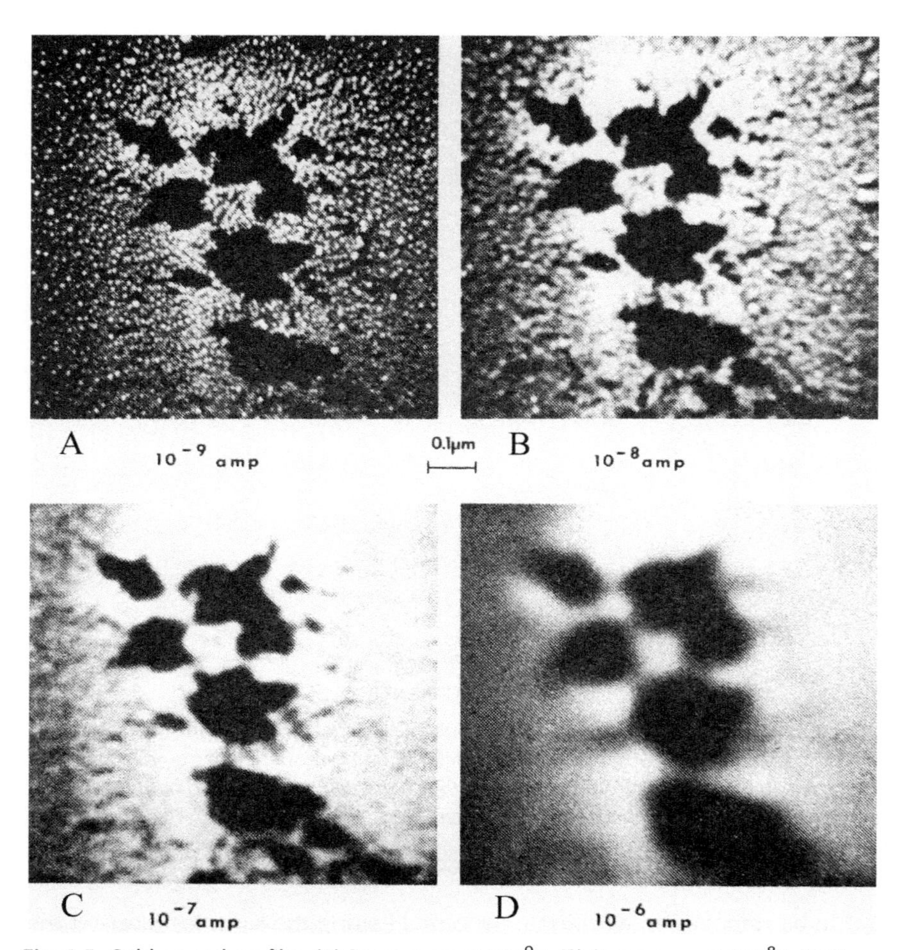

Fig. 8.5 Gold on carbon film. (A) Beam current 10^{-9}A, (B) Beam current 10^{-8}A, (C) Bean current 10^{-7}A, (D) Beam current 10^{-6}A. *(Reprinted from Broers A.N. 1979. Kohler illumination and brightness measurement with lanthanum hexaboride cathodes. J. Vac. Sci. Technol. A, 16, 1692–1698, with the permission of AIP Publishing).*

heating the sample in this way for a few seconds was quite remarkable. Samples such as catalase and graphitized carbon could be examined for about 5 min in STEM mode without any sign of contamination. Since 1970, contamination has been studied in depth, but the mechanisms are still speculative. Perhaps heating reduces contamination because the mobility of the contaminants on the surface is reduced by removing liquids. In any case, knowing this prompted Xiaodan Pan and me many years later in Cambridge to try heating samples of SiO_2, upon which we were writing patterns by direct electron bombardment. Bombardment enhanced the etch rate of SiO_2 in hydrofluoric acid but contamination was blocking the etch and

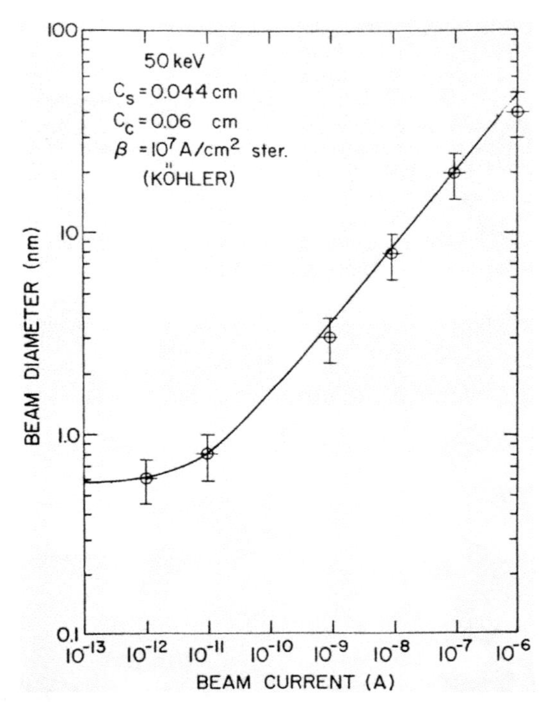

Fig. 8.6 Beam diameter versus beam current for the short focal length final lens electron probe operating at 50 keV in the Kohler illumination mode and a gun brightness of 10^7 A/cm^2.ster. *(Reprinted from Broers A.N. 1979. Kohler illumination and brightness measurement with lanthanum hexaboride cathodes. J. Vac. Sci. Technol. A, 16, 1692–1698, with the permission of AIP Publishing).*

had to be removed in a separate process. Heating the samples reduced the contamination rate to a point where this was no longer necessary (Pan and Broers, 1993).

The ability to examine samples in STEM mode without contaminating them was particularly valuable when examining nanostructures.

Surface microscopy with low-loss scattered electron

As I have discussed, scanning electron microscopes have mainly been used in the surface imaging mode in which secondary electrons excited by the primary electrons are drawn to an Everhart Thornley scintillator detector, and the signal from the detector is used to modulate the brightness of the image. To collect the secondary electrons efficiently, the sample has to be kept out of the magnetic field of the final probe-forming lens, meaning that this lens has to have a relatively large focal length and consequently high aberrations (C_S = 1.6 cm, C_C = 1 cm). Despite this, as

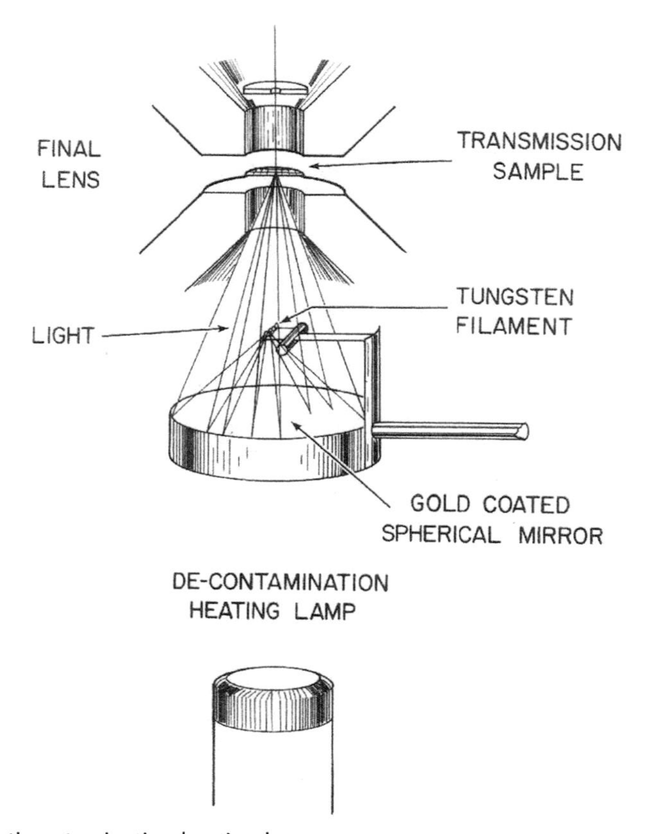

Fig. 8.7 Anti-contamination heating lamp.

already discussed, the ultimate resolution in the surface images was not set by the electron optics but by the size of the region from which secondary electrons emerge from the sample. The minimum beam diameter of the IBM SEM had been about 3 nm and the size of the region from which secondary electrons emerge appears to be between 5 and 10 nm.

Wells et al. (1973) had for several years been exploring the use of high energy electrons scattered from the sample surface with little energy loss to generate the signal for surface images, rather than secondary electrons. One of the motivations for building the new short focal length probe was to see if it would be possible, with the sample placed in the middle of the short focal length lens, to collect and detect these low-loss electrons and take advantage of the smaller electron beam. Low-loss electrons are beam electrons that have been scattered through less than about 20 degrees and have lost little energy. Provided one was examining a portion of the sample close to the edge of the sample holder, a significant fraction of these electrons should pass around the edge and be focused back parallel to

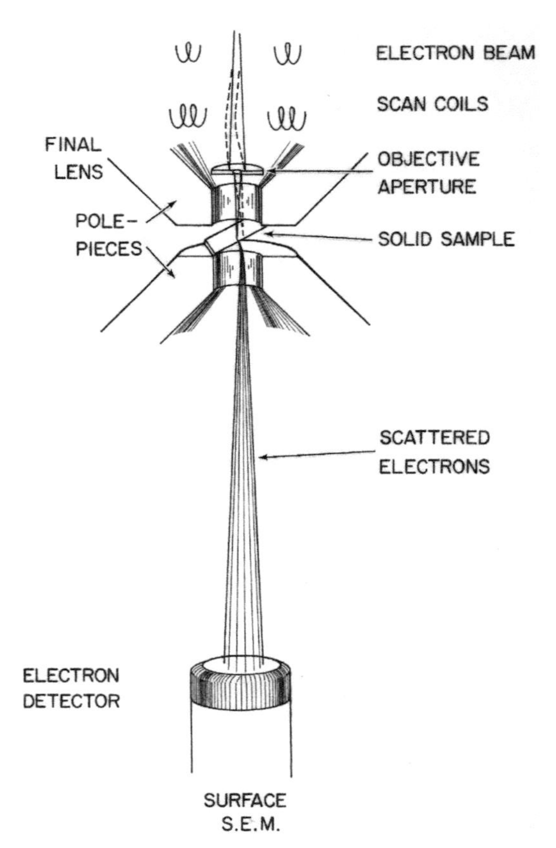

Fig. 8.8 Sample and beam configuration used for low-loss surface microscopy.

the axis of the lens by the second half of the lens field. They could then be detected with an on-axis scintillator below the lens.

The general configuration for low-loss imaging is shown in Fig. 8.8. Electrons scattered through small angles with little energy loss are redirected back onto the axis by the lower half of the lens field and strike the detector. Those scattered through larger angles with higher energy loss are over-focused by the lens field and miss the detector. They are not shown in the diagram. However, the most efficient method for detecting electrons scattered with little energy loss is not to select them by the angle they are scattered through but to place an energy filter between the lens and the detector.

Because I had wanted to measure the energy spread in the beam created by Boersch Effect I had already designed a retarding field energy filter that could be used to select and detect the low-loss electrons emerging from the final lens (Broers et al., 1975). This detector/filter is shown in Fig. 8.9. It consisted of two, flat, 5 cm meshes of 90% transparency

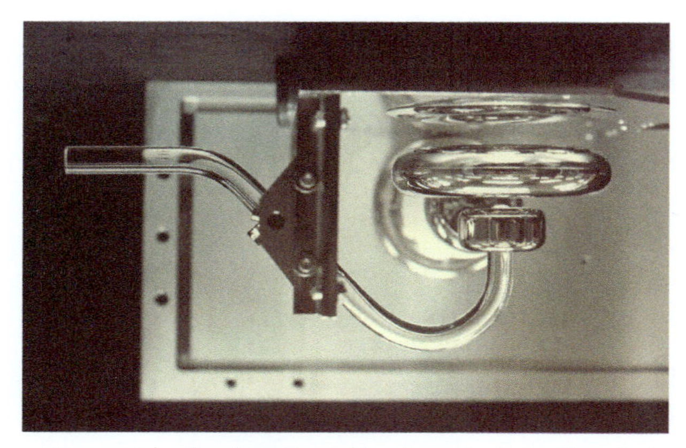

Fig. 8.9 Apparatus used for the detection of low-loss electrons. The electrode between the lens and the detector is held at a voltage close to the beam voltage.

(25 μm wire on a 0.5 mm pitch) spaced 2 cm apart. The first mesh was 10 cm below the lens gap and held at ground potential and the second at a potential close to the cathode potential, up to 75 kV. The high-voltage mesh is surrounded by a "donut" to suppress unwanted electrical discharges. The meshes were parallel to each other to within $0^{\circ}2'$ and perpendicular to the beam within $0^{\circ}3'$. A quartz light pipe coated with plastic scintillator and a thin layer of aluminum was used as the electron detector. A 500 MΩ voltage divider was used to hold the scintillator at 80% of the potential applied to the high-voltage mesh. The resolution of the filter was about 300 V when the beam width at the first mesh was several millimeters, which was typical for low-loss microscopy.

As shown in Fig. 8.10, a movable aperture plate was located immediately above the first mesh. It contained apertures varying in diameter from 20 μm to 1 cm. The 1 cm aperture was used for low-loss surface microscopy.

The smaller apertures were used to eliminate scattered electrons and provide contrast for transmission microscopy. A dark-field stop was also included.

All specimens were prepared on 1 mm × 10 mm cleaved silicon strips, which were mounted with a glancing angle of incidence of 20 to 30 degrees as shown in Fig. 8.8. Satisfactory images could be obtained up to 150 μm from the edge of the sample. The electron beam was initially focused and adjusted using a transmission specimen that was located at the same level as the edge of the silicon strip. It was then focused on the sharp edge of the silicon strip. After this the strip was advanced and the focus current and the photo-multiplier voltage increased to look for details on the upper surface of the strip. This procedure worked well and we were immediately able to obtain images with satisfactory contrast and signal-to-noise with a beam current of about 10^{-11} Amp. Fig. 8.11 shows an image of a 0.5 μm

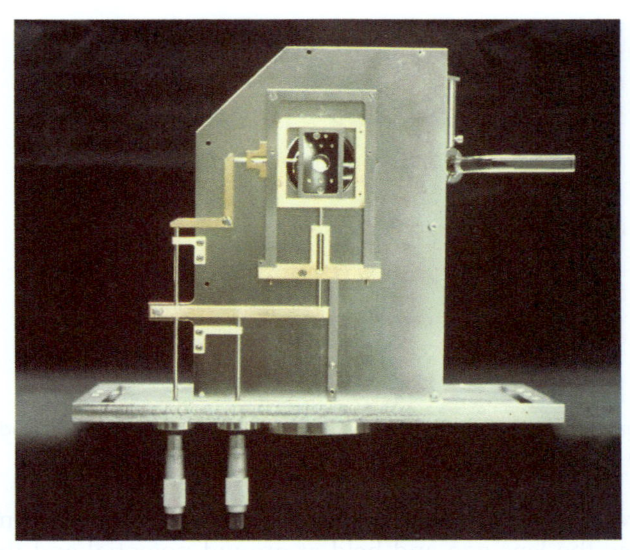

Fig. 8.10 Top view of retarding energy filter showing the mechanism for locating a variety of apertures between the lens and the detector. The energy filter was built by John Sokolowski.

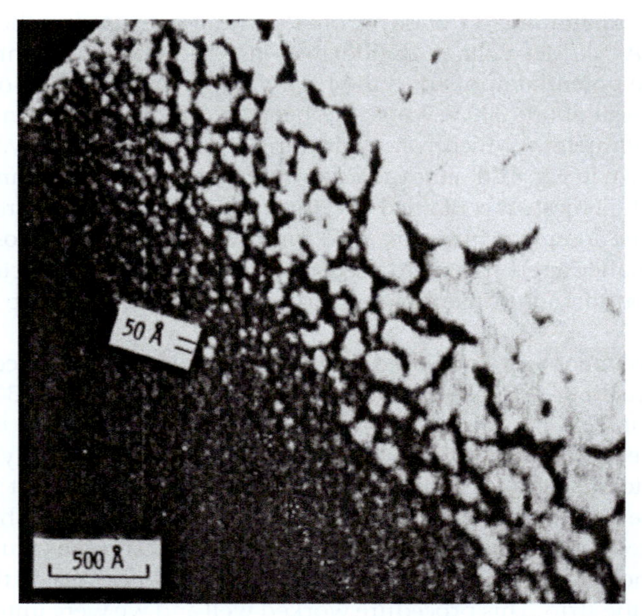

Fig. 8.11 AuPd dots on a 0.5 μm latex sphere. *(Reprinted from Wells O.C., Broers, A.N., and Bremer, C.G. 1973. Method for examining solid specimens with improved resolution in the scanning electron microscope. Appl. Phys. Lett. 23, 353–355 with the permission of AIP Publishing).*

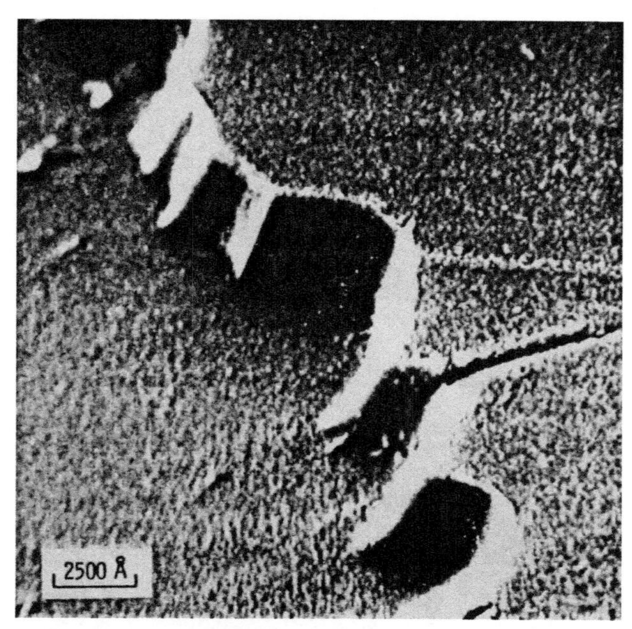

Fig. 8.12 Etched step in 400 nm SiO_2 layer using the energy filter to reject electrons that had lost more than 230 eV at 23 kV. *(Reprinted from Wells O.C., Broers, A.N., and Bremer, C.G. 1973. Method for examining solid specimens with improved resolution in the scanning electron microscope. Appl. Phys. Lett. 23, 353–355 with the permission of AIP Publishing).*

diameter latex ball partially covered with a thin layer of gold and Fig. 8.12 a silicon oxide step on a silicon wafer. The edge sharpness was about 1.5 nm and point-to-point separations between gold particles down to about 3 nm were visible. These images were obtained with a beam current of 7×10^{-12} A and a beam voltage of 65 kV. The detector used for these images was a quartz light pipe 1 cm in diameter coated with a plastic scintillator and a thin layer of aluminum. It was placed approximately 14 cm below the center of the lens gap.

Figs. 8.13 and 8.14 show how the energy filter improves the low-loss image by rejecting electrons that have penetrated the sample and lost more than 230 eV. The beam voltage was 23 kV and the beam current was approximately 10^{-11} Amp. Fig. 8.15 shows the 25 nm period band of the superlattice resolution test sample shown in Fig. 7.12.

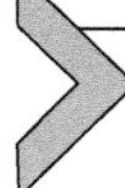

Low-loss scanning electron microscopy of biological specimens

To test this new form of scanning electron microscopy of surfaces, a study of bacterial viruses and other biological samples was launched in

Fig. 8.13 Silicon sample coated with 50 Å of gold–palladium. (A) was obtained without the filter and (B) with the filter rejecting slower electrons (Broers et al., 1975).

collaboration with Barbara Panessa and Joseph Gennaro Jr. at New York University (Broers et al., 1975; Panessa-Warren and Broers, 1979). Until the development of this new type of low-loss scanning electron microscopy it had not been possible to examine the surface of bulk biological samples with a resolution comparable to that of transmission electron microscopes.

Low Loss SEM offered this resolution. Specimen size and low magnification operation were restricted compared to the standard SEM because the sample was placed in the small lens gap, but this did not prove a

Fig. 8.14 Aluminum contacts of an integrated circuit. Upper image without the filter and lower image with the filter.

significant problem for examining small biological samples such as viruses, and the images, like those of the standard SEM, were easily interpreted in three dimensions.

Sample coating

A major issue with the low-loss technique is that it is imaging the thin metal coating rather than the biological tissue itself. The primary reasons for coating the samples are to increase the number of scattered electrons needed to form the low-loss image and to improve electrical and thermal

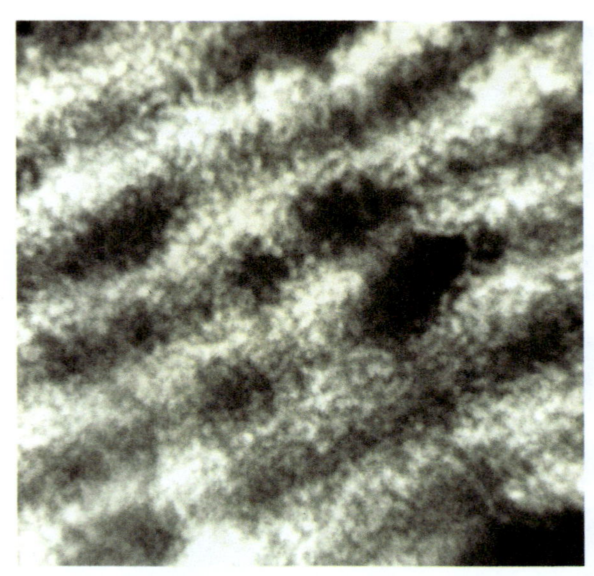

Fig. 8.15 GaAS-GaAsP superlattice SEM resolution test sample. Ridge period 25 nm. *(Reprinted from Broers A.N. 1979. Kohler illumination and brightness measurement with lanthanum hexaboride cathodes. J. Vac. Sci. Technol. A, 16, 1692–1698, with the permission of AIP Publishing).*

conductivity. At the same time, the coating must be thin enough not to obscure minute surface detail. We were not able to make the coating uniform enough to reveal details below 20–30 Å. A similar issue exists with the staining of samples for transmission electron imaging where most of the contrast is due to scattering of electrons by the stain rather than by the biological tissue. Replicas that are peeled off and examined in transmission are also limited by the structure of the replica material itself.

In the standard SEM, coatings are also used to increase the number of secondary electrons and to avoid sample charging, because it interferes with their collection. However, because the resolution is set by the size of the area from which secondaries are emitted, which is 50 Å to 150 Å, the grain size of the coating is seldom visible and doesn't limit the resolution.

It is possible with low voltage beams, and environmental sample chambers to examine biological samples in the standard SEM without drying the samples, but only at relatively low resolution.

Our dried samples were coated on a rotary tilt holder with a thin layer of carbon followed by platinum-palladium or gold–palladium. Total layer thicknesses were between 20 Å and 50 Å. Charging problems are less severe in low-loss imaging because the primary and scattered electrons are high-energy electrons and are, therefore, less influenced by the relatively small surface fields created if portions of the sample charge up. Similarly, layers of carbonaceous contamination that build up on the surface after prolonged examination at high magnification, have little effect on the image. The T4 coliphage shown in Fig. 3 was used for demonstrating the resolution of the technique for several weeks without the image being significantly changed. I will discuss the coating problem further at the end of this section.

Sample staining

We found that samples treated with heavy metal stains, and coated with a thin layer of metal, exhibit a differential backscatter emission, giving rise to bright and dark areas within the specimens that indicate the stain localizations. Dividing Staphylococcal cells treated with 2% ethanolic-phosphotungstic acid (PTA) during dehydration showed increased signal from the periphery of the daughter cells and a marked decrease in the signal in the central region (Figs. 8.18–8.20). Similarly, some T4-coliphages treated with 2% aqueous filtered uranyl acetate after fixation and another staining bath after acetone dehydration (1% uranyl acetate in absolute acetone) are seen with capsids, some of which appear solid, while others are transparent. The latter may have lost their nucleic acid, whereas those with solid heads contain stained DNA which scatters more electrons and makes them appear brighter.

At about the same time as we were examining these biological samples with low-loss microscopy with a short focal length final lens SEM, Kondo, and Hasegawa at JEOL were also using a short focal length lens to obtain images with secondary electrons. Their results showed improvement over standard SEM results but did not have the clarity of the low-loss images at high magnifications.

The bacteriophage images were obtained with a beam current of 2×10^{-11}amp. The current reaching the detector varied between 5×10^{-13} amp and 1.5×10^{-12} amp.

Micrographs of phage obtained in STEM compare closely with those obtained with high-resolution TEM. The periodic structure in the tail of the 3C phage shown in Fig. 8.18A is resolved as is the structure in the tail of the T4 phage shown in the low-loss surface image, see Fig. 8.17.

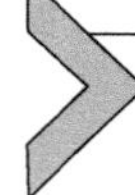

Preparation and examination of 3C and T4 Bacteriophage

Log phase *Staphylococcus aureus* (strain 3C), grown on nutrient agar, was inoculated with 3C phage and incubated at 30 °C. Samples were taken at 1-, 2-, and 5-h intervals after inoculation and were prepared for STEM and low-loss SEM. A small area of the inoculated plate was scraped with a transfer loop, and the specimens were placed on a Formvar–carbon-coated copper grid. The residual stain was drained off the grid, and the specimen was stained with 2% aqueous, filtered uranyl acetate. Specimens for low-loss SEM were taken at the same time and were placed on an oxidized silicon sliver ($9 \times 1 \times 0.4$ mm) and fixed in 1.5% glutaraldehyde in 0.1 M cacodylate buffer (pH 7.4) for 1 to 5 h at room temperature. The slivers were then washed in 0.1 M cacodylate buffer, post-fixed in 1% aqueous osmium tetroxide at room temperature and washed in distilled water. To increase the backscatter yield some of the samples were bathed in 2% aqueous uranyl acetate for 1 h (ref). The residual stain was removed by washing with distilled water. The specimens were then dehydrated in acetone, dried by the critical point method (ref), and coated with a thin layer of carbon-platinum-palladium or carbon-gold–palladium.

T4 coliphages were prepared by inoculating *Escerichia coli* (strain B) grown on nutrient agar (to which tryptophan had been added) with T4 bacteriophage and grown at 37°C. Samples were taken at 10 and 30 min after inoculation and prepared as described for 3C phage.

We first examined T4 coliphage, as shown in Figs. 8.16 and 8.17. The metal coating was 50 nm for Fig. 8.16 and 25 nm for Fig. 8.17. Fig. 8.16 shows the head, collar (white arrows), tail, and end plate of a T4 coliphage coated with 50 Å of gold–palladium. Small projections (S) arising from the end plate are similar in location and size to the spikes, which function in phage absorption. Rod-like structures (black arrows) were frequently observed extending from the end plate or lying against the phage tail and projecting up into the phage head. These may be tail fibers that are partially obscured by the metal coating. Fig. 8.17 shows T4 coliphages coated with 25 Å. The thinner coating reveals the structure in the tail, and the central core and six-fold symmetry of the end plate.

Fig. 8.18 shows images of 3C staphylococcal phage. Fig. 8.18A is a STEM image of a 3C phage clearly showing the head (h), narrow neck (n), helical tail (t), and knob-like end plate (ep). This specimen has been stained with 2% uranyl acetate. Fig. 8.18B is a low-loss image of 3C phages that reveals similar morphology. A typical cylindrical head (h), and a collapsed head (arrow), are characteristic of phages that have lost their nucleic acid. At higher magnification in Fig. 8.18C and D, the narrow neck region (n) joining the cylindrical head to the tubular tail (t) is apparent. A repeating periodic structure can be seen in the tail.

Barbara Panessa-Warren and I also examined the attachment of the phage 3C phage to *S. aureus* as shown in Figs. 8.19 and 8.20. Fig. 8.19 shows a 3C phage attached to *S. aureus*. This sample was not stained, and the semi-transparency of the phages reveals the hollow, tubular construction of the tail at the point of insertion into the thick carbohydrate surface layer (arrows). Head (h); neck (n).

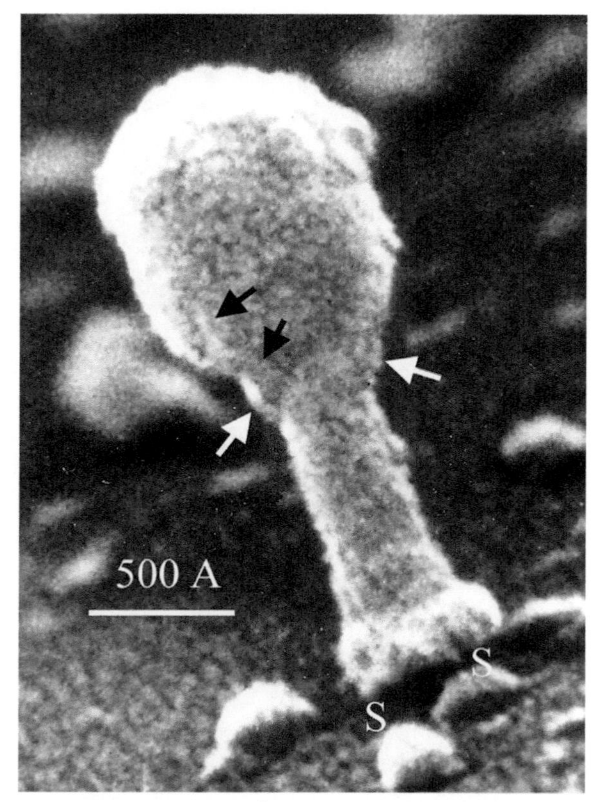

Fig. 8.16 T4 coliphage coated with 50 Å C-Au-Pd Used with permission of AAAS. *(From Broers A.N., Panessa, B.J., and Gennaro, J.F. Jr. 1975. High resolution scanning electron microscopy of bacteriophage 3C and T4. Science, 189, 637–639, permission conveyed through Copyright Clearance Center, Inc.).*

Fig. 8.19 shows the attachment of a 3C phage to *S. aureus*. Thin projections (arrows) anchor the phage tail into the carbohydrate-surface-coat of the bacterial cell. The collapsed, transparent configuration of the head (h) probably indicates that the head is empty and devoid of nucleic acid. A clear central region corresponding to the rod canal can be seen within the phage tail.

Figs. 8.21–8.23 show *S. aureus* bacteria at three stages of cell division. Fig. 8.21 shows the bacterium at the beginning of cell division. A raised ridge, corresponding to the transverse septum can be seen encircling the *S. aureus* cell (arrows). This sample was treated with a heavy metal stain to produce differential electron scattering, resulting in visualization of the nuclear area (n) in each area.

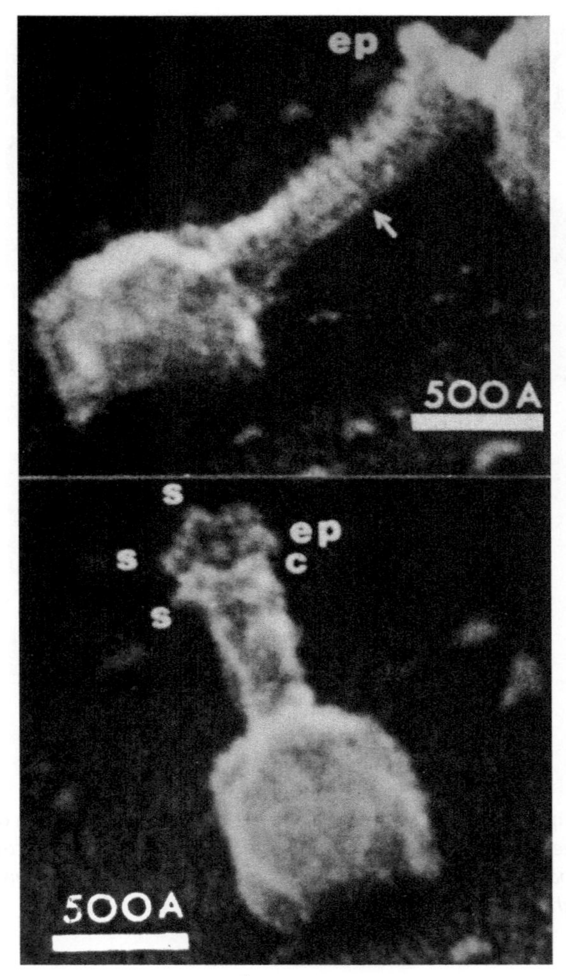

Fig. 8.17 T4 coliphage coated with 25 Å C-Au-Pd Used with permission of AAAS. *(From Broers A.N., Panessa, B.J., and Gennaro, J.F. Jr. 1975. High resolution scanning electron microscopy of bacteriophage 3C and T4. Science, 189, 637–639, permission conveyed through Copyright Clearance Center, Inc.).*

Figs. 8.19–8.24 are reproduced with permission from the Illinois Institute of Technology, Symposia Office, 10, West 35th Street, Chicago, Illinois 60616, USA.

Fig. 8.22 shows a bacterium at a later stage of division. The dividing cell wall produced a groove (arrows) that encircled the cell. Strands of surface carbohydrates stretched across the gap. Fig. 8.23 shows a bacterium during the last stage of cell division. Only a small region of the cell wall remains to unite the two daughter cells.

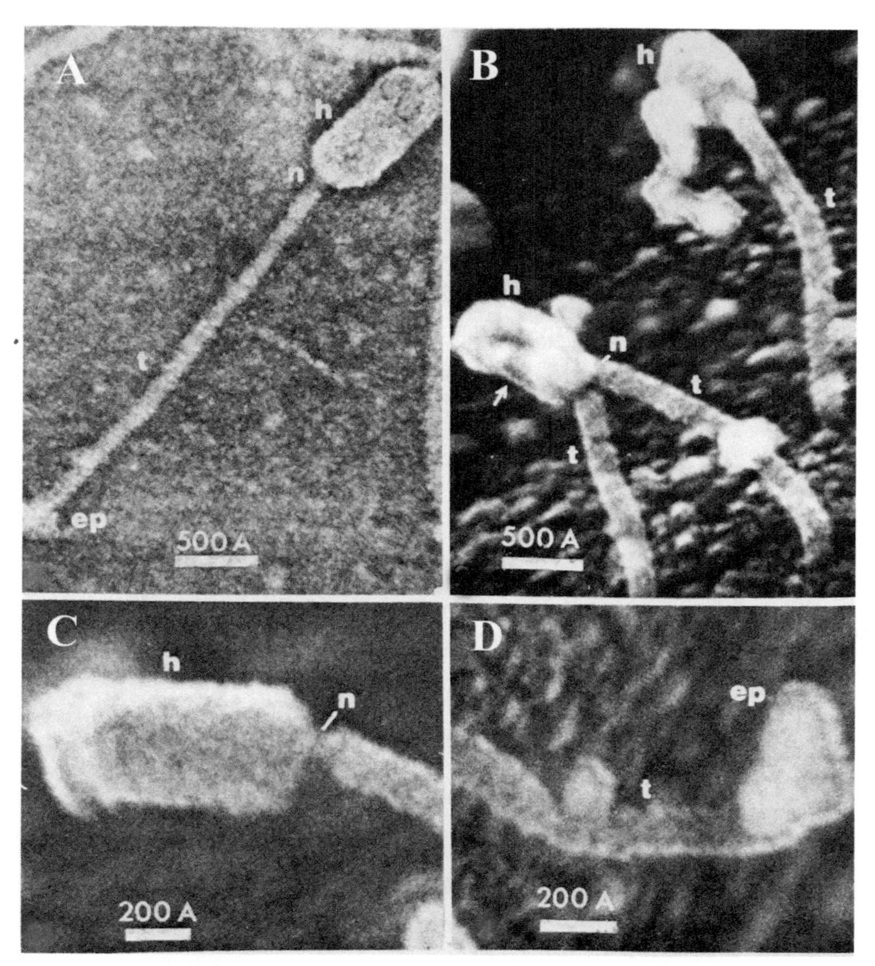

Fig. 8.18 3C staphylococcal phage coated with 30 Å C-Au-Pd. Used with permission of AAAS. *(From Broers A.N., Panessa, B.J., and Gennaro, J.F. Jr. 1975. High resolution scanning electron microscopy of bacteriophage 3C and T4. Science, 189, 637–639, permission conveyed through Copyright Clearance Center, Inc.).*

Fig. 8.24 shows an unstained *S. aureus* bacterium coated with C-Au-Pd. It does not exhibit a darkened nuclear region.

Low-loss imaging of bacterial virus T7

Barbara Panessa-Warren also prepared the DNA containing T7 coliphage for low-loss imaging, thereby eliminating the need for negative staining, three-dimensional reconstruction, or metallic replicas. T7 coliphage was widely used for studies in microbial genetics and molecular

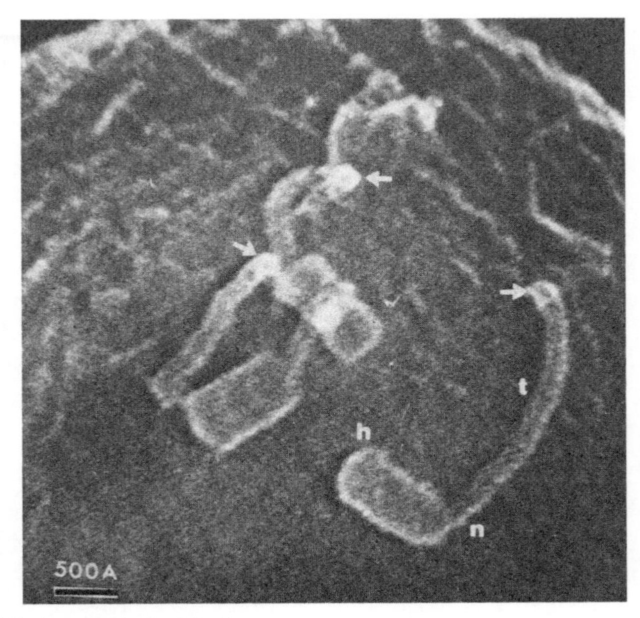

Fig. 8.19 3C phage attached to *S. aureus*.

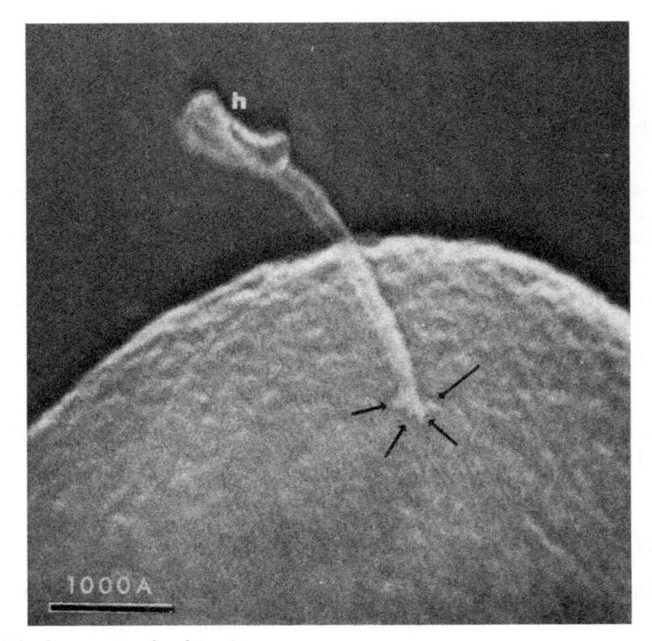

Fig. 8.20 3C phage attached to *S. aureus*.

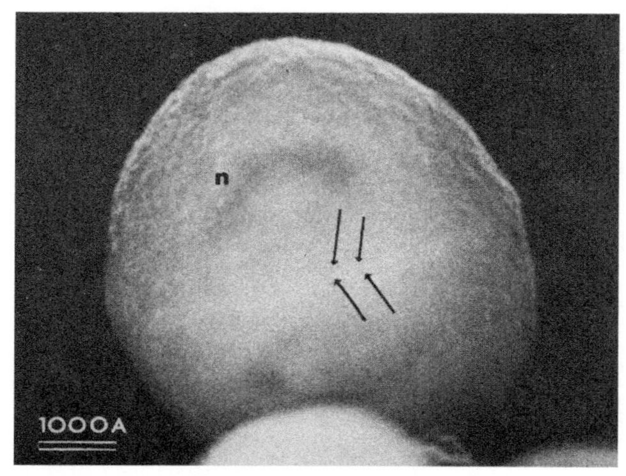

Fig. 8.21 *S. aureus* bacterium at the beginning of cell division.

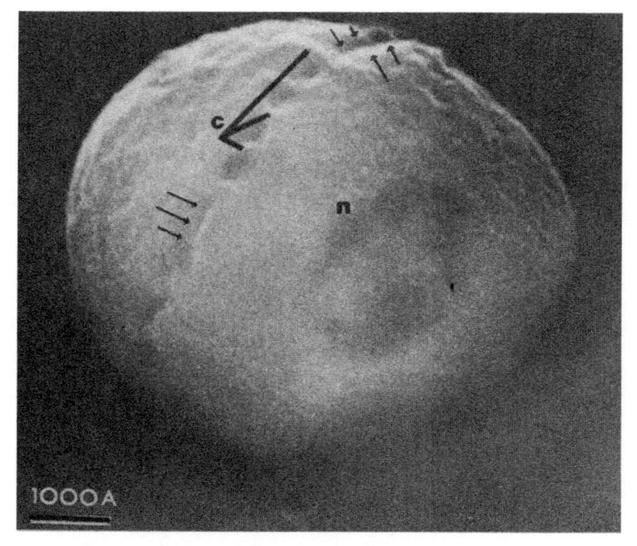

Fig. 8.22 *S. aureus* bacterium at a later stage of division.

biology. It was first described in 1945 by Dumeree and Fano (1945) and its physical properties were extensively examined by Davidson and Friefelder (1962). It had been ascertained that its head was icosahedral (Luria and Darnell, 1967) and was 47.0–70.0 nm in diameter, and that it had a small rudimentary tail (15 nm × 10 nm) which lacked an endplate (Laskin and LeChevalier, 1973; Tikonenko, 1970; Luftig and Haselkorn, 1968). All studies employed conventional TEM with negative

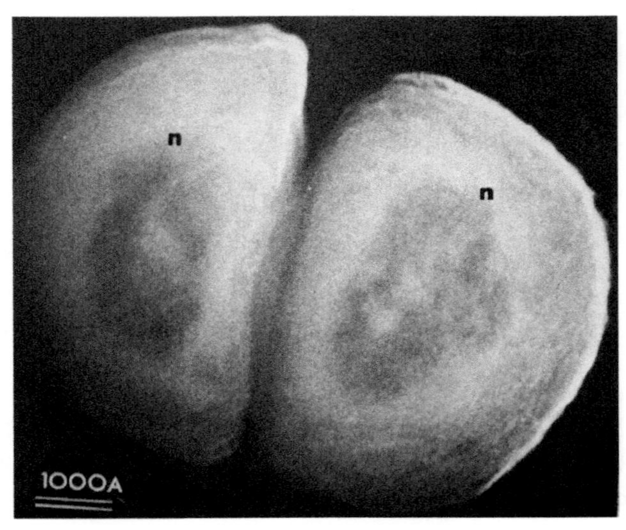

Fig. 8.23 *S. aureus* bacterium in the last stage of division.

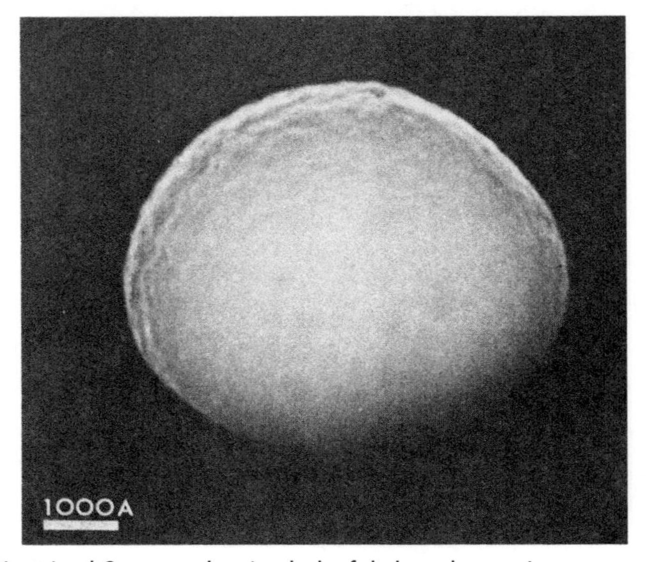

Fig. 8.24 Unstained *S. aureus* showing lack of dark nuclear region.

staining. Secondary electron SEM was difficult because of the small size of the phage.

As before, the specimens were prepared on oxidized silicon slivers. These agar-coated slivers were pressed into sterile nutrient agar plates so that the agar-coated surface was level with the agar on the sterile plates. The plates were then inoculated with E.coli B bacteria. After 3–6 h at 37OC,

T7 coliphage were added directly to the agar slivers and incubated for 10–180 min. Plaques were observed around the slivers after 180 min.

At 6, 8, 10, or 180 min. After the addition of the virus, the slivers were removed and placed in a droplet of 1.5% glutaraldehyde on 0.1 M cacodylate buffer (pH 7.2), post-fixed in 1% aqueous osmium tetroxide for 1 h at room temperature, washed in distilled water, dehydrated in acetone, and dried by the critical point method, as with the 3C phage. The samples were then rotary-metal coated at 10^{-5} mmHg (using liquid N_2baffles) with 3–7 nm gold–palladium. The coating was applied in small successive increments, which produced a uniform coating with a structure of less than 2 nm.

Isolated cleaned phage were prepared by centrifugation with polyethylene glycol and resuspended in Tris buffer (0.001 M $MgSO_4$, 0.001 M Tris, 0.001 M NaCl). A droplet of these clean viruses (4×10^{10} virus particles/mL) was placed on a clean SiO_2 to which 4 droplets of 1.5% glutaraldehyde were added. The sliver was then placed in an airtight, micro-petri dish and maintained at 4 °C overnight. These samples were washed, dehydrated, dried, and coated as previously described.

Negatively stained conventional preparations were made, for comparison, by air-drying the phage at room temperature on formvar–carbon-coated copper grids. All samples examined by STEM were stained with 2% aqueous uranyl acetate (Fig. 8.25).

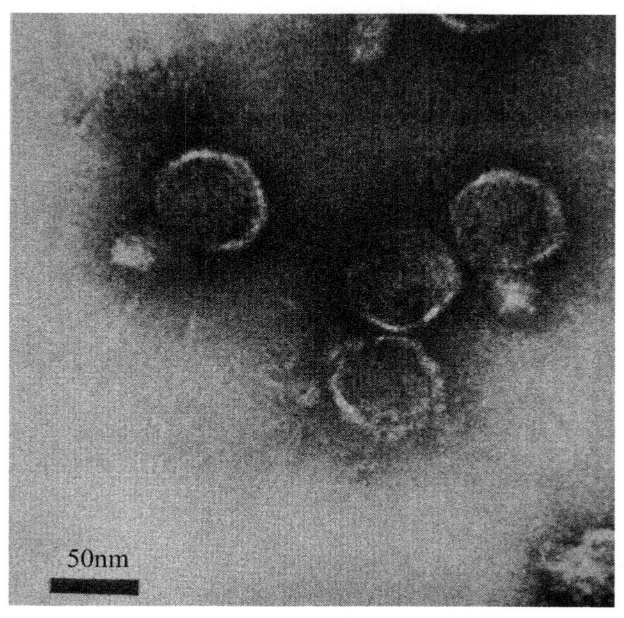

Fig. 8.25 T7 viruses were air dried on formvar–carbon grids, stained with 2% uranyl acetate & examined in STEM mode (Panessa-Warren and Broers, 1979).

As before, the low-loss images were obtained with a beam current of about 2×10^{-11}A and a beam energy of 45 keV. The surface of the bulk samples was positioned at a 20°glancing angle relative to the incident beam. In general, phage or bacteria chosen for detailed examination were oriented approximately normal to the beam. STEM samples were examined at a lower beam current of about 10^{-12}A and a beam diameter of about 0.5 nm. The samples were not tilted, and an aperture below the lens was used that subtended a solid angle approximately equal to the incident beam solid angle.

Results

T7 phage examined by surface low-loss microscopy after fixation, dehydration, and critical point drying showed uniform capsids measuring 50–60 nm in diameter (Figs. 8.26 and 8.29), and no variation in shape was observed. Some SEM micrographs showed a single protuberance suggestive of a tail (Figs. 8.27, 8.29C and D); however, these tail-like prominences were never clearly delineated, nor could any endplate, collar, or tail fibers be discerned (Figs. 8.26–8.29D).

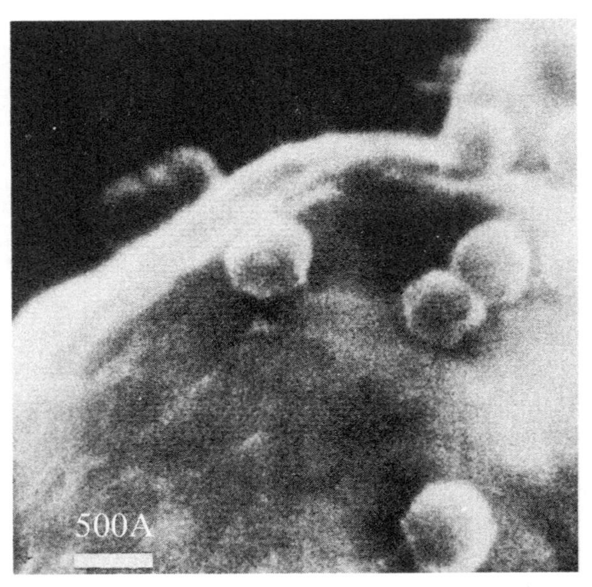

Fig. 8.26 This specimen was fixed for 8 min after the addition of T7 virus. Adsorbed virus particles show a random distribution on the bacterial surface. Although the grain size of the metal coating in this image is barely apparent (less than 2 nm),no, tails, or tail fibers are evident (Panessa-Warren and Broers, 1979).

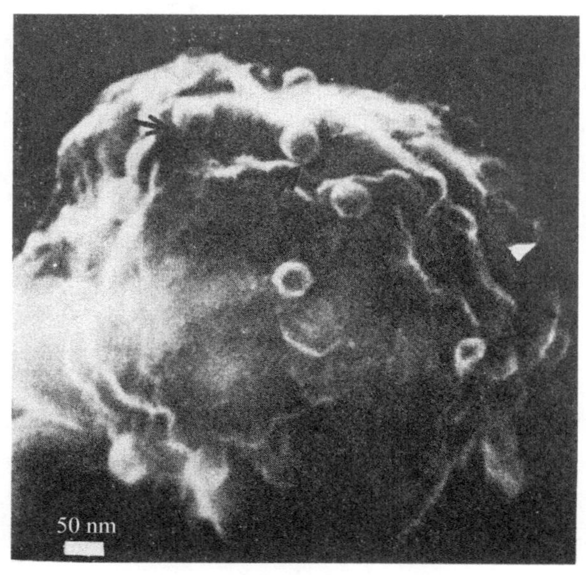

Fig. 8.27 T7 phage adsorbed in a *E. coli* bacterium which seem to be about to lyse. Note the uniformity and six-sided symmetry of the adsorbed capsids. Micrograph with lysed and damaged bacterial cells frequently show the capsids with a single protuberance (arrows) extending to the bacterial surface, which may be indicative of the tail (Panessa-Warren and Broers, 1979).

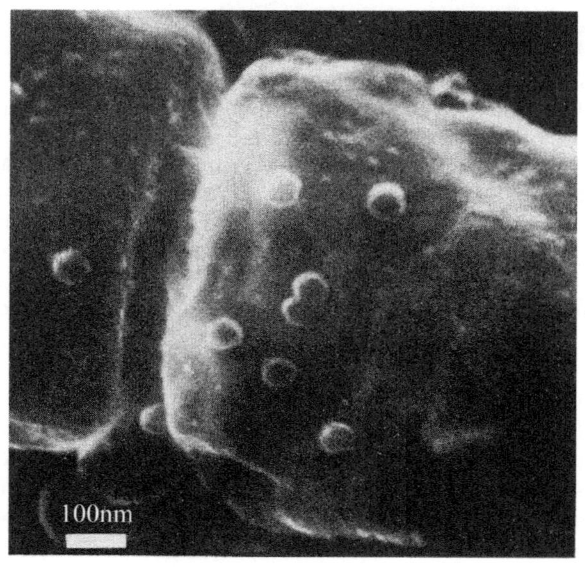

Fig. 8.28 T7 phage particles adsorbed to a dividing *E.coli* bacterium show a random distribution (Panessa-Warren and Broers, 1979).

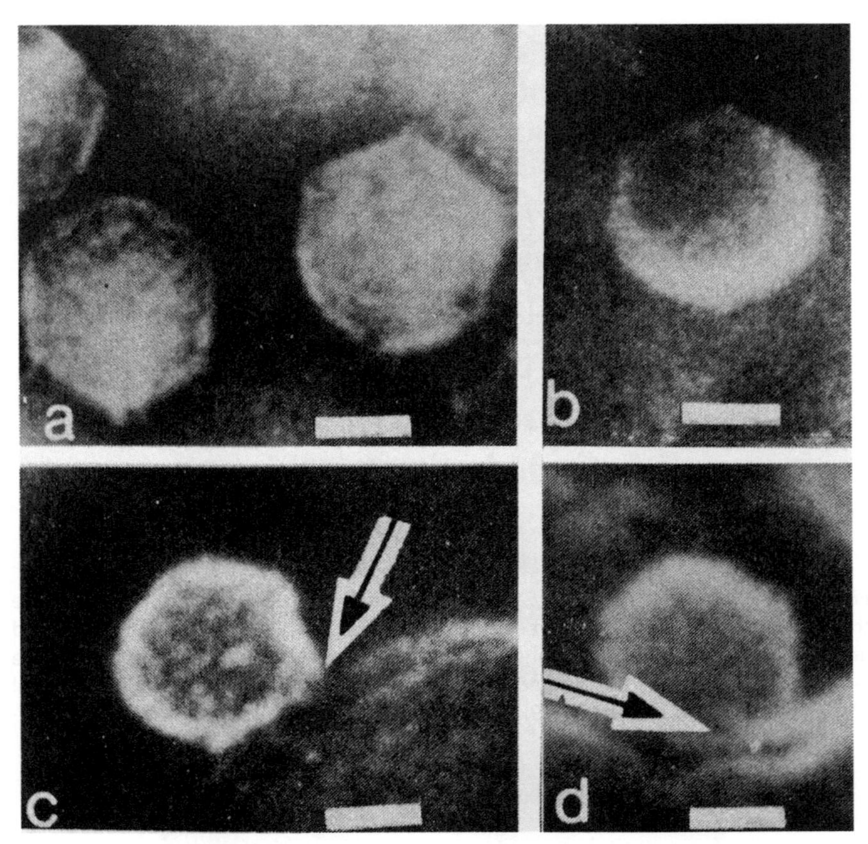

Fig. 8.29 (A) STEM image of air-dried uranyl-acetate-stained T7-phage which clearly show the angular capsid. Upon closer examination, there appears to be a repeating pattern or periods on the capsid, suggestive of capsomers (B) Surface micrographs of adsorbed phage show the same morphology as that seen in the STEM images (A). The virtually structureless coating on this adsorbed phage capsid reveals no capsid ultrastructure or capsomer delineation. Adsorbed phage particles which show protuberances (arrows) attached to the bacterial cell wall. Note the large grain size (3 nm) of the metal coating. This is a common artifact when a coating is applied by one deposition, or when platinum palladium is used (Bar = 25 nm) (Panessa-Warren and Broers, 1979).

The grain structure in the metal coating was less than 2 nm on many of the samples (Figs. 8.26–8.29B) and it should have been possible, therefore, to see the capsomeres if any surface morphology existed. It could be that the tail fibers were destroyed in sample preparation or were obscured in the particular sample configurations examined. The phages predominately appear as isometric icosahedral, sitting directly on the bacteria (Figs. 8.26–8.28). The tail appears to be partially buried in the cell wall, and there was no evidence of a contractile helical tail sheath or a tubular non-contractile tail, as is found with other T-coliphages (Broers and Sokolowski, 1975).

Cleaned, centrifuged samples of coliphage prepared with polyethylene glycol and suspended in Tris buffer were also examined by surface microscopy and showed considerable damage (Fig. 8.28). Capsids were often flattened or polygonal and appeared to have thickened (20–50 nm wide) tails with long strands extending from the base of the tail onto the specimen sliver. From the size and placement of these strands, they may represent extruded nucleic acid.

STEM and TEM examination of air-dried uranyl acetate-stained T7 coliphages reveals hexagonal capsids ranging in size from 45–67 nm in diameter. Many of the capsids were found to have invaginations of the planar surface or were laterally flattened, indicating specimen damage. What appeared to be a rudimentary tail (15–40 nm in width) extended approximately 1.5–2.0 nm in length from the base of the capsid.

Discussion

The high-resolution low-loss method of examining biological samples has shown that there is little variation in the size of the T7 capsid provided the phage are meticulously prepared (fixation, dehydration, critical point drying, and coating). The controversy that exists in the literature concerning the size of the capsid (measurements ranging from 47–70 nm in diameter) may represent variations resulting from the specimen preparation. We found similarly large ranges of capsid measurements when specimens were air-dried and uranyl-acetate-stained, as well as when the phages were subjected to repeated washing and centrifugation. Fixed samples prepared by critical point drying did not show these marked variations in capsid size.

The suggestion of a repeating pattern seen in the empty capsids (devoid of DNA) by STEM (uranyl-acetate-stained phages), may be indicative of the capsomeres (Fig. 8.29A). In this case the junction of the T7 capsomeres may be morphologically smooth. When negatively stained, the viral capsomeres may appear as distinct units due to biochemical-compositional differences at the junctions. Of the capsomeres (resulting in the uptake and accumulation of the stain), rather than be outlined with stain due to topography.

During phage absorption, the tail appears to be buried in the bacterial cell wall (Fig. 8.26). Absorbed phage seem quite rigid and withstand tissue processing without becoming detached or broken. After examining many absorbed virus samples, it appears that the phage adsorb randomly. Occasionally, they were grouped in clusters. But more frequently, the viruses were scattered over the entire bacterial surface. Viral-host

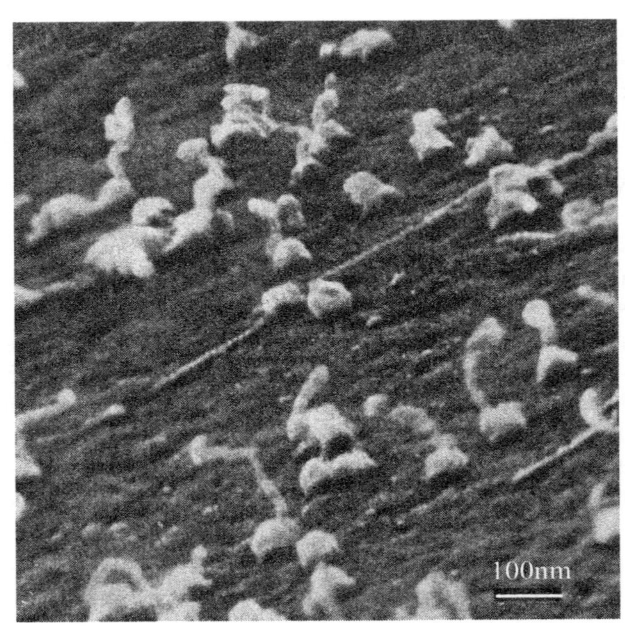

Fig. 8.30 Coliphage which have been isolated and cleaned by routine centrifugation procedures are known to be less virulent. This micrograph clearly shows clean but damaged capsids. The filamentous, twisted, strands extending from individual capsids appear to be extruded DNA (Panessa-Warren and Broers, 1979).

preparations that were examined never showed any repeating patterns of viral attachment, or any preference for specific regions on the bacterial surface wall (Figs. 8.26–8.28).

Phage examined after successive washes and centrifugation appear severely damaged. Collapse of capsids as well as tail deformation were observed. The reported decrease in virulence of phage cleaned in this manner (Gould, n.d.) may, therefore, be associated with observed morphological damage, or to the partial or complete loss of nucleic acid from the capsid (as suggested by the appearance of the filamentous extruded material seen in Fig. 8.30).

Conclusions

This examination of T7 bacteriophage reinforced our conclusion that high-resolution, low-loss SEM is an effective tool that complements transmission electron microscopy by permitting three-dimensional morphological imaging at comparable magnifications. In this case, viruses were grown on the specimen holders and examined intact. Although the specimens were delicate and rather difficult to handle, we saw little evidence of sample damage. This may be attributed to the fact that the average incident radiation dosage using SEM can be less than in a typical TEM. Specimen preparation was found to be our greatest limitation in obtaining clean, high-resolution images. Occluding biological layers of material had to be removed without altering viral or bacterial morphology. The metal coating posed a serious problem, in that the periodicity of the coating could alter surface detail, and too little coating failed to scatter enough incident electrons for adequate imaging. Further work remains to be done to find increasingly better biological preparation methods that will produce samples of sufficient quality to exploit the high-resolution capabilities of this type of microscopy.

References

Broers, A.N., 1973a. High resolution scanning electron microscopy of surfaces. In: Anderson, C.A. (Ed.), Microprobe Analysis. John Wiley and Sons, New York.

Broers, A.N., 1973b. High-resolution thermionic cathode scanning electron microscope. Appl. Phys. Lett. 22, 610–612.

Broers, A.N., Sokolowski, J., 1975. Energy filter for high resolution low-loss scanning electron microscopy. In: Bailey, C.W. (Ed.), Proc. 33rd Annual Meet. Electron Microscopy Soc. of Am, pp. 148–149.

Broers, A.N., Panessa, B.J., Gennaro Jr., J.F., 1975. High resolution scanning electron microscopy of bacteriophage 3C and T4. Science 189, 637–639.

Davidson, P.F., Friefelder, D., 1962. The physical properties of T7 bacteriophage. J. Mol. Biol. 5, 635.

Dumeree, M., Fano, V., 1945. Bacteriophage – resistant mutants in Escherichia coli. Genetics 30, 119.

Ennos, A.E., 1953. The origin of specimen contamination in the electron microscope. Brit. J. Appl. Phys. 4, 101.

Ennos, A.E., 1954. The sources of electron-induced contamination in kinetic vacuum systems. Brit. J. Appl. Phys. 5, 27–31.

Gould D.: Personal communication.

Hillier, J., 1948. Investigation of contamination in the electron microscope. J. Appl. Phys. 19, 226.

Laskin, A., LeChevalier, H. (Eds.), 1973. Handbook of Microbiology, Vol. 1, Organismic Microbiology. CRC Press, Cleveland.

Luftig, R., Haselkorn, R., 1968. Comparison of blue-green algae virus LPP-1 and the morphologically related viruses G-3 and coliphage T7. Virology 34, 674.

Luria, S.E., Darnell, J., 1967. General Virology, second ed. Wiley, New York.

Pan, X., Broers, A.N., 1993. Improved electron beam pattern writing in SiO_2 with the use of a sample heating stage. Appl. Phys. Lett. 63, 1441–1442.

Panessa-Warren, B.J., Broers, A.N., 1979. High resolution SEM of T7 bacteriophage. J. Ultramicrosopy 4, 317–322.

Smith, K., 1956. PhD thesis, University of Cambridge.

Tikonenko, A., 1970. Ultrastructure of Bacterial Viruses. Plenum, New York.

Wells, O.C., Broers, A.N., Bremer, C.G., 1973. Method for examining solid specimens with improved resolution in the scanning electron microscope. Appl. Phys. Lett. 23, 353–355.

Further reading

Broers, A.N., 1979. Kohler illumination and brightness measurement with lanthanum hexaboride cathodes. J. Vac. Sci. Technol. A 16, 1692–1698.

CHAPTER NINE

Microfabrication in the STEM

Contents

The first fabrication experiments in the STEM were made with contamination resist and ion milling. The samples had to be pre-contaminated in a vacuum system pumped with an un-trapped diffusion pump filled with Dow Corning silicone oil type 504 because the new probe was ion-pumped, and the background contamination rate was too slow to be useful in a fabrication process.

Fig. 9.1 shows 8 nm wide and 10 nm thick AuPd wires on carbon membranes fabricated in this way (Broers et al., 1976). These wires were to be compared with the 40 nm lines formed in Cambridge in 1963. The improvement in resolution was attributed to the 10 times smaller electron beam and the use of a thin carbon substrate that effectively eliminated exposure due to electrons back-scattered from the substrate.

Because these wires were narrower than 10 nm and more than five times smaller than we had made before, one of my co-authors,

Advances in Imaging and Electron Physics, Volume 231
ISSN 1076-5670
https://doi.org/10.1016/B978-0-443-31462-9.00009-5

Fig. 9.1 Gold-palladium wires on a 10 nm thick carbon film. Wires are 8 and 15 nm wide and 10 nm thick. Contamination resist was written at line charge densities of 10^{-6} C/cm and 1.3×10^{-6} C/cm at 45 kV. *(Used with permission of IOP Publishing Ltd. from Broers, A.N., Molzen, W.W., Cuomo, J.J., Wittels, D., 1976. Electron beam fabrication of 80 Å structures. Appl. Phys. Lett. 29, 596–598 permission granted through Copyright Clearance Center Inc.).*

Jerry Cuomo, suggested that we call them nanostructures rather than microstructures and the process used to make them nanolithography rather than microlithography. After all, we were making structures that were less than 10 nm in size. I thought at the time that re-naming might be helpful to distinguish what we were doing from the practicable processes used for the mass manufacture of chips. So we used it in the title of an IBM Research Report that was later published in *J. Vac. Sci. Technol.* in 1979 (Molzen et al., 1978, 1979).

We were not aware that Taniguchi, in a lecture in 1974, had already used the prefix nano when he referred to nanotechnology, which he described as "mainly consisting of the processing of separation, consolidation, and deformation of materials by one atom or one molecule" inferring that "nano" referred to dimensions below 1 nm. We were certainly not doing things on that molecular scale.

However, 4 years later, the prefix nano first appeared in the title of a contribution at the 3-Beams conference in 1983 (Komoru et al., 1983) that referred to "a maskless etching process with a resolution of about 20 nm". It had, therefore, gone from 1 to 20 nm by the early eighties. Today "nano" has been devalued completely and is used to describe almost anything small. We have continued to use it for structures at the limits of conventional resist processing, that is, between 10 and 20 nm.

The limit to the size of the wires shown in Fig. 9.1 was set by the thickness of the contamination needed to mask the gold film. This naturally depended on the etch rates of the metal and the contamination. 20 nm thick contamination lines were adequate to shield 22.5 nm AuPd films and 50 nm of contamination to shield 22 nm of niobium. The contamination thickness was measured on test samples by shadowing test samples at an oblique angle with a heavy metal.

Imperfections in the metal film also limited resolution in this instance. A subsequent unpublished study by Parikh and myself concluded that backscattering from the thin substrate could not account for the cusp formed at the intersections of the lines shown in Fig. 9.1. This suggested that, once again, resolution was being set by the range of the secondary electrons in the metal film and the growing contamination layer and shows why it is difficult with this method to produce metal structures smaller than about 5 nm even if the grains structure in the metal would allow it.

Figs. 9.2 and 9.3 show conductors fabricated on Si_3N_4 substrates using contamination resist. Fig. 9.2 shows a 12 nm wide gold wire connecting to gold pads on a membrane substrate. Fig. 9.3 is an SEM image of two similar lines showing the cones of contamination that built up while the

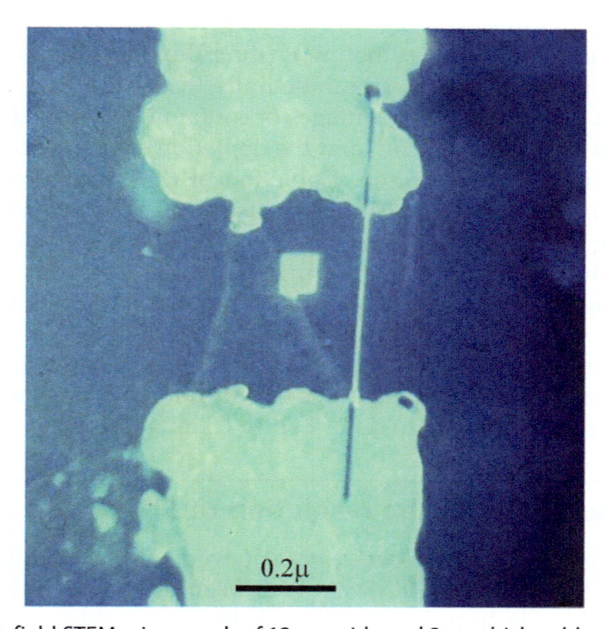

Fig. 9.2 Dark-field STEM micrograph of 12 nm wide and 2 nm thick gold stripe between two gold pads on a Si_3N_4 membrane. Wire was formed with contamination resist and ion etching. Pads were fabricated with conventional electron beam lithography using the lift-off process.

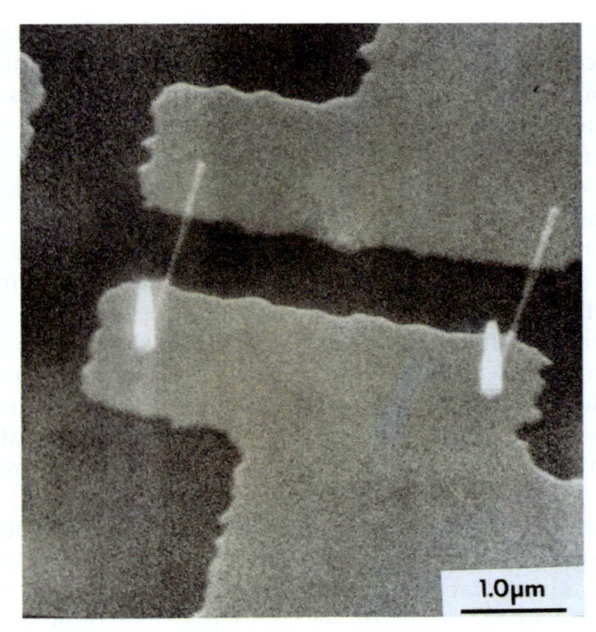

Fig. 9.3 Contamination cones and lines formed on a AuPd coated Si_3N_4 membrane. The cones formed while the electron beam was stationary at the end of each line.

electron beam was stationary before writing the lines. It shows that the aspect ratio in the resist pattern can be more than 3 to 1.

The most significant disadvantages of contamination resist are its low sensitivity of about 10^{-1} C/cm^2, and the depletion effect that occurs when dense patterns must be written. Depletion of the contamination arises because contaminant molecules become used up in the area immediately surrounding the electron beam, and resist thickness becomes inversely proportional to pattern density. In principle, depletion can be minimized by providing an in-situ vapor source of contaminants, but we never tried this.

This contamination-resist process was used to make most of the devices in our study of the properties of thin-film structures with dimensions in the sub-0.1 μm range described in Chapter 10.

Measurement of the resolution of PMMA

After exploring what could be achieved with contamination, we moved to PMMA resist. PMMA had been the standard resist for high-resolution electron beam lithography since 1968 when Haller, Hatzakis, and Srinivasan published their original paper. The exposure dose for PMMA (5×10^{-5}–2×10^{-4} C/cm^2) was too high for many semiconductor direct exposure applications, but its resolution was higher than the faster resists. My experiments in Cambridge in 1964 with an electron beam with a diameter of about 20 nm had only produced linewidths on about 0.25 μm.

It had, for example, been used in 1970 to make the surface wave transducers described in Chapter 6, which had aluminum fingers 0.15 μm wide. 60 nm metal wires were also made using lift-off on thin window substrates in 1968. In addition to PMMA's use with electron beams, exposure with 4.5 nm x-rays produced features of 17.5 nm (Flanders, 1980).

Our first experiments in the new STEM in 1978 showed that it was possible to produce 25 nm linewidths in dense metal patterns with PMMA. These patterns were formed in Au—Pd layers on silicon nitride window substrates. Ion etching was used to remove the unwanted metal. PMMA was three orders of magnitude more sensitive than contamination resist and did not suffer from the depletion effect described above so could produce complex patterns without difficulty.

The minimum beam diameter, as described in Chapter 8, was 0.5 nm for a current of 10^{-12}A and could be maintained below 10 nm for currents up to 10^{-8}A. Diameters of less than 2 nm were used for all of the initial experiments with PMMA described here and an accelerating voltage of 56 kV. The electron column was slaved to a flying-spot scanner to generate complex patterns.

The test samples were prepared by evaporating 22.5 nm of PdAu (60:40) onto silicon wafers containing 60 nm thick silicon nitride membrane windows, which were fabricated as described in Chapter 6.

The wafers were then spin-coated at 2000–5000 rpm with 2010 PMMA diluted 50:1 in chlorobenzene, and the resist baked at 150 °C for 1 h before electron beam exposure.

The resist was measured to be 110 nm thick by Tolansky interference microscopy. Exposure was made at a charge density of 5×10^{-4} C/cm^2, and the resist developed in methyl-isobutyl ketone:isopropyl alcohol (1:3) for times between 15 and 45 s. The samples were etched with 1 kV argon ions at a current density of 0.1 mA/cm^2 for about 90 s. This was enough to remove the unprotected PdAu (60:40) while not damaging the resist or the SI_3N_4 window.

Fig. 9.4 is a STEM image of an etched 22.5 nm thick PdAu pattern supported on a 60 nm thick Si_3N_4 window substrate. The smallest lines and spaces in this pattern were nominally 25 nm wide with a center-to--center spacing of 50 nm. The pattern seems to have been slightly under etched so that the exposed lines are about 20 nm wide and the spaces between the lines about 30 nm wide.

In describing the resolution of a microfabrication process, it is necessary to distinguish between its ability to produce isolated structures and densely packed structures where the spaces between lines are comparable in size to the linewidths. It is generally possible to achieve smaller dimensions in isolated structures by taking process biases to extremes.

For example, lines in resist patterns can be artificially narrowed by suitable manipulation of exposure and development conditions. Similarly, over-etching can be used to reduce linewidths when larger spaces between lines can be tolerated. It is possible, therefore, to produce dimensions below 25 nm with PMMA when only isolated lines are needed or when larger spaces between lines can be tolerated. See, for example, Figs. 9.4 and 9.5, which show slightly under-etched and slightly over-etched samples in which the most closely spaced metal (black) lines are 20 and 30 nm and the spaces 30 and 20 nm. Fig. 9.6 shows loss of adhesion with a zone plate pattern fabricated with the PMMA lift-off process when the line width reached 20 nm. This did not appear to be due to the undercutting of the resist because no improvement was found when the resist thickness was reduced from 110 to 60 nm.

In these experiments we could not resolve 15 nm lines on 30 nm centers but, as discussed above, could produce 15 nm lines or spaces by overexposure or underexposure.

When PMMA is overexposed by a factor of 10, it acts as a negative resist. That is, the exposed regions become less soluble in certain solvents than the unexposed resist. To examine the resolution of this mode, open raster patterns such as that shown in Fig. 9.7 were exposed at a line charge density of 4×10^{-8} C/cm. A blanket exposure of 5×10^{-4}/cm^2 was used to expose the resist surrounding the raster pattern. This allowed positive and negative images to be developed with the same development procedure. The sample was etched with 1 keV argon ions at a current density of 0.1 mA/cm^2 for 90 s.

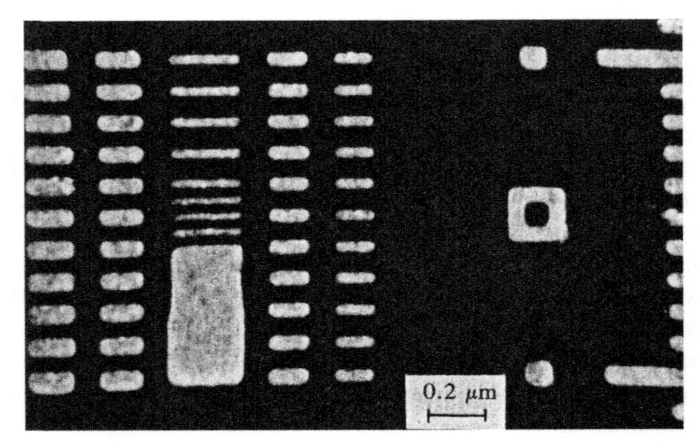

Fig. 9.4 Bright-field STEM image of a pattern ion-etched in 22.5 nm thick PMMA layer on a 60 nm thick Si_3N_4 thick membrane. Minimum lines (black) are 30 nm and spaces 20 nm wide. PMMA was exposed at 5×10^{-4} C/cm^2 with a 56 kV 1 nm diameter electron beam. *(Reprinted from Broers, A., Harper, J.M.E., Molzen, W.W., 1978a. 250A linewidths with PMMA electron resist. Appl. Phys. Lett. 33, 392–394 with the permission of AIP Publishing).*

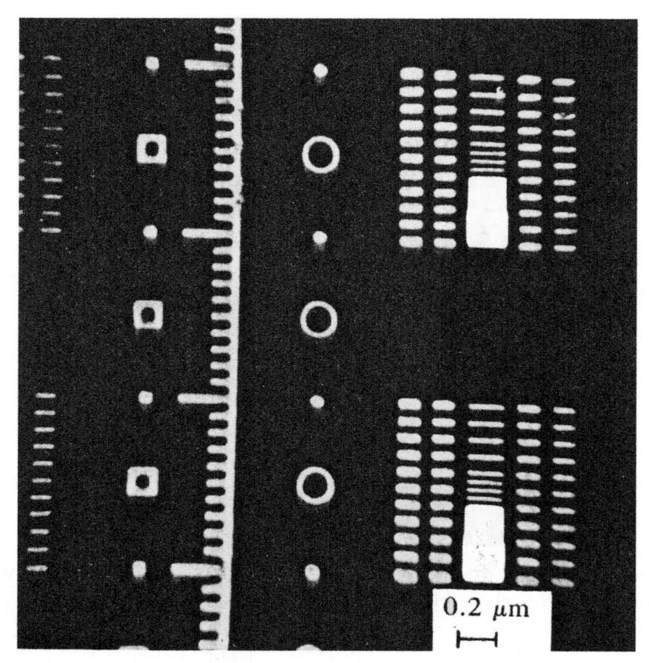

Fig. 9.5 Bright-field STEM image similar to Fig. 9.4 but ion etched for slightly longer so that the minimum lines and spaces were 20 and 30 nm. *(Reprinted from Broers, A., Harper, J.M.E., Molzen, W.W., 1978a. 250A linewidths with PMMA electron resist. Appl. Phys. Lett. 33, 392–394 with the permission of AIP Publishing.).*

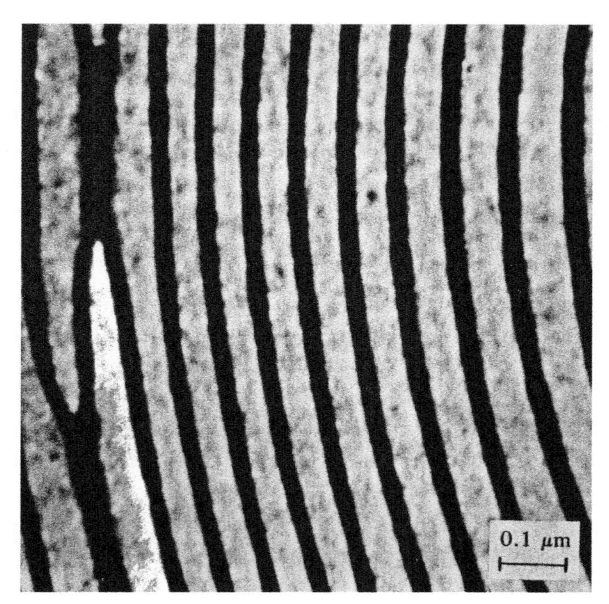

Fig. 9.6 Bright field STEM image of portion of a concentric ring zone plate in 22.5 nm thick AuPd film on a Si_3N_4 membrane showing loss of resist adhesion when linewidth reached 20 nm.

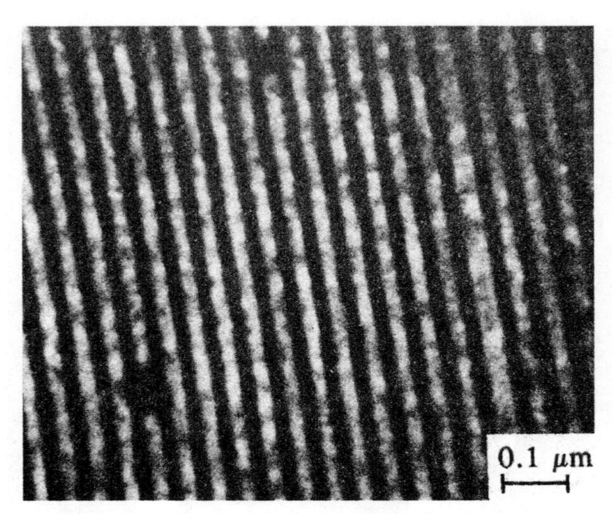

Fig. 9.7 15 nm AuPd lines on 50 nm centers on a 60 nm thick SI 3 N_4 membrane produced by ion etching with argon ions. A line charge density of 4×10^{-8} C/cm^2 was needed to expose the resist in the negative mode. *(Reprinted from Broers, A., Harper, J.M.E., Molzen, W.W., 1978a. 250A linewidths with PMMA electron resist. Appl. Phys. Lett. 33, 392–394 with the permission of AIP Publishing).*

For purely negative patterns, a resist solvent such as chlorobenzene or acetone can dissolve away the unexposed resist, thus avoiding the need for blanket exposure.

Conclusions

These preliminary experiments showed conclusively that the exposed region in the resist was not confined to the area bombarded by the primary electrons. There was a region surrounding the beam that dissolved away in the developer or remained after development for a negative resist, that was approximately 15 nm wide. This may result from exposure by secondary electrons excited by the primary electrons or due to the molecular nature of the resist and the development process.

It became clear to me at this time in 1978 that we needed to measure this distribution and find a mathematical model for it that could be used to calculate the resulting exposure for any shape or combination of shapes. Philip Chang had already done this for exposure due to electron scattering, both forward scattering in the resist and backscattering from the substrate. The next section describes how we measured this distribution.

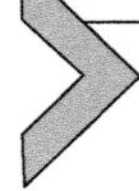

Measurement of the resolution contrast function for PMMA as an electron resist

In the 1970s the usual way to describe the "resolution" of a fabrication process was merely to say how small a structure it could make.

However, as just discussed this was only accurate when applied to a specific structure. A more useful way was to refer to the smallest center-to-center spacing that could be achieved in an infinite array of lines and spaces, but even this was specific and could not be used to predict the outcome of exposing any shape or combination of shapes.

A point spread function for the process that modeled the delocalization created by the remote excitation of secondary electrons and the subsequent straggling of the secondaries into the resist was needed. The function also needed to consider the ratio of the dissolution rate of the exposed and unexposed resist.

Chang showed that it was possible to accurately model the forward and backward scattering of electrons in the imaging medium by assuming that these distributions were Gaussian. Their widths and the width of the electron beam were then added in quadrature with the beam distribution to gain the overall distribution. Their widths for thick resists, and samples thick compared to the primary electron penetration range, these distributions become very significant and must be considered in calculating the exposure distribution needed to produce a given shape.

For modeling purposes, the point spread function could also be assumed to be Gaussian and be added in quadrature with the other distribution to allow the ultimate resolution of the resist process to be determined.

To measure this distribution with the high-resolution probe, we used a relatively high accelerating voltage of 50 kV, thin resist layers (<0.1 μm) to reduce forward electron scattering, and thin (<0.1 μm) membrane substrates to effectively eliminate back-scattering. The beam diameter used for exposing the test patterns and for subsequently examining the developed resist patterns was less than 1 nm in diameter (Broers, 1981).

Sample preparation

The membrane substrates were fabricated by selectively etching holes through the silicon wafers, as described in Chapter 6. We used 30 and 60 nm thick membranes, but the 30 nm membranes were generally too fragile to be practicable; almost half of them broke during processing. Because no loss in resolution was observed with the 60 nm membranes, which was expected from electron scattering data, they were used for most of the experiments.

After window fabrication, the silicon wafers were diced into chips 2.6 × 1 mm. For electron beam exposure and subsequent examination in the scanning transmission microscopy mode (STEM), the chips were mounted in a sample holder designed for standard 3 mm electron microscope grids. Each chip contained two windows.

The membrane substrates were coated with PMMA using a conventional spin-coater with a vacuum chuck designed to avoid applying atmospheric pressure across the membranes. Very dilute resist (1:50, 2010

PMMA:chlorobenzene) and relatively slow spinning rates (2000–3000 rpm) produced layers with thicknesses between 30 and 100 nm. As mentioned below the precise thicknesses were made from STEM images of the developed samples after shadowing the samples with a thin layer of metal. Resist thickness was uniform to better than 10 nm over the window areas. Resist layers were baked at 160 °C for 1 h to remove residual solvent.

A 5 nm thick layer of AuPd was deposited on the underneath of the membranes to provide detail for focusing the electron beam. The structure in this layer provided high contrast in the bright field STEM image, making it easy to check that the beam diameter was less than 1 nm. The number of primary electrons backscattered from this layer was less than 1 in 500 (Wells, 1974), and few of the low-energy secondary electrons formed in the AuPd layer would have penetrated the Si_3N_4 membrane to the resist. So the AuPd later would have had little effect on resist exposure.

Image spreading due to the beam's finite angular width was about 2 nm (resist thickness 30 nm, substrate thickness 60 nm, beam half-angle 10^{-2} rad), assuming that the beam was focused on the bottom of the membrane. The AuPd layer had sufficient conductivity to avoid sample charging during exposure.

Exposure

The test pattern used to measure the exposure distribution in PMMA is shown in Fig. 9.8. The pattern was written under the control of the flying

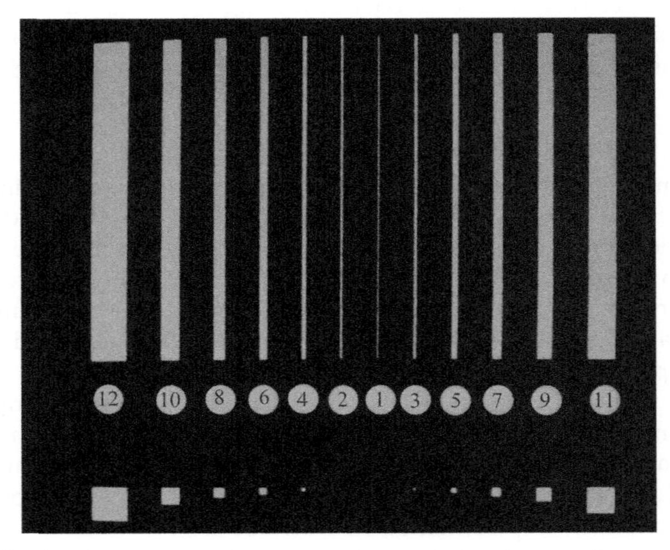

Fig. 9.8 Nominal linewidths written by the sub-nm electron beam in the exposed resist pattern (1) 4.4 nm, (2) 7.5 nm, (3), 11 nm (4) 14 nm, (5) 20 nm, (6) 26 nm, (7) 33 nm, (8) 38 nm, (9) 53 nm, (10) 62 nm, (11) 96 nm, and (12) 120 nm.

spot scanner. The image shown in Fig. 9.8 was obtained while the mask was scanned using the beam current arriving at the sample to modulate the brightness. The exact widths of the exposed lines were measured from the width of the "beam-on" pulses on this video signal. This eliminated any errors that might have arisen due to the finite resolution of the scanner.

The size of the pattern was adjusted so that the linewidths were as follows: (1) 4.4 nm; (2) 7.5 nm; (3) 11 nm; (4) 14 nm; (5) 20 nm; (6) 26 nm; (7) 33 nm; (8) 38 nm; (9) 53 nm; (10) 62 nm; (11) 96 nm; (12) 120 nm.

Beam current was measured using a Faraday cup located below the sample. Because of the small area of the pattern and the relatively slow scan of the STEM system, the beam current had to be kept at a level of about 2×10^{-12} A. The minimum frame time was 10 s. The pattern was typically 2.5 μm × 2.5 μm for a minimum linewidth of 2.5 nm.

Before each pattern was exposed, the beam was offset to an area adjacent to the exposed area, and the beam focused on the AuPd layer underneath the membrane. The sample was mechanically moved approximately 20 μm between exposure sites. The vertical accuracy of the sample stage was such that focus adjustment after movement was unnecessary in most cases.

Samples were developed for about 50 s in standard PMMA developer: one part methyl isobutyl ketone to three parts isopropyl alcohol. Our measurements indicated that the developer dissolved the unexposed resist at a rate of about 25 nm/min, so the initial resist thickness was approximately 20 nm thicker than that measured for the developed patterns.

After development, 5 nm of AuPd was vapor deposited onto the top of the resist at an incidence angle of 45°. This metal layer clearly outlined the resist pattern, and the shadow length was an accurate measure of the resist thickness.

Determination of exposure distribution for thin resist on a thin window substrate

Eleven different exposure times were used for each experimental run. The shortest time gave an exposure below that needed to open up the largest shapes. The longest time required the resist to develop through to the substrate in the site of the narrowest line.

If the exposure distribution is Gaussian, then using the "reciprocity principle" of Chang (1975), the exposure dose Q_S received at the center of the infinitely long rectangle of width w can be shown to be given by

$$Q_W = Q_0 \, erf(w/2\sigma) \tag{9.1}$$

where Q_0 is the exposure in the center of an infinitely large shape, and σ is the standard deviation of the distribution.

Fig. 9.9 shows three bright field STEM micrographs of the developed resist patterns. The 20-s exposure was slightly greater than was needed to

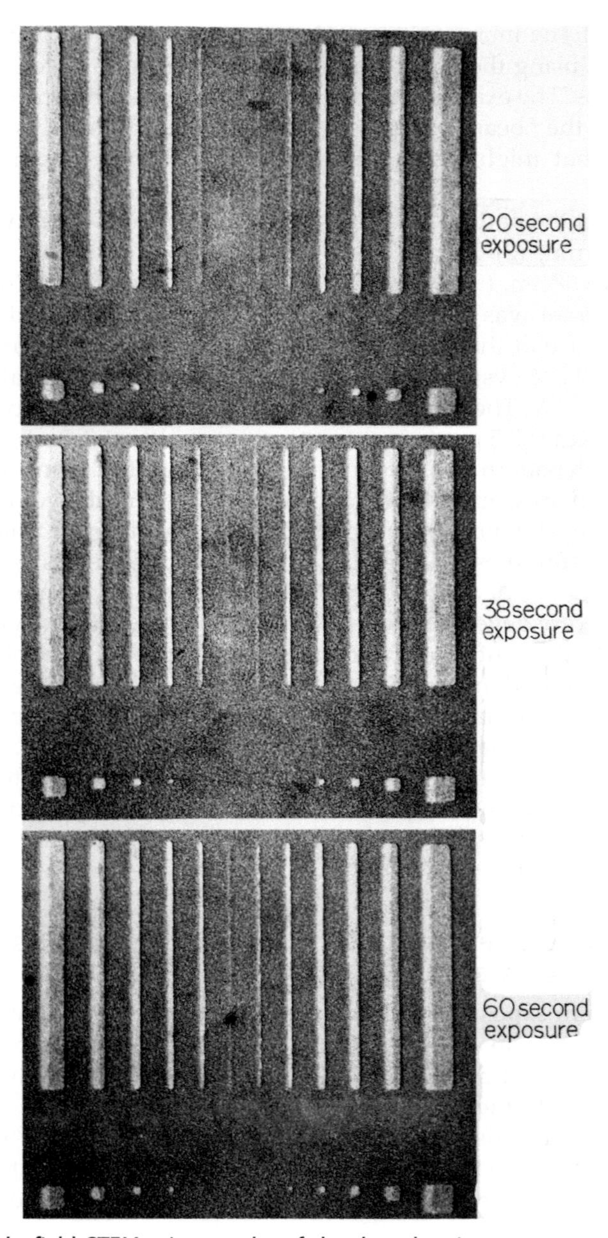

Fig. 9.9 Bright-field STEM micrographs of developed resist test patterns exposed at three levels. The 20 s exposure was slightly greater than was necessary to open up the shapes that were larger than the exposure distribution. The 60 s exposure was sufficient to open up line (2) that had a written linewidth of 7.5 nm.

develop the lines that were larger than the exposure distribution. The narrowest line that opened up in this case was (5), which had a nominal width of 20 nm. However, the width of the line had become 25 nm by the time the resist was removed through to the substrate. The 38-s exposure was sufficient to open up line 3 with a written width of 11 nm but this had grown to 15 nm before all of the resist had been removed.

The standard deviation σ of the exposure distribution was calculated from patterns of this type exposed at progressively greater doses. σ was calculated from the nominal linewidth (w) of the narrowest line that had developed through to the substrate and the ratio of the exposure needed to open this line to the exposure needed to develop through to the substrate in the sites of the large shapes (Q_W/Q_0). If the distribution were truly Gaussian, then the same σ would have been found for each test exposure. In practice, this was not the case. Heavier exposures, in which the narrowest lines had developed through to the substrate, generally yielded σ values higher than those deduced from lower exposure doses. In a series of five separate experiments, the average values of σ varied from 11 to 14 nm. The resist thickness was about 30 nm.

Fig. 9.10 plots the fractional exposures received for different linewidths. The fractional exposure was determined from the exposure times

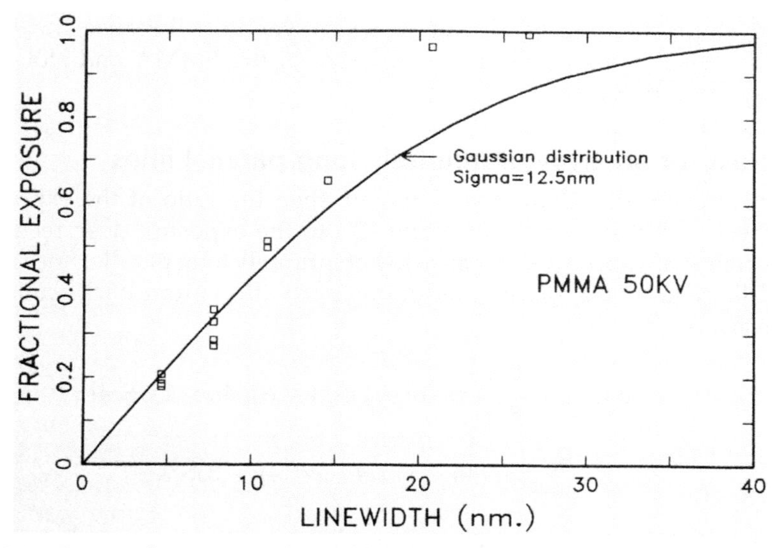

Fig. 9.10 Fractional exposure vs linewidth for PMMA resist. *(Used with permission of IOP Publishing from Broers, A.N., 1980. Resolution limits of PMMA resist for electron beam exposure. In: Bakish, R. (Ed.). 9th Int. Conf. on Electron & Ion Beam Sci. & Technol., pp. 396–406. Electrochemical Soc., Princeton, NJ permission conveyed through Copyright Clearance Center, Inc.).*

needed to open up different linewidths. For example, it took about five times longer to open the 4.4 nm nominal linewidth line than it did to open the large shapes.

Fig. 9.10 also shows (continuous line) the fractional exposure that would be obtained with a resist with a Gaussian distribution $\sigma = 12.5$ nm. An additional experiment in which identical exposures were made on three different resist thicknesses (50, 80, and 100 nm) showed, as expected from electron scattering data, that there was little dependence on resist thickness for thicknesses below 100 nm. The small increase for the thicker resist layers was probably due to the longer development time. As the exposure was increased to open the narrowest lines, the linewidth of the wider lines increased. The rate of increase can also be used to estimate the σ of the spread function in the resist. However, this will not be discussed here.

If the exposure is due to low-energy secondary electrons alone, and the excitation of secondaries is isotropic and random, it would appear reasonable for the distribution to be Gaussian. The deviation observed could result from a dependence of the development rate on linewidth that was large compared to the exposure distribution. In this case, the narrowest line that developed through to the substrate was line (5), which was 20 nm wide, but the developed width was 25 nm. The diameter of the 50 kV electron beam used to write the patterns was less than 1 nm.

Development may be slower for narrower lines because the lines become close in size to the molecular size of the PMMA and slow the development process.

Contrast for an array of infinitely long parallel lines

If the exposure distribution is Gaussian, then the ratio of the exposure dose received at the center of a line (Q_l) to the exposure dose received at the center of a space (Q_S) for an array of infinitely long parallel lines with equal linewidth and spacing, assuming zero effect from back-scattered electrons, is given by

$$\frac{Q_l}{Q_S} = \frac{erfS/4 + \sum_{n=1}^{n=m} [erf(4n+1)S/4 - erf(4n-1)/S/4]}{\sum_{n=1}^{n=m} [erf(4n-1)S/4 - erf(4n-3)S/4]} \tag{9.2}$$

where S is the center-to-center spacing between the lines, and m is the number of lines taken into account on either side of the line under consideration. It is also useful to define what we shall call the Resist Contrast Function (RCF), K, where

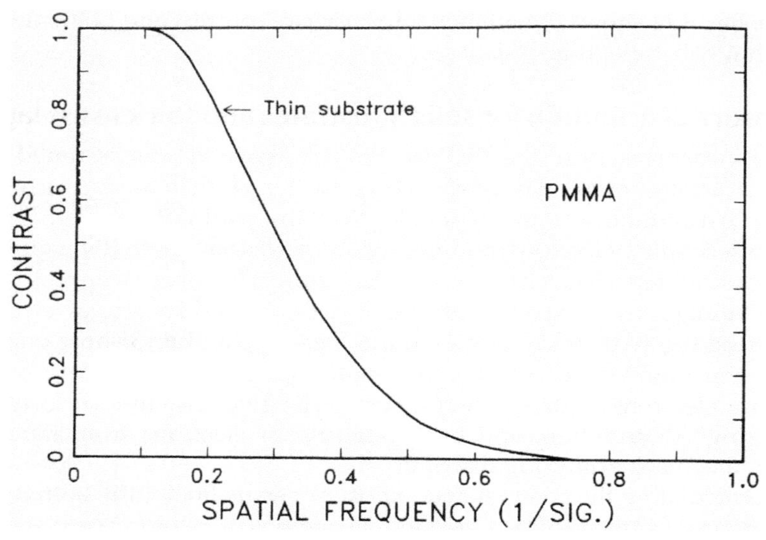

Fig. 9.11 Contrast vs spatial frequency for PMMA resist.

$$K = \frac{(Q_1 - Q_S)}{(Q_1 + Q_S)} \tag{9.3}$$

This definition is similar to that of the modulation transfer function (MTF) for an optical system, except that in the optical case, the pattern distribution is generally assumed to be sinusoidal rather than the square wave we have assumed for the electron beam case. This is discussed in the next section. Fig. 9.11 shows the RCF for a resist that exhibits a Gaussian exposure distribution. For a thin substrate, 60% contrast ($K=0.6$) corresponds to a spatial frequency (ν) of 0.28 σ^{-1}. If it is assumed that the distribution for PMMA is Gaussian and the $\sigma=12.5$ nm, then 60% contrast is obtained for $\nu=0.28/12.5\times10^{-7}=2.24\times10^5$ lines/cm (44.6 nm center to center spacing).

It is assumed that the electron beam diameter is zero and that lateral electron scattering in the resist is negligible. 60% is the order of contrast normally required for adequate definition in a developed resist pattern. Broers et al. (1978a) reported a minimum center-to-center spacing of 50nm for a 110nm thick PMMA layer on top of a 20nm layer of AuPd on a Si_3N_4 membrane. This is in good agreement with the results reported here.

The experimental data indicate that a Gaussian distribution with a sigma (σf) of 12.5 nm can approximate the exposure distribution. This spread was assumed to arise from the straggling of the secondaries into the resist, although the size of the PMMA molecules may also have played a role. Fig. 9.11 shows the relationship between contrast and linewidth for this thin substrate, thin resist case.

Exposure distribution for solid substrates and thick resist layers

In the measurement of the RCF for PMMA above, it was assumed that electron scattering could be ignored because the width of the forward scattering distribution was much smaller than the width of the distribution due to secondary electrons, and the substrate was so thin that the number of backscattered electrons was in effect zero. However, when electron beam lithography is used to fabricate devices on solid substrates, electron scattering has to be taken into account. Figs. 9.12 and 9.13 show qualitatively the mechanisms of electron scattering. Lateral scattering of the primary electrons as they penetrate the resist gives rise to the "forward-scattering" distribution, and backscattering of electrons from the substrate, the "back-scattering distribution.

In calculating the contrast and exposure versus linewidth, both distributions are assumed to be Gaussian with sigmas of σ_f for forward scattering and σ_b for backscattering. Contrast for a pattern of equal lines and spaces is given by

$$K = \frac{Q_L - Q_S}{Q_L + Q_S} \tag{9.4}$$

where Q_L is the dose at the center of a line, and Q_S is the dose at the center of a space.

$$\begin{aligned} Q_L = {} & erfS/4\sigma_f + \sum_{n=1}^{n=m}\left[erf(4n+1)S/4\sigma_f - erf(4n-1)S/4\sigma_f\right] \\ & + \eta\left\{erfS/4\sigma_b + \sum_{n=1}^{n=m}\left[erf(4n+1)S/4\sigma_b - erf(4n-1)S/4\sigma_b\right]\right\} \end{aligned} \tag{9.5}$$

$$\begin{aligned} Q_S = {} & \sum_{n=1}^{n=m}\left[erf(4n-1)S/4\sigma_f - erf(4n-1)S/4\sigma_f\right] \\ & + \eta\left\{\sum_{n=1}^{n=m}\left[erf(4n-1)S/4\sigma_b - erf(4n-3)S/4\sigma_b\right]\right\} \end{aligned} \tag{9.6}$$

S is the center-to-center spacing between the lines, and m is the number of lines taken into account on either side of the line under consideration. η is the ratio of the exposure due to backscattered electrons to the exposure due to incident electrons, and $(1+\eta)$ is the exposure dose in the center of a large shape.

Contrast (K_G) for an isolated space is assumed to be given by

$$K_G = \frac{Q_P - Q_G}{Q_P + Q_G} \tag{9.7}$$

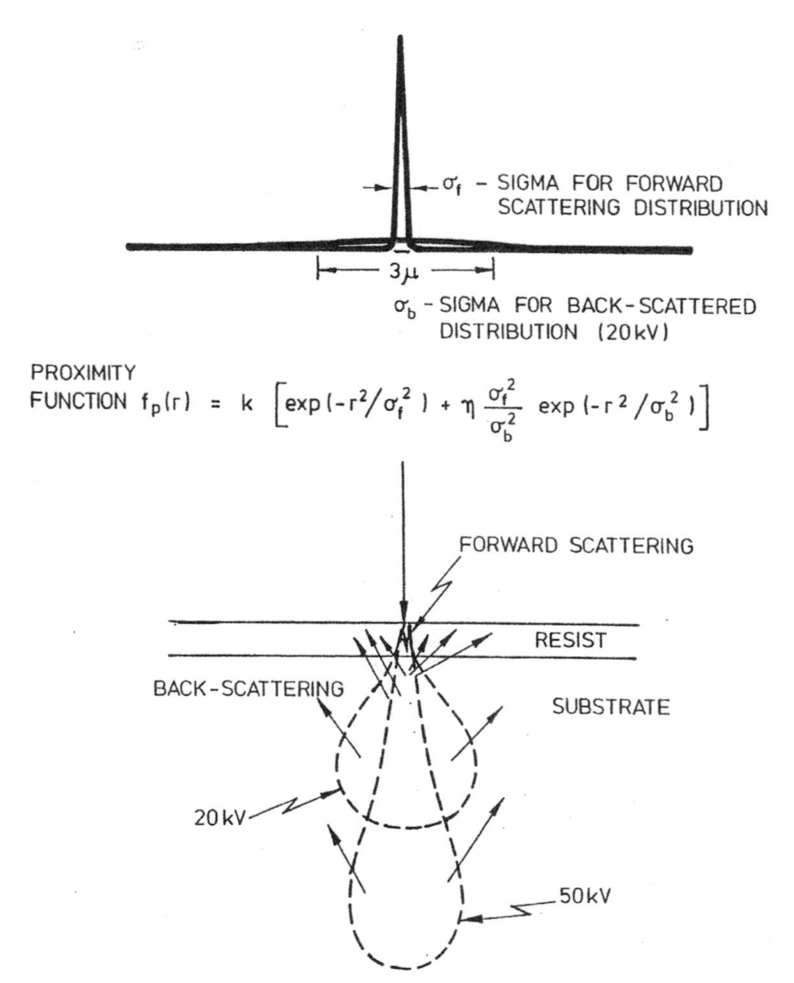

Fig. 9.12 Electron scattering in electron beam exposure of a resist on a solid substrate. The curve at the top of the figure shows the exposure distribution due to the incident and backscattered electrons. *(Reprint Courtesy of IBM Corporation ©).*

where Q_P is the exposure dose received at a point half the gap dimension away from the gap, and Q_G is th exposure dose received at the center of the gap.

$$Q_P = 1 + \eta - 0.5\left(erf\frac{3l}{2\sigma_f} - erf\frac{l}{2\sigma_f}\right) - \frac{\eta}{2}\left(erf\frac{3l}{2\sigma_f} - erf\frac{l}{2\sigma_f}\right) \tag{9.8}$$

$$Q_G = 1 + \eta - erf\frac{l}{2\sigma_b} - erf\frac{l}{2\sigma_b} \tag{9.9}$$

where l (nm) is the gap width.

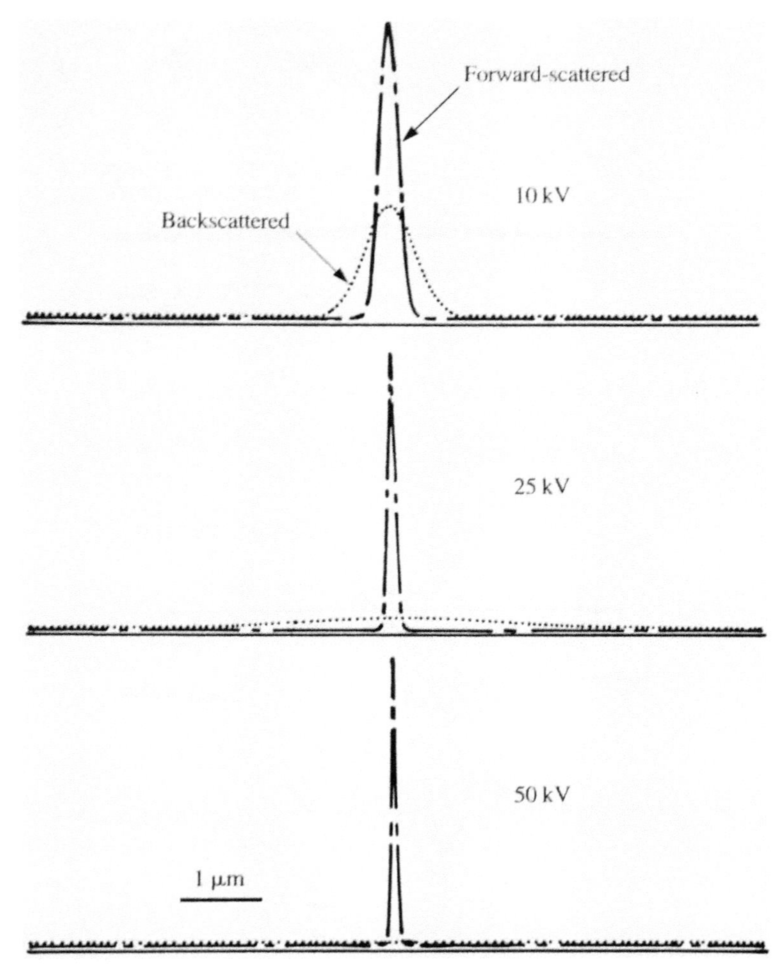

Fig. 9.13 Exposure distributions for a 1 μm thick resist layer on a silicon substrate for 10 kV (σ_f=0.3 μm, σ_b=0.8 μm, η=1), 25 kV (σ_f=0.12 μm, σ_b=3 μm, η=0.86), and 50 kV (σ_f=0.07 μm, σ_b=8 μm=0.5) electrons. *(Reprint Courtesy of IBM Corporation ©).*

Fig. 9.14 shows the contrast vs linewidth for lines and spaces and for isolated spaces for 25 kV electrons. Both thick and thin substrates and resist thicknesses of <0.1, 0.5, and 2 μm are included. It is important to note that with a thick substrate, the contrast is the same for 0.05 μm lines as it is for 1 μm lines, provided the resist is thin (<0.1 μm). This is because the backscattered distribution is much wider than either the secondary distribution or the forward-scattered distribution.

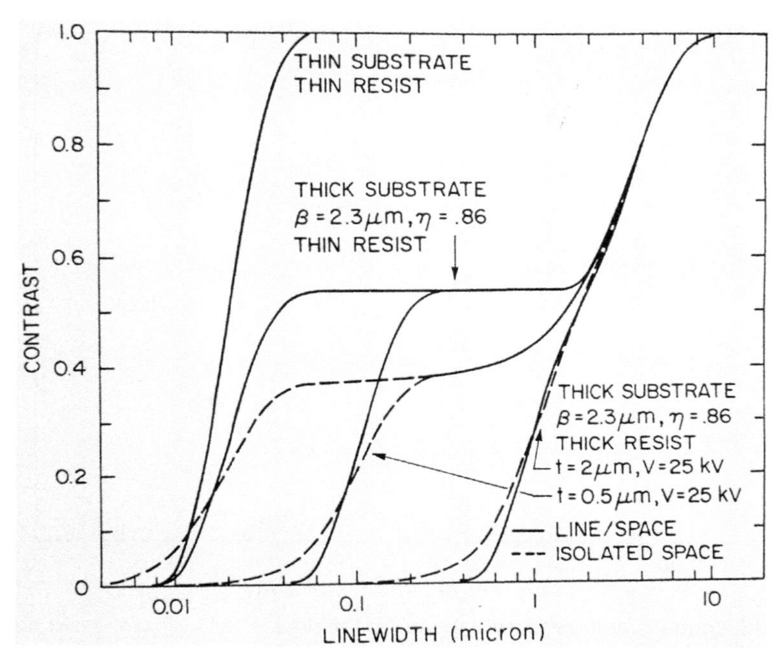

Fig. 9.14 Contrast vs linewidth for equal lines and spaces and isolated spaces and isolated spaces. *(Reprinted with permission from Broers, A.N., 1984. Practical and fundamental aspects of lithography. Materials for Microlithography. American Chemical Society. Chapter 2. ©1984 American Chemical Society).*

The fraction of the exposure due to backscattered electrons is reduced for multilayer resist processes where the image is formed in a thin top layer that is backed by a thicker underlying layer. The underlying resist is of lower atomic weight than silicon and consequently backscatters fewer electrons. The contrast for the multi-layer siloxane process described by Parasczak (1983) is shown in Fig. 9.15. Contrast is improved for all linewidths.

For relatively thick resist (>0.25 μm) and lower scattering potentials (<50 kV), lateral scattering of high energy electrons determines resolution. It is possible again to approximate the effective spreading by a Gaussian distribution, and the standard deviation of the distribution (σ_f) can be obtained from (Greeneich, 1979). The data in this figure are determined from Monte Carlo simulations but are in good agreement with experimental measurements reported in Chung and Tai (1978). We have used the expression to approximate lateral scattering when calculating contrast (Figs. 9.15–9.17) and aspect ratio (Fig. 11.20, X-ray exposure). Eq. (9.10) closely fits the Monte Carlo data up to the point that the number of collisions suffered by an electron passing through the resist exceeds 100.

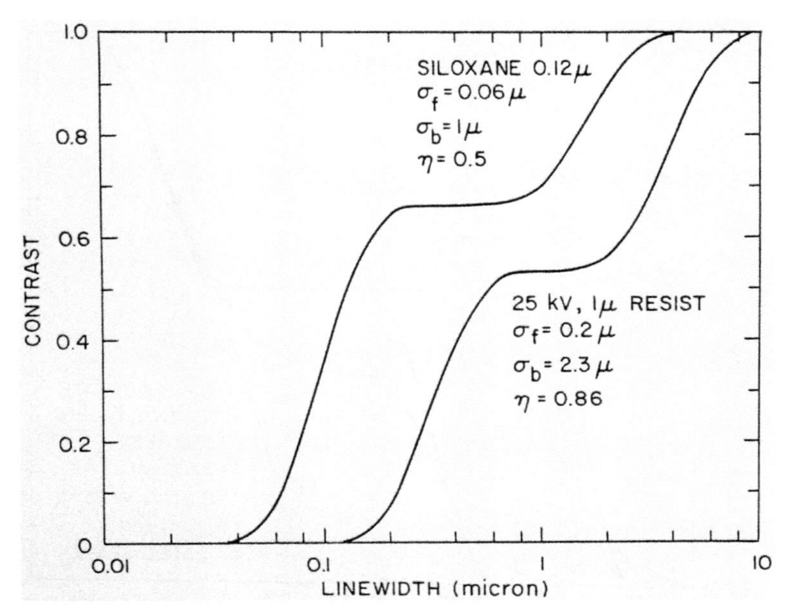

Fig. 9.15 Contrast for lines-space patterns for the double layer siloxane resist process Parasczak (1983) is higher than for the single layer case for all linewidths. *(Reprinted with permission from Broers, A.N., 1984. Practical and fundamental aspects of lithography. Materials for Microlithography. American Chemical Society. Chapter 2. ©1984 American Chemical Society).*

$$\sigma_f = \left\{ \frac{9.64.z(\mu\text{m})}{V(\text{kV})} \right\}^{1.75} \tag{9.10}$$

In general, higher accelerating potential offers higher resolution and higher contrast. Higher energy electrons scatter less in the resist and backscatter from the substrate over a larger area. The overall effect is that contrast is increased for all linewidths except those close to the width of the backscattered distribution, and contrast at these relatively large line widths is so high that this is not significant. Fig. 9.16 shows the improvement to be gained in going from 25 to 50 kV. The effective resist sensitivity is reduced at higher voltages because less of the electron energy is dissipated in the resist, but the electron brightness is increased in proportion to the voltage, and this increases the beam current, partially compensating for the loss. The overall improvement obtained at higher electron energy should make it easier to correct proximity effects and to write high-resolution patterns in thick resist layers.

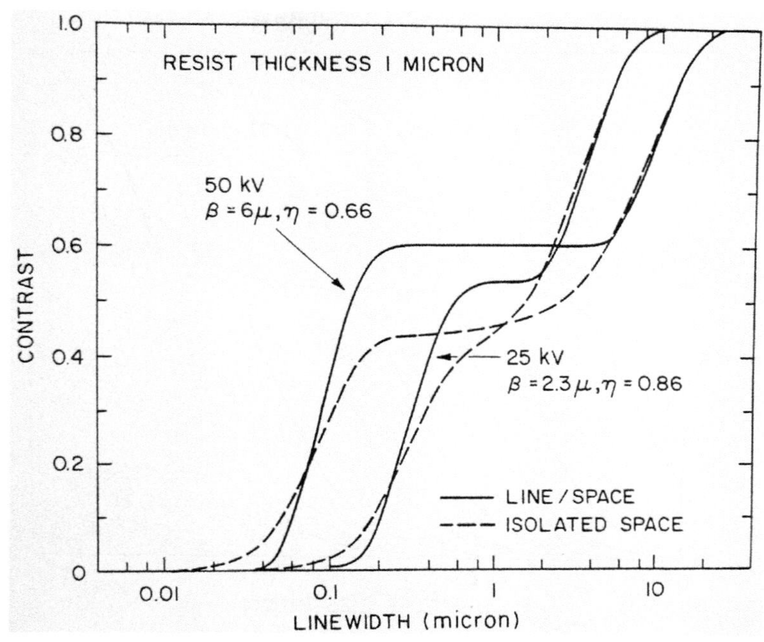

Fig. 9.16 Contrast versus linewidth for 25 and 50kV electrons exposing a 1 μm thick resist layer on a silicon substrate. Improved proximity effect has been reported by Neill and Bull (1980). Backscattering for 50kV electrons was obtained by extrapolation from data given in Neill and Bull (1980) and Parikh and Kyser (1979). *(Reprinted with permission from Broers, A.N., 1984. Practical and fundamental aspects of lithography. Materials for Microlithography. American Chemical Society. Chapter 2. ©1984 American Chemical Society).*

Normalized aperture exposure

The resist contrast function described above is useful in estimating ultimate resolution but does not provide a good indication of the width of the "process window" for a given lithography process. An array of lines may be clearly resolved, but only for exposure and development conditions at which the isolated lines and apertures do not open. Increasing exposure and/or development to open isolated shapes will overdevelop the line/space array. The width of the process window can be better estimated from the ratio of the exposure at the center of a small aperture to the exposure in a large shape. This is shown in Fig. 9.17 for electron and optical cases.

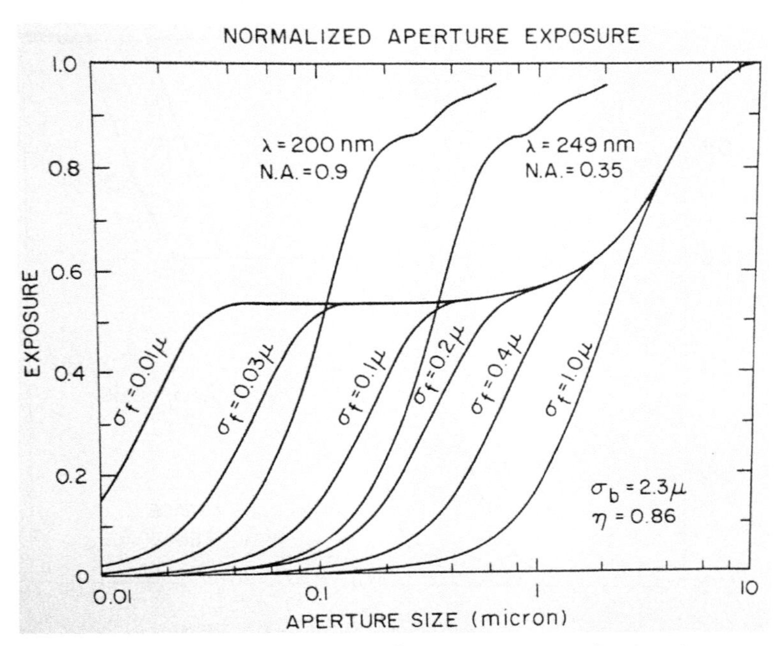

Fig. 9.17 Exposure received at the center of an aperture, normalized to the exposure at the center of a large shape, for electron and optical cases. *(Reprinted with permission from Broers, A.N., 1984. Practical and fundamental aspects of lithography. Materials for Microlithography. American Chemical Society. Chapter 2. ©1984 American Chemical Society).*

Fig. 9.17 shows, for example, that the exposure at the center of a 1 μm × 1 μm aperture is only 55% that at the center of a large area for a 25 kV electron beam exposing a 1 μm thick resist layer (σ_f=0.25 μm). With a high numerical aperture step and repeat camera lens, the aperture exposure is >80% of that of large areas. This example neglects the secondary influences such as depth-of-field, resist contrast, etcetera, but realistically describes the ease of finding exposure and development conditions that are satisfactory for the full range of pattern shapes.

Fig. 9.18 shows the more favorable case of electron beam exposure of the multilayer siloxane resist. Here, the exposure due to backscattered electrons is reduced from 86% to 50% of the primary exposure, and the imaging layer is very thin. Exposure in the 1 μm × 1 μm aperture exceeds 70% and is greater than the exposure for the optical case below 0.4. It is clear that it will be much easier to find acceptable exposure and development conditions with the siloxane process. The penalty is that more processing steps are required making the process more expensive.

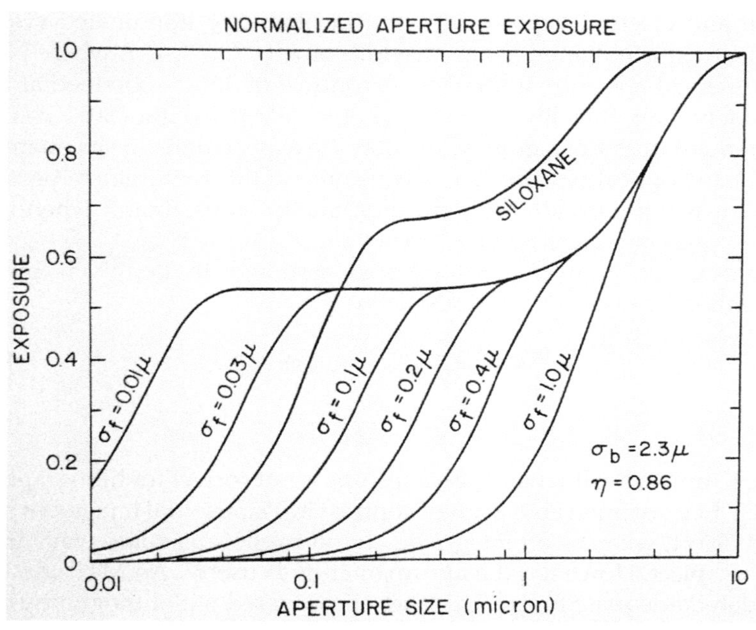

Fig. 9.18 Aperture exposure for multi-layer siloxane electron beam resist. *(Reprinted with permission from Broers, A.N., 1984. Practical and fundamental aspects of lithography. Materials for Microlithography. American Chemical Society. Chapter 2. ©1984 American Chemical Society).*

Contrast and resolution for optical exposure

Both reflecting and refracting lenses have been used in optical microcircuit cameras. In the early 1970s, the Perkin Elmer Micralign cameras that replaced contact and proximity printers used mirror lenses. The first generation Misaligns had a resolution of about 2 μm. A period followed when refracting lenses with more than a dozen elements were used to reduce dimensions from about 2 to 0.5 μm. To further increase resolution, however, it was necessary to reduce the wavelength of the light below that at which the materials available for refractive elements were sufficiently transparent, and today, the highest resolution cameras all use mirrors. These require complex multi-layer coatings and have to be fabricated with mechanical tolerances that approach atomic dimensions.

In practice, 60% modulation had been considered necessary in the optical image for satisfactory resist exposure. However, it is frequently more appropriate to also require that at least 30% contrast is maintained at half the nominal linewidth, as is discussed in the next section comparing optical and electron exposure. For an incoherently illuminated system, 60% contrast is obtained for a linewidth of $\lambda/(1.28\times$ (NA)) where NA is the numerical aperture of the lens. A contrast of 30% is obtained at approximately half this linewidth ($\lambda/(2.36\times$ (NA)), so the 30% contrast requirement does not significantly alter the expectations for incoherently illuminated optical systems. The significance of this requirement becomes apparent when partially coherent illumination is used, and when comparing optics and electron beam exposure.

Contrast at a given resolution is assumed to be given by the Modulation Transfer Function (M.T.F.)

$$MTF = \frac{2}{\pi}(\phi - \cos\phi\sin\phi) \tag{9.11}$$

$$\phi = \cos^{-1}\frac{\lambda}{4l.(\mathrm{NA})} \tag{9.12}$$

where L (nm) is the linewidth. Strictly, this is not correct for lithography as the M.T.F., as defined above, gives contrast in a sinusoidal image of a sinusoidal object, whereas an integrated circuit mask is a square wave transmission object. However, the approximation is useful. An MTF of 60% is considered adequate for typical cases encountered with lithography cameras and 80% for cases where the image size control of a tenth of the minimum linewidth is required.

Higher contrast can be obtained for relatively large linewidths by using partially coherent illumination; that is by arranging that the image of the source only partially fills the pupil of the projection lens. 30% to 50% filling of the pupil has proven optimum and, for example, can increase contrast from 60% for the incoherent case, to more than 80% (ref). Partial coherence increases the depth of field but increases exposure time. Too high a degree of coherence produces undesirable interference effects between the lines.

The depth of the field depends on substrate reflectivity, the degree of partial coherence, and the minimum feature size. In practice, however, the classical depth of field for the incoherent case $\pm(\lambda/2(\mathrm{NA})^2$ gives a reasonable approximation. Two-layer resist processes in which the image is formed in a thin and flat resist layer on top of a much thicker planarizing layer alleviate the need for a large depth of field and make it easier to form high resolution, high aspect ratio, resist patterns (Lin, 1979) (King, 1981). Satisfactory results can be obtained at contrast levels as low as 40%.

Fig. 9.19 shows the optical case for an N.A. = 0.15 and a wavelength of 405 nm. 60% contrast for the incoherent case is obtained at a linewidth of about 2.2 μm. Partial coherence increases contrast for linewidths above about 1 μm. Fig. 9.20 shows the case for a refractive lens with an N.A. = 0.35 operating with UV light with a wavelength of 250 nm. This was regarded as the high-resolution case to be obtained in the early 1990s. The linewidth for 60% contrast is about 0.5 μm.

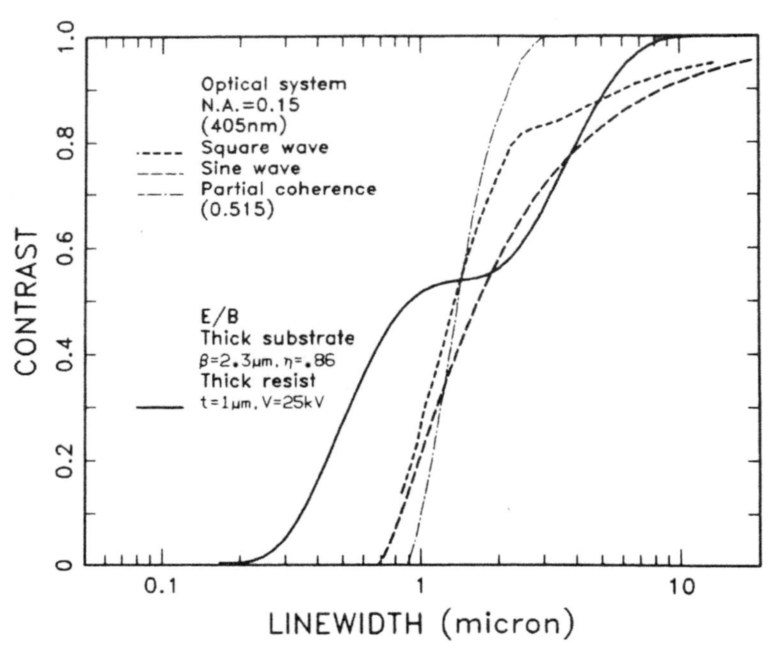

Fig. 9.19 Contrast vs linewidth for electron beam and optical projection. System parameters are those common for microcircuit technology circa 1980. *(Used with permission of IEEE from Broers (1981) permission conveyed through Copyright Clearance Center, Inc.).*

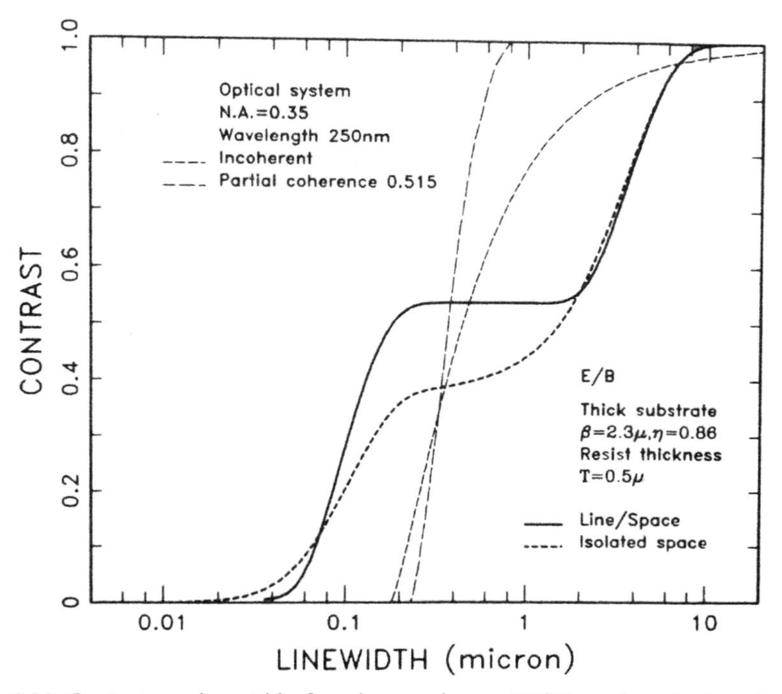

Fig. 9.20 Contrast vs linewidth for electron beam (25 kV) and optical projection lithography. System parameters common for the mid to late 1980s. *(Used with permission of IEEE from Broers (1981) permission conveyed through Copyright Clearance Center, Inc.).*

Contrast for partially coherent illumination is obtained from Offner (1979). Too high a degree of coherence increases interferences between lines, rounds corners on patterns, and increases exposure time. A major advantage of partial coherence is that it effectively increases the depth of focus (King, 1981). Fig. 9.20 shows the case for an N.A. of 0.35 and a wavelength of 250 nm.

The depth of field for the incoherent case is given approximately by $\pm\lambda/2(NA)$ (Bowden and Thomson, 1978) but depends on the substrate reflectivity, the degree of coherence of the illumination, and the minimum feature size (King, 1981). As with electron exposure two-layer resist processes in which the image is formed in a thin, flat resist layer on top of a much thicker planarizing layer alleviate the need for a large depth of field and make it easier to form high-resolution high-aspect ratio resist patterns (Hatzakis et al., 1980; Lin, 1979; Tai et al., 1979). These resist systems make it possible to obtain satisfactory results at contrast levels considerably lower than 60%.

Contrast for optical and electron beam exposure

The relationship between the theoretical contrast function and the minimum reproducible linewidth is difficult to establish for both optical and electron beam exposure because it depends on variations in resist contrast, substrate reflectivity or scattering power, resist thickness, and many other variables. Several workers have modeled the exposure and development process for specific cases, taking many of these variables into account (Dill, 1975; Greeneich, 1980) but few have made direct comparisons between optical and electron beam exposures. Here I make such comparisons, but caution that many factors must be considered in interpreting the data. In particular, it is not sufficient to consider just the contrast at the minimum required linewidth. The contrast at these higher spatial frequencies determines the edge slope on the resist and the sharpness of corners etc., In practice, it appears that at least 30% contrast has to be maintained at a linewidth half that of the nominal minimum linewidth. There is a considerable difference between electron beam and optical exposure in this respect. For electron beam, contrast is constant over a broad range of linewidths. This is because there is a large difference between the widths of the secondary electron and backscattered electron distributions. In cases where the forward-scattering spread function is wider than the secondary distribution, the forward-scattering distribution is still generally much narrower than the backscattered distribution. In the region of constant contrast, the backscattered electrons in effect *fog* the image, but the image remains relatively sharp. A definition of 60% is unnecessarily pessimistic because the resist process can be adjusted to be insensitive to the background fog.

Fig. 9.19 compares electron beam and optical exposure for operating conditions in standard use in 1980. For the partially coherent case, the contrast for optics exceeds that for electron beam for linewidths above about 1.25 μm, but below 1.25 μm, the contrast falls so rapidly that it is zero at 0.9 μm. The realistic capability can again be predicted by taking twice the linewidth at which 30% contrast is obtained. This yields a minimum linewidth of 2.2 μm for the optics and 1 μm for the electron beam case.

Fig. 9.20 compares electron beam and optical exposure for operating conditions in the mid to late 1980s. It shows that a 0.35 NA ($\lambda = 250$ nm) optical system has higher contrast for linewidths above 0.5 μm than a 25 kV electron beam exposing 0.5 μm thick resist on a thick substrate. At 1.25 μm linewidth, the contrast of the optical system is 85% compared to 54% for the electron beam. However, for reasons just discussed the exposure transition at the boundary between a line and a space is sharper for the electron beam case than it is for the optical system even though the nominal contrast is lower, Contrast for the electron beam does not fall to 30% until a linewidth of 0.15 μm is reached which suggests that the smallest useful linewidth is 0.3 μm. In practice, the contrast with electron beam resists was high enough for linewidths approaching 0.25 μm to be produced in semiconductor processes by the 1980s, but this required the application of proximity correction where the exposure dose is adjusted according to the density of the pattern. For the optical system, linewidth at 30% contrast is 0.3 μm, suggesting a minimum useful linewidth of 0.6 μm. Optical lithography was not capable of 0.25 μm linewidth in the 1980s. This had to await further reductions in the wavelength of the exposing UV light.

In Chapter 10, I return to semiconductor lithography in the 1980s and 1990s and conclude this chapter with a description of the processes that have been discovered that allow electron beams to pattern materials directly without the need for development. Some of them offer higher resolution than may be obtained with conventional resists.

Direct fabrication with electron beams

If only the size of the electron beam mattered, it would have been easy to make devices and structures with dimensions below 10 nm, but as I have discussed at length, electron scattering and the excitation of secondary electrons in the region surrounding the beam conspire to make it difficult or impossible.

In 1989, Donald M. Eigler and Erhard K. Schweizer at IBM showed that single atoms could be placed with the probe of a scanning tunneling microscope (Eigler and Schweizer, 1990). This was an amazing accomplishment, but this process is too slow to be useful for manufacturing devices. It has even been difficult to use it to make devices that are useful in the laboratory.

What is needed to make electron beam lithography practicable for dimensions below 10 nm is a process that is sensitive only to the beam electrons, and itself has a sub-10-nanometer resolution. The beam electrons could be confined to a beam diameter of 0.3 to 0.5 nm, or even lower, if the aberration correction techniques that have been applied to electron microscopes were applied. Exposure speed is also not a concern for making devices for scientific exploration. The issue is the process and its ability to make structures in given materials and be able to make electrical contact to them.

One process that appeared to potentially have these characteristics was the direct sublimation under electron bombardment of a variety of materials, including NaCl, MgF_2, AlF_3, LiF, and Al_2O_3. Work on this process started following a discussion I had with John Matthews over a cup of coffee in the Yorktown laboratory in 1977. John was a distinguished electron microscopist from South Africa who had installed the second electron microscope in South Africa. In the 1960s, he started a collaboration with IBM Research and eventually, in January 1970, joined IBM Research as Head of the Thin Film Research Group. One of his interests was the epitaxial growth of face-centered cubic metals, especially gold, on clean sodium chloride crystals. A difficulty he encountered was that the sodium chloride crystals would sublimate away under electron bombardment. He could cut a path through NaCl crystals scattered on a carbon film with the illuminating beam in his TEM, as shown in Fig. 9.21. He knew I was looking for ways to make structures with electron beams and suggested I try to measure the resolution of this process with my 0.5 nm electron beam. Naturally, I was keen to do this but was not very hopeful that it would be a high-resolution

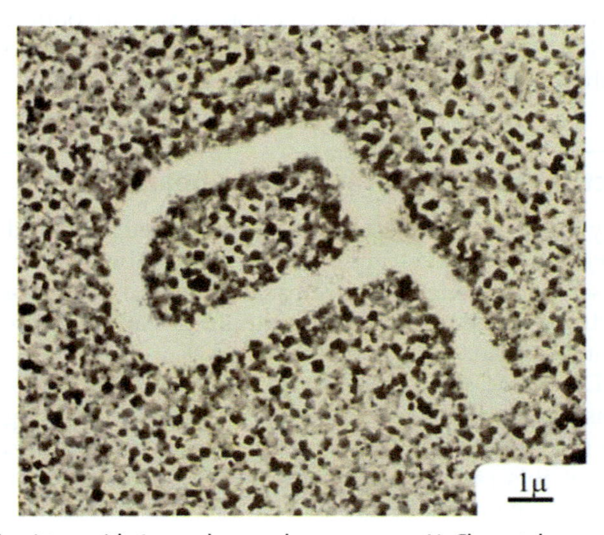

Fig. 9.21 "P" written with 1 μm electron beam among NaCl crystals.

process. I thought that the crystals might just melt or disintegrate, although NaCl does not melt until 800°C, and John had shown that there was no residue, it simply disappeared.

I soon discovered that the STEM beam, in effect, drilled holes in the crystal and did this in a few seconds. I left the beam stationary on a NaCl crystal and watched the transmitted electron signal grow back to the level that it had been off the crystal in a few seconds. When I switched the STEM scan on again, I took the micrographs shown in Figs. 9.22 and 9.23 (Broers et al., 1978a). The larger holes in Fig. 9.21 were about 10 nm in diameter, and the smaller ones were 5 nm in diameter. The beam voltage was 50 kV, the beam current was 10^{-11} A, and the beam diameter was about 0.7 nm. The crystal was 0.25 μm thick, and the convergence half-angle of the beam was 10^{-2} rad. Assuming that the beam was focused on one face of the crystal, the beam would have formed a conical-shaped hole, with the base of the cone being about 5 nm in diameter. This would explain the difference between the apparent size of the hole and the beam diameter and suggest that the resolution of the process was better than 5 nm.

We continued these experiments to show that it was possible to use a patterned MgF_2 film to mask a gold film from ion milling and produce slots less than 10 nm in width in a 20 nm thick gold film (see Fig. 9.24). However, the films did not stand up well to a variety of dry or wet etching techniques, and the exposure doses that are required were very heavy ($0.1\,C/cm^2$), and we were unable to use the method to make superconducting switches and SQUIDs.

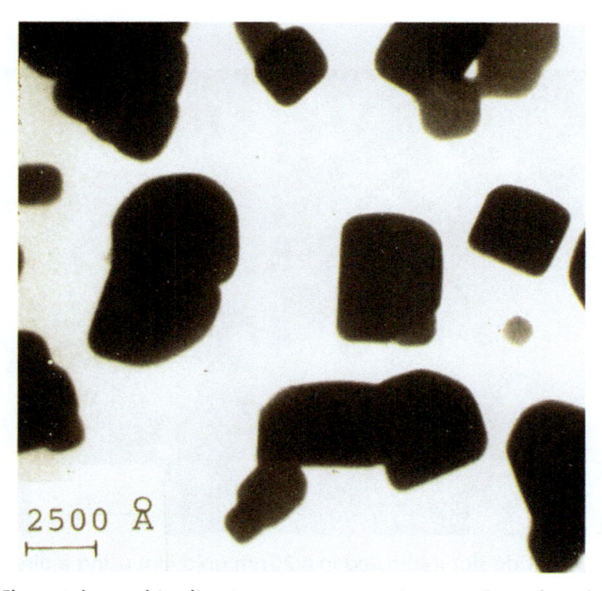

Fig. 9.22 NaCl crystals used in direct exposure experiments. Crystal to the right of the center of the micrograph is also shown in Fig. 9.24.

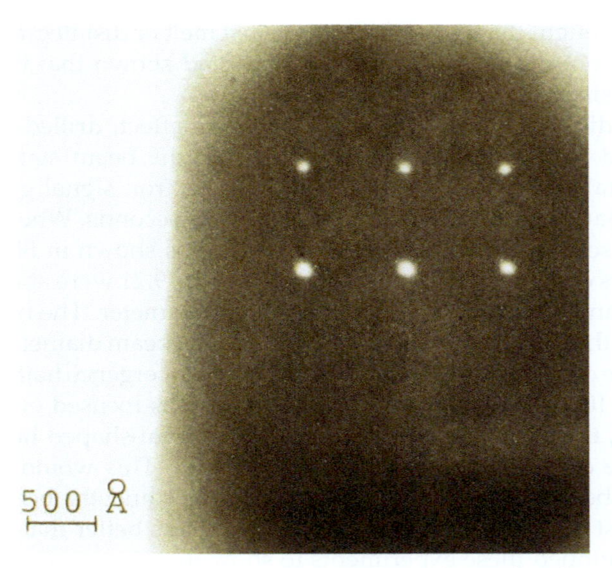

Fig. 9.23 Holes 5 nm in diameter in 0.25 μm thick NaCl crystal with 1 nm diameter electron beam. *(Reprinted with permission from Broers, A.N., 1984. Practical and fundamental aspects of lithography. Materials for Microlithography. American Chemical Society. Chapter 2. ©1984 American Chemical Society).*

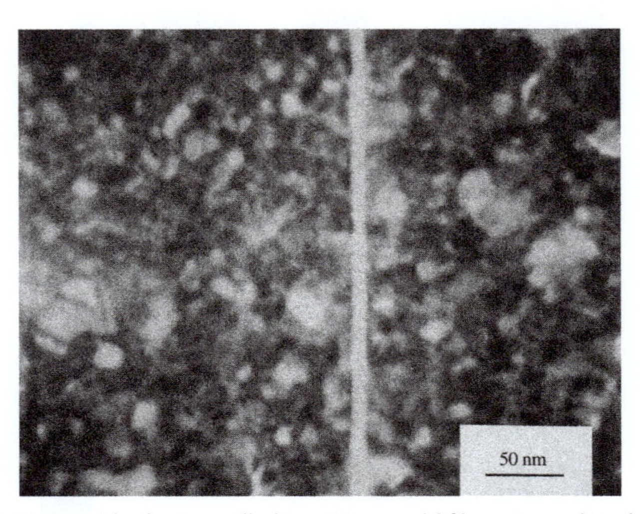

Fig. 9.24 A 10 nm-wide slot ion milled in a 20 nm gold film using a directly patterned MgF_2 as a mask.

Isaacson and Muray (1981) confirmed that resolution below 5 nm is possible with direct sublimation by writing structures down to 1.5 nm in size in thin NaCl and AlF_3 films. Kratschmer and Isaacson (1986) used one of the patterned AlF_3 films to mask a silicon nitride film against dry etching but the smallest structures produced in the nitride were about 20 nm in size.

Mochel et al. (1983) discovered that the sublimation process works with Al_2O_3. They produced holes that were 1 nm in diameter. Both Mochel et al. and Isaacson et al. used EELS (Electron Energy Loss Spectroscopy) to analyze the process and showed that, in general, a metal-rich deposit remains after the sublimation process. The details of the sublimation mechanism remained unclear, however, and it has yet to be shown that the process can produce devices of practical importance.

Fig. 9.25 shows some of the ways device structures might be made with direct sublimation lithography. I do not know whether the method has been used successfully to fabricate devices.

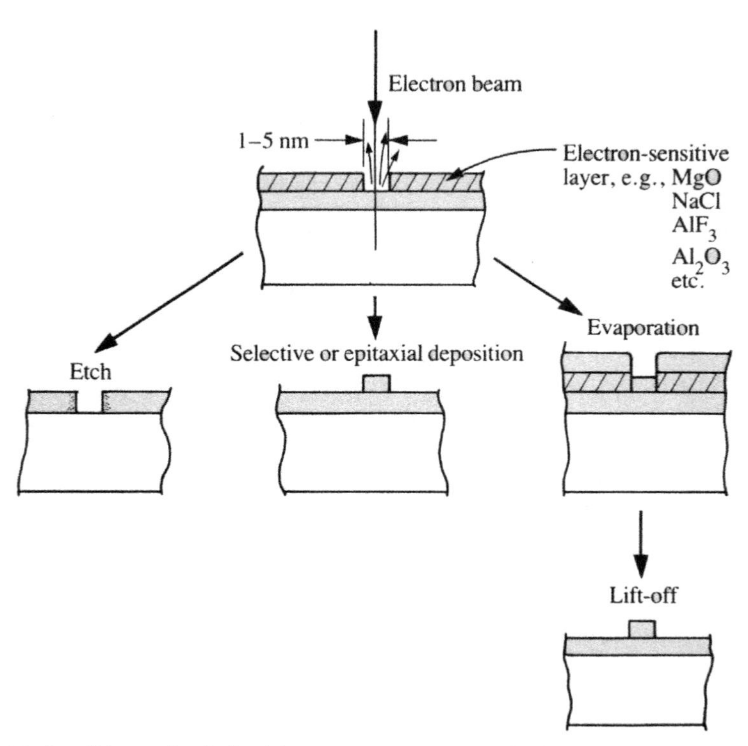

Fig. 9.25 Possible methods for fabricating structures with direct-sublimation electron beam lithography. Selective deposition methods could include plating. *(Reprint Courtesy of IBM Corporation ©).*

Exposure of multi-layer Langmuir–Blodgett films

In the early 1980s, I worked with Mel Pomerantz to see whether an electron beam could be used to pattern multi-layer Langmuir–Blodgett films (Broers et al., 1978b; Broers and Pomerantz, 1983). We discovered that it was possible to write lines at much lower doses ($\sim 10^{-4}$ C/cm^2) than were required for the other direct sublimation processes just described where exposure doses above 0.1 C/cm^2 were needed.

There were earlier reports of the use of Langmuir–Blodgett films for high-resolution writing. Zingsheim (1977) showed that films of cadmium arachidate, after exposure to an electron beam with a diameter of 0.5 nm, could be a template for decoration by large molecules of evaporated silver. The resolution was about 10 nm. He suggested that contamination is not the mechanism behind these effects. He looked for changes in the film surface morphology by an angled evaporation shadowing technique but did not observe any. Barraud et al. (1980) have studied the optimization of polymerized Langmuir–Blodgett films for high resolution and contrast. Even using thick substrates, they achieved a resolution of about 60 nm. The thickness of the substrate is important because electrons backscattered from the substrate also expose the film, but their directions are diffuse. The effect is to lower the contrast of the written image by partially exposing the film adjacent to the primary beam.

Experimental methods

The Langmuir–Blodgett films used in our experiments were prepared by methods that have been reviewed by Pomerantz (1980) and Pomerantz et al. (1978). Briefly, the bath of distilled water contained 10^{-3} M $MnCl_2$, and the pH was adjusted to about 7 by the slow addition, with stirring, of dilute KOH. Stearic acid dissolved in hexane (1 mg/mL) was spread on the surface, and the solvent was allowed to evaporate for about 30 min. Under these conditions, it has been shown that manganese stearate ($Mn(H_3C–(CH_2)_{16}–COO)_2$, abbreviated as $MnSt_2$) is formed stoichiometrically to within a few percent experimental error. We used $MnSt_2$ because we were most familiar with it from work on two-dimensional magnets, but other metal ions or carboxylic acids would probably give similar results as we shall describe. The surface pressure applied was about 30 dyne/cm. The rate of dipping was about 2 cm/min. The substrates used were thin amorphous graphite films on electron microscope grids 3 mm in diameter and of 200 mesh. Typically, about 100 layers were deposited, i.e., 50 unit cells of head-to-head molecules with a total thickness of about 250 nm. The substrates were so thin (about 20 nm) that they produced a minimal number of backscattered electrons, ensuring maximum contrast in the written patterns.

A 50 keV electron beam with a diameter of about 1 nm and a current between 10^{-10} and 10^{-11} A was used to write the lines. Lower current beams were used to examine the results of the writing in the scanning transmission bright field mode. The scanning rate during the writing was about 10 μm/s. This produces an electron dose of about 10^{-8} C/cm. Block areas became exposed at about 5×10^{-3} C/cm^2, which was significantly lower than the dose needed to expose a useful thickness of contamination resist.

Experimental results

The effect of a raster scan of lines on 138 layers of $MnSt_2$ films is shown in Fig. 9.26.

It was found that where the beam has scanned, there is less absorption, as observed in the micrographs obtained with the lower-intensity beam. The film had not been developed in any way; it was observed immediately after writing. The width of the lines was about 10 nm, and their spacing was about 20 nm. Another indication of the high resolution is shown in Fig. 9.27, in which the intersection of lines is displayed. There is little or no rounding of the corners. This shows that exposure is confined to a region with a diameter of less than 10 nm, and this region may be even smaller than 10 nm.

Fig. 9.28 shows a line written by the beam during its fly-back after writing a frame of lines. This gives an indication of the speed with which lines

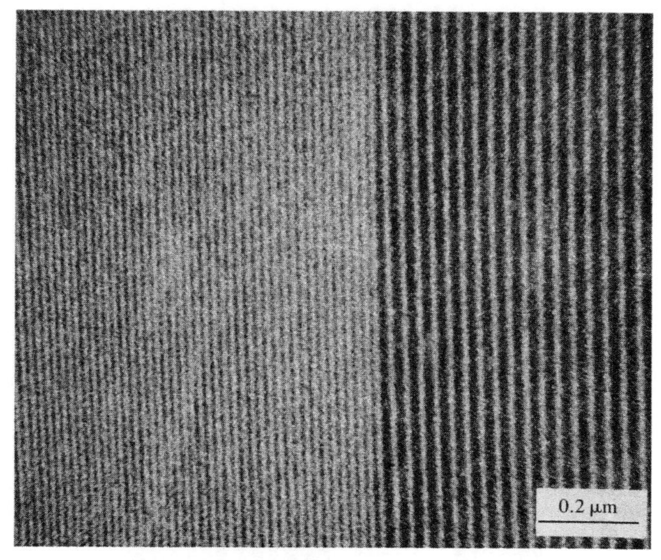

Fig. 9.26 Bright-field STEM micrograph of lines written in a stack of manganese-stearate Langmuir–Blodgett films. *(Reprinted with permission from Broers, A.N., 1984. Practical and fundamental aspects of lithography. Materials for Microlithography. American Chemical Society. Chapter 2. ©1984 American Chemical Society.).*

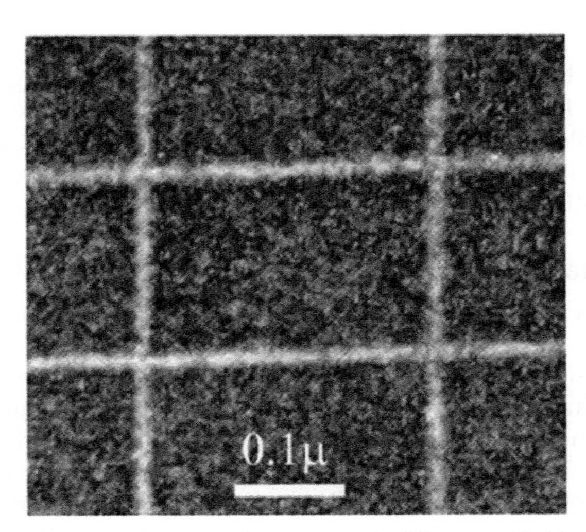

Fig. 9.27 Intersecting lines written in the same stack of films used in Fig. 12.6. *(Used with permission of Elsevier Science and Technology Journals from Broers, A.N., Pomerantz, M., 1983. Rapid writing of fine lines in Langmuir-Blodgett films using electron beams. Thin Solid Films 99, 323–329 permission conveyed through Copyright Clearance Center, Inc.).*

Fig. 9.28 Exposure showing trace of the electron beam during fly-back between frames. *(Used with permission of Elsevier Science and Technology Journals from Broers, A.N., Pomerantz, M., 1983. Rapid writing of fine lines in Langmuir-Blodgett films using electron beams. Thin Solid Films 99, 323–329 permission conveyed through Copyright Clearance Center, Inc.).*

can be written in the MnSt2 layers. Although not measured, the fly-back speed of the beams was faster than the line scan rate.

To determine the profile of the exposed lines, we performed an angled evaporation shadowing experiment. A layer of gold was evaporated onto the sample at an angle of 45° in the plane normal to the lines. Fig. 9.29 shows this sample and the buildup of gold on the side of the lines facing the source of the evaporant. We concluded that the electron beam produced a trough in the film, and the absence of metal at the bottom of the trough allowed us to determine that the depth of the trough must be at least 50 nm. The aspect ratio (depth/width) of the troughs was, therefore, more than 5.

From experiments (conducted in collaboration with S. Herd) on films deposited onto the substrates with holes in them, we observed that the film tended to shrivel when exposed to the electron beam, but it did not disappear entirely. After extended exposure, no further change was observed. Connecting these two experiments, it appeared that the electron beam caused the film to distort (shrink) and reach some final contracted state that is resistant to further change. The large aspect ratio found in the shadowing experiments suggests that the contraction may be quite large or is associated with some loss of material.

Another interesting effect is shown in Fig. 9.30. This shows a film in which lines had been drawn and then repeatedly examined in an area that appears white. The area being examined was abruptly increased, and the micrograph was recorded. We previously interpreted this figure as

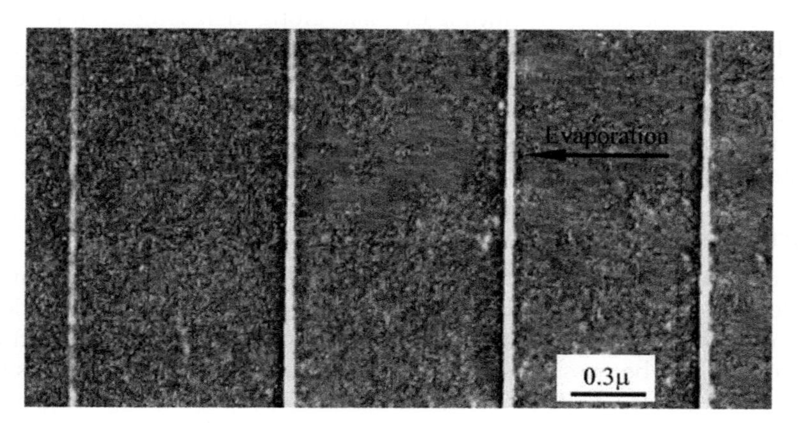

Fig. 9.29 Slots in stack of Langmuir Blodgett films shadowed to reveal the profile of the lines. The metal was shadowed from the right. *(Used with permission of Elsevier Science and Technology Journals from Broers, A.N., Pomerantz, M., 1983. Rapid writing of fine lines in Langmuir-Blodgett films using electron beams. Thin Solid Films 99, 323–329 permission conveyed through Copyright Clearance Center, Inc.).*

Fig. 9.30 Square in the middle of the micrograph formed when the lines were being repeatedly examined by the electron beam. *(Used with permission of Elsevier Science and Technology Journals from Broers, A.N., Pomerantz, M., 1983. Rapid writing of fine lines in Langmuir-Blodgett films using electron beams. Thin Solid Films 99, 323–329 permission conveyed through Copyright Clearance Center, Inc.).*

showing that the heavily exposed lines had become less transmitting (denser) because of the subsequent low-intensity exposure during the examination. It seemed that, compared with that of the lines in the least-exposed regions, the contrast of the lines in the white area has reversed. A closer examination of the original micrographs shows, rather, that the contrast along the lines is the *same* along their lengths; it is independent of subsequent exposure. However, the figure does suggest that the lower exposure used in the central square is more efficient in reducing the contrast than either the intense beam that drew the lines or the yet lower-density beam used for the micrograph. This suggests that the loss of material is not a monotonic function of beam density.

Discussion

Our experiments cast a new light on the well-known fact that Langmuir–Blodgett films are sensitive to electron beams. This property is a handicap when attempts are made to use electron diffraction to study the structure of films. It may be possible to turn it to advantage because it facilitates the fabrication of structures on a scale of tens of nanometers. Zingsheim (1977) had already indicated one use for such prepared surfaces in organizing large molecules. It might be useful to use the structurally modulated surface to align liquid crystals, which had been done (Flanders et al., 1978) on a scale of 300 nm but not smaller. Such a surface might also be useful as a substrate for graphoepitaxy, in which the topography of the surface is used to create interesting crystal growth (Geis et al., 1979).

It is also possible to envisage the preparation of diffraction gratings for far-UV radiation by angle evaporation of metal onto arrays of peaks cut by the electron beam. Such gratings might be used not only as dispersive elements for wavelength analysis and separation but also to generate monochromatic UV light by the Smith-Purcell effect (1953). This involves passing an electron beam close to a corrugated surface. The beam takes on the spatial wavelength of the corrugation and radiates a corresponding frequency. The practical use of such gratings will require that they be of fairly large area; so far, we have made only small-area gratings of the order of $1\,\mu m^2$.

Thus, we see that Langmuir–Blodgett films, in conjunction with electron beam writing, may provide another medium for fabricating very small structures.

Our experiments did not establish a minimum linewidth for the method. The lines we measured were about 10 nm, but there is evidence that we may have been delivering more dose than was necessary and that we might have produced narrower lines if we had used lower doses. The process appeared so sensitive to electron bombardment that it was difficult to avoid significant exposure when examining the samples in STEM mode. It would have been better to examine them in a conventional TEM, where the instantaneous exposure dose can be much lower.

We were not successful in using these Langmuir–Blodgett stacks as masks for standard dry or wet etching processes so, as with the direct sublimation methods, the process did not prove useful for device fabrication.

Radiation damage lithography

We realized in 1978 that we could probably fabricate structures smaller than the electron-resist interaction range by processes that use electrons whose energy is great enough to introduce radiation damage in the sample. For crystalline materials such as silicon, this would require an energy above about 150 keV. Several potential methods for converting the damaged areas into structures are shown in Fig. 9.25. The damage might enhance or retard the etch rate of the material for dry or wet etch processes, it might change other properties such as the critical temperature for superconductors, or it might affect the integrity of epitaxial films grown on the substrate. In the case of epitaxial films, damaged areas might etch at a different rate than defect-free areas or exhibit different electrical properties (conductivity, etc.). An example of the latter might be that regions of a superconducting film deposited on damaged areas of a single crystal substrate would exhibit normal conductivity and, therefore, act as weak links between the superconducting regions grown on the un-damaged areas.

Electrons with an energy of less than half the damage threshold energy might also be used for radiation damage lithography so that each electron creates only a single damage event. The damage should then be localized within the beam diameter, which can be as small as 0.3 nm. Jones et al. have conducted some preliminary experiments confirming that the basic process is feasible.

One reason for acquiring the 400 kV electron microscope in Cambridge, described in Chapter 12, was to further explore the use of electron probe-induced radiation damage to fabricate sub-10 nm structures and devices.

References

Barroud, A., Rosilio, C., Ruaudel-Teixier, A., 1980. Thin Solid Films 65, 31.

Bowden, M.J., Thomson, L.F., 1978. Solid State Technol. 15, 72.

Broers, A.N., 1981. Resolution limits of PMMA resist for exposure with 50 kV Electrons. J. Electrochem. Soc. 128, 166.

Broers, A.N., Pomerantz, M., 1983. Rapid writing of fine lines in Langmuir-Blodgett films using electron beams. Thin Solid Films 99, 323–329.

Broers, A.N., Molzen, W.W., Cuomo, J.J., Wittels, D., 1976. Electron beam fabrication of 80 Å structures. Appl. Phys. Lett. 29, 596–598.

Broers, A., Harper, J.M.E., Molzen, W.W., 1978a. 250A linewidths with PMMA electron resist. Appl. Phys. Lett. 33, 392–394.

Broers, A.N., Cuomo, J., Harper, J., Molzen, W., Laibowitz, R., Pomerantz, M., 1978b. High resolution electron beam fabrication using a STEM. In: Sturgess, J.M. (Ed.), Electron Microscopy. vol. III, pp. 343–354.

Chang, T.H.P., 1975. Proximity effect in electron beam lithography. J. Vac. Sci. Technol. 12, 1271–1275.

Chung, M.S.C., Tai, K.L., 1978. Electron and ion beam science and technology. In: Bakish, R. (Ed.), Proc. 8th Int. Conf. on Electron and Ion Beam Sci. and Tech. Electrochemical Soc. Inc., Princeton, NJ, p. 242.

Dill, F.H., 1975. Optical lithography. IEEE Trans. Electron Dev. 22, 440.

Eigler, D.M., Schweizer, E.K., 1990. Positioning single atoms with a scanning tunnelling microscope. Nature 344, 524–526.

Flanders, D.C., 1980. Appl. Phys. Lett. 36, 93.

Flanders, D.C., Shaver, D.C., Smith, H.I., 1978. Alignment of liquid crystals using submicrometer periodicity gratings. Appl. Phys. Lett. 32, 597.

Geis, M.W., Flanders, D.C., Smith, H.I., 1979. Crystallographic orientation of silicon on an amorphous substrate using an artificial surface-relief grating and laser crystallization. Appl. Phys. Lett. 35, 71.

Greeneich, J.S., 1979. J. Vac. Sci. Technol. A 16, 1749.

Greeneich, J.S., 1980. In: Brewer, G.R. (Ed.), Electron Beam technology in Microelectronic Fabrication. Academic.

Hatzakis, M., Canavello, B.J., Shaw, J.M., 1980. Single-step optical lift-off process. IBM J. Res. Dev. 24 (4), 452–460.

Isaacson, M., Muray, A., 1981. In situ vaporization of very low molecular weight resists using 1/2 nm daimeter electron beams. J. Vac. Sci. Technol. 19, 1117–1120.

King, M.C., 1981. Principles of optical lithography. In: Einsbruch, N.G. (Ed.), VLSI Electronics Microstrucvture Science. vols. 1–2. New York Academic Press.

Komoru, M., Hiroshima, H., Tanoue, H., Kanayama, T., 1983. Maskless etching of a nanometer structure by focused ion beams. J. Vac. Sci. Technol. B1 (4), 985–989.

Kratschmer, E., Isaacson, M., 1986. Nanostructure fabrication in metals, insulators and semiconductors using self-developing metal inorganic resist. J. Vac. Sci. Technol. B4, 361–364.
Lin, B.J., 1979. SPIE Proc., 174.
Mochel, M.E., Humphreys, C.J., Mochel, J., Eades, J.A., 1983. Cutting of 20 Å holes and lines in metal-ß-aluminas. In: Proc. 41st Annual Meeting of the Electron Microscopy Soc. of America. San Francisco Press, San Francisco, pp. 100–101.
Molzen, W.W., Broers, A.N., Cuomo, J.J., Harper, J.M.E., Laibowitz, R., 1978. Materials and Techniques Use in Nanostructure Fabrication. IBM Research Report 7339.
Molzen, W.W., Broers, A.N., Cuomo, J.J., Harper, J.M.E., Laibowitz, R., 1979. Materials and techniques used in nanostructure fabrication. J. Vac. Sci. Technol. 16, 269–272.
Neill, T.R., Bull, C.J., 1980. In: Kramer, R.P. (Ed.), Microcircuit Engineering 80. Delft University Press, Delft, the Netherlands, p. 45.
Offner, A., 1979. Photogr. Sci. Eng. 23, 374.
Parasczak, J., 1983. J. Vac. Sci. Technol. B 1, 1372.
Parikh, M., Kyser, D.F., 1979. Energy deposition functions in electron resist films on substrates. J. Appl. Phys. 50, 1104.
Pomerantz, M., 1980. In: Dash, J.G., Ruvalds, J. (Eds.), Phase Transitions in Surface Films. Plenum, New York.
Smith, S.J., Purcell, E.M., 1953. Phys. Rev. 92, 1069.
Tai, K.L., Sinclair, R.G., Vadimsky, R.G., Moran, M., Rand, M.J., 1979. Bilevel high resolution photolithographic technique for use with wafers with stepped and/ or reflecting surfaces. J. Vac. Sci. Technol. 16, 1977–1979.
Wells, O.C., 1974. Scanning Electron Microscopy (Chapter 3). McGraw-Hill, New York.
Zingsheim, H.P., 1977. In: Johari, O. (Ed.), Scanning Electron Microscopy. vol. I. Chicago Press, Chicago, IL, p. 357.

CHAPTER TEN

Nano-devices fabricated with the short focal length electron probe

Contents

Having used the high-resolution electron probe for scanning electron microscopy and to explore the resolution limits of electron beam lithography, I entered major collaborations with Bob Laibowitz, Praveen Chaudhari, Richard Voss, Cory Umbach, Roger Koch, and others to make nanoscale devices. These were mainly designed to gain a better understanding of superconductivity and to explore the potential of superconducting electronic switches. My collaborators were all working in the IBM Research laboratory and had world-leading expertise with superconductors. IBM was building what was called a Josephson computer at the time that was to harness the high speed and low power of superconducting diodes.

Most of the operational sub-tenth-micron devices were made on thin window substrates. The thin window made it possible to make electrical contact to the devices and to examine them using scanning transmission electron microscopy. Earlier attempts to make contact to devices on carbon films supported by standard electron microscope grids had failed. The windows also eliminated exposure from backscattered electrons, although this was not a serious problem when making isolated wires because the total pattern area was small compared to the area from which the backscattered electrons emerged.

Advances in Imaging and Electron Physics, Volume 231
ISSN 1076-5670
https://doi.org/10.1016/B978-0-443-31462-9.00010-1

At first, we found the window membranes fragile and difficult to handle, but soon found that composite membranes of SiO_2 and Si_3N_4, where the compressive stress in the oxide film was balanced by tensile stress in the nitride, were relatively robust. The membranes were smooth and flat allowing wafers with arrays of windows to be spin-coated with resist without difficulty. Electron beam lithography was used to expose the patterns for the arrays of contact pads. Hundreds of twin window substrates complete with contact pads were made on a single silicon wafer. The wafer was then diced up to produce the individual substrates.

Josephson Effect in Nb Bridges

Led by Bob Laibowitz, we started in 1978 by making devices to explore the physical nature of electrical conductivity in Josephson nanobridges.1978 (Broers and Laibowitz, 1978; Laibowitz et al., 1979). These were niobium wires about 25 nm wide, 20 nm thick, and about 0.8 μm long. The length was set by the pad separation. Fig. 10.1 shows the general configuration for the Nb nanostripe devices.

At several stages in the fabrication process, the samples were examined by STEM. A typical final device is shown in Fig. 10.2. The width of the stripe is 25 nm and the irregular shape of the edges is due to the large grain size in the niobium film combined with the characteristics of the lift-off technique used to form the contact pads. This was the narrowest Nb strip that we successfully tested at the time.

Four-terminal I-V measurements were obtained on these Nb samples with and without microwaves. Resistance measurements as a function of temperature were also made in the region of the transition temperature.

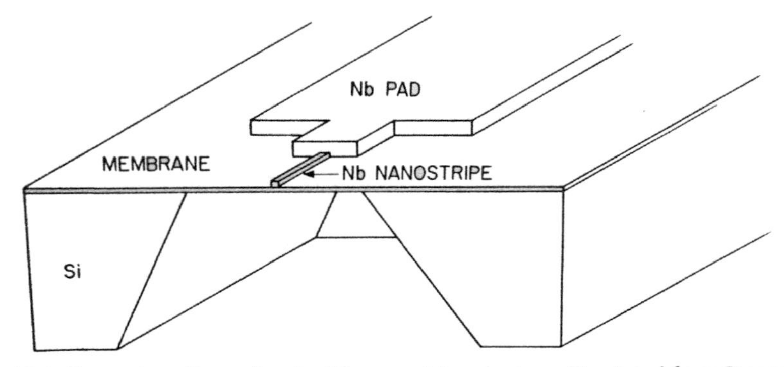

Fig. 10.1 General configuration for Nb nanostripe devices. *(Reprinted from Broers A.N., and Laibowitz, R.B. 1978. High resolution electron beam lithography and applications to superconducting devices. in: B.S. Deavor Jr. et al. (Ed.). Future Trends in Superconducting Electronics. Am. Instit. Phys, N.Y. 289–297 with the permission of AIP Publishing).*

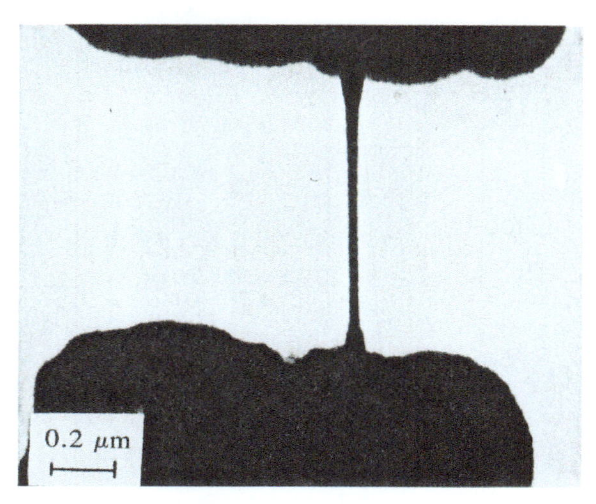

Fig. 10.2 STEM scanning transmission electron micrograph of a 25 nm wide and 20 nm thick Nb stripe between niobium contact pads. *(Reprinted from Broers A.N., and Laibowitz, R.B. 1978. High resolution electron beam lithography and applications to superconducting devices. in: B.S. Deavor Jr. et al. (Ed.). Future Trends in Superconducting Electronics. Am. Instit. Phys, N.Y. 289–297 with the permission of AIP Publishing).*

Because contact to the nanostripes was made through pads of several widths we often saw a complex R vs temperature curve due to the variations in the series resistances of the pads. An example of such a curve is shown in Fig. 10.3, which also shows the sample dimensions. The highest transition temperature was usually that of the large connecting pads; a second transition probably caused by the intermediate pads was often seen, and finally, as had been observed in past work, the weak link itself generally had the lowest transition temperature. The bank T_c in Fig. 10.3 is somewhat lower than bulk Nb, presumably due to impurity contamination. Higher bank T_cs have been obtained. It is interesting to observe that above the bank T_C the resistance still slightly decreased with decreasing temperature, which may have been the first observation of fluctuation phenomena in a one-dimensional Nb stripe. Sample geometries where the strip resistance was a much larger part of the total resistance were needed to effectively study such phenomena.

The I-V curves for these structures generally show intermediate states when their zero voltage critical current is exceeded. This might be expected as these are long bridges in which heating, impurities, or defects can cause such effects (Laibowitz et al., 1978). A full description of the switching behavior of these nanobridges had to await further studies in which the sample geometry was controllably varied. Such experiments could only be accomplished using high-resolution techniques. An example of an I-V characteristic with and without microwave radiation is shown in Fig. 10.4 The device was biased up to its first intermediate state.

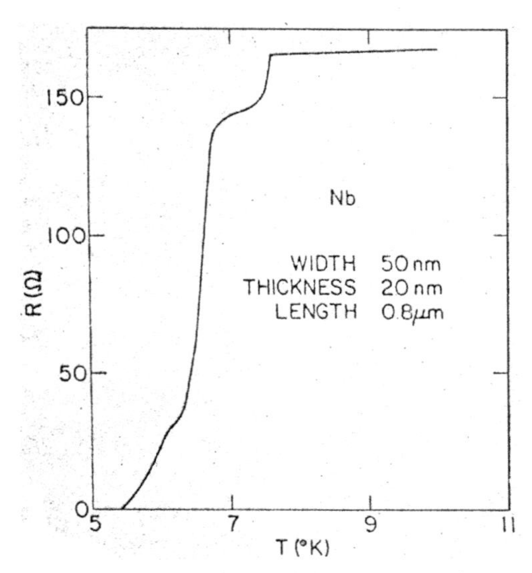

Fig. 10.3 Resistance vs temperature for a Nb line measured at a current of 0.11 μA. *(Reprinted from Broers A.N., and Laibowitz, R.B. 1978. High resolution electron beam lithography and applications to superconducting devices. in: B.S. Deavor Jr. et al. (Ed.). Future Trends in Superconducting Electronics. Am. Instit. Phys, N.Y. 289–297 with the permission of AIP Publishing).*

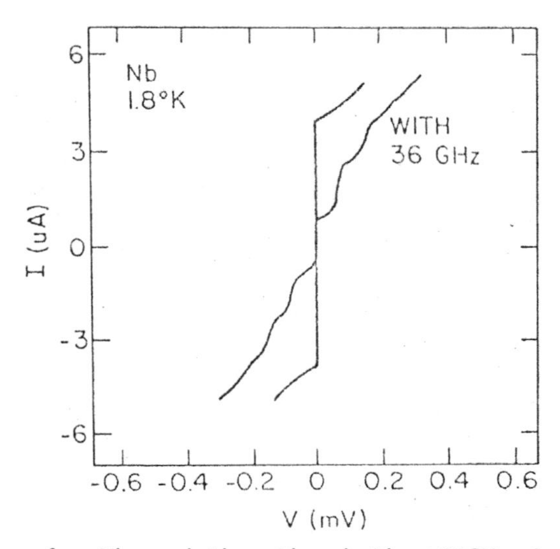

Fig. 10.4 I-V curves for a Nb nanobridge with and without 36GHz microwaves applied. *(Reprinted from Broers A.N., and Laibowitz, R.B. 1978. High resolution electron beam lithography and applications to superconducting devices. in: B.S. Deavor Jr. et al. (Ed.). Future Trends in Superconducting Electronics. Am. Instit. Phys, N.Y. 289–297 with the permission of AIP Publishing).*

Three microwave-induced steps can be observed and their spacings correspond to the Josephson relation $2ev = h\nu$. Steps had not been generally observed above the first transition, but in the nanobridge studied here, they were observed over the entire range below the bridge T_C. This result and the high normal temperature of these nanobridges (greater than about 100°C) had the potential to enable applications using bridges to become practicable.

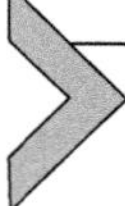

Ratio of Superconducting Transition Temperatures in Granular Nb Films (Laibowitz et al., 1980)

As pointed out, the superconducting transition temperature of the various components of the superconducting devices depended on their size and their crystallographic structure. It had been well-known for many years that granular superconductors in the form of thin films with grain dimensions $L_o = 30\text{-}200$ Å frequently showed a step structure in the resistance in the transition to the superconducting state (Broers and Laibowitz, 1978; Laibowitz, 1973). The experimental data indicated that the transition occurred in two distinct parts, and the size of the low T_c foot structure seems to vary from sample to sample. In this section, I describe experimental data we obtained for thin granular Nb films with theoretical equations emerging from a microscopic theory of a granular superconducting system (Patton et al., 1979). The theory predicts that the individual grains become superconducting at a temperature T_{co} which, for grains not too small compared to the coherence length ξ, will be close to the bulk superconducting temperature. The whole material, however, does not show long-range superconducting behavior until a lower temperature T_c where the resistance goes to zero. T_c will depend on the magnitude of the intergrain Josephson coupling energy, which from the microscopic theory is related to the inverse of the intergrain resistance (Ambegaokar and Baratoff, 1963; Deutscher et al., 1974; Patton et al., 1979). With certain simplifying assumptions, the theory predicts that the ratio of the two transition temperatures T_c and T_{co} for a sample with all three dimensions large compared to L_o is given by the equations (Patton et al., 1979).

$$T_c/T_{co} = R_o/\left(R_o + R_{ij}\right) \tag{10.1}$$

$$R_o = \left[\frac{\pi^2}{7\zeta(3)} z \frac{h}{e^2}\right] \approx 1.17z\frac{h}{e^2} \approx z.4500\Omega \tag{10.2}$$

R_{ij} is the average intergrain resistance, and z is the average number of nearest neighbors seen by a given grain.

The granular Nb films were formed by vapor deposition from an electron-beam heated source and were generally around 300 Å thick. The films were deposited on window substrates to facilitate detailed

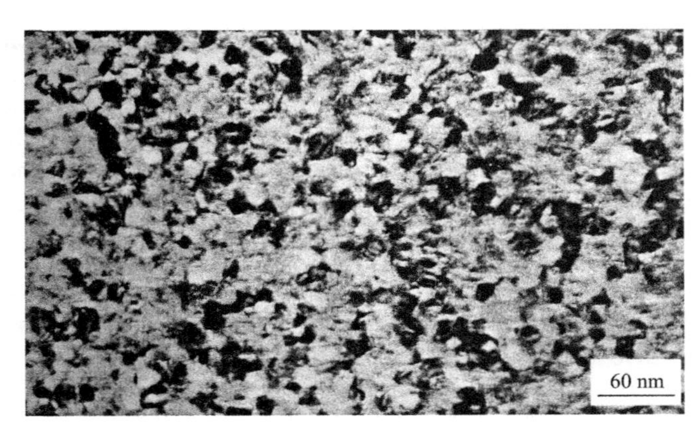

Fig. 10.5 TEM micrograph of 30 nm thick Nb film. *(Reprinted from Laibowitz R.B., Broers, A.N., Stroud, D., Patton, B.R. 1980. Ratio of superconducting transition temperatures in granular Nb films. AIP Conf. Proc. 58, 278–281, with the permission of AIP Publishing).*

examination by TEM. Fig. 10.5 is a TEM micrograph of a 300 Å thick film. The granularity is apparent with grain sizes varying from 30 Å to greater than 200 Å.

Ultra-narrow thin film wires were then fabricated from the granular Nb films using the previously described techniques. Typical conductor dimensions were length 0.12 to 1 μm, width 25 to 75 nm, and thickness 30 nm as shown in Table 10.1. The resistive transition for such a stripe is shown in Fig. 10.6. The two transitions are visible; however, the detailed shape of such curves generally can vary from sample to sample. The bridge fabrication process does contaminate the samples and in a few

Table 10.1 Ptoperties of nanobridges of various dimensions.

R(Ω)	l(μm)	w(nm)	t(nm)	ρ(μΩcm)	R_{ij}(Ω)	T_c(K)	T_c/T_{co}
137	0.28	110	30	164	1090	5.9	1.49
138	0.28	110	30	164	1100	5.8	1.52
146	0.33	110	30	148	990	6.3	1.40
595	1.0	40	30	71	470	6.7	1.31
306	1.0	50	30	46	310	7.4	1.19
19	0.35	300	100	165	1100	6.6	1.41
245	1.0	60	27	36	240	6.3	1.19

Reprinted from Laibowitz R.B., Broers, A.N., Stroud, D., Patton, B.R. 1980. Ratio of superconducting transition temperatures in granular Nb films. AIP Conf. Proc. 58, 278–281 with the permission of AIP Publishing.

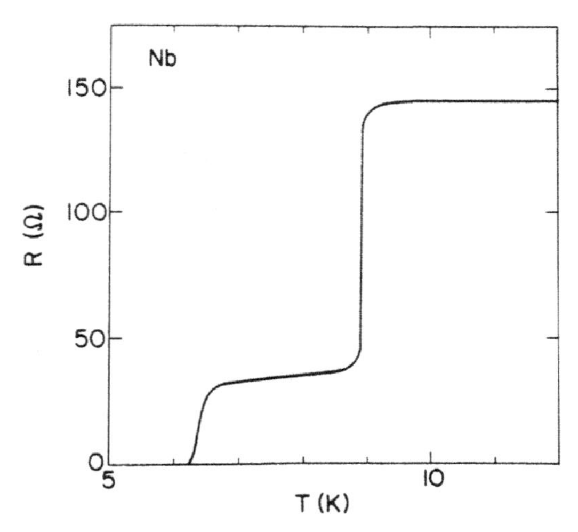

Fig. 10.6 Resistance vs temperature for a Nb line (third in row in Table 10.1). *(Reprinted from Laibowitz R.B., Broers, A.N., Stroud, D., Patton, B.R. 1980. Ratio of superconducting transition temperatures in granular Nb films. AIP Conf. Proc. 58, 278–281 with the permission of AIP Publishing).*

cases, the sample remains resistive even to temperatures as low as 1.5 K. The fabrication process also leaves a protective layer of carbon on the bridge which often makes it difficult to obtain clear micrographs of the final grain structure.

To apply Eq. (10.1) to the data, the average intergrain resistance must be extracted from the measured resistivity. If the films are assumed to consist of high-conductivity grains coated with a high-sensitivity skin, then the effective resistivity ρ of the film is best calculated by means of the Clausius-Mossotti-Maxwell-Garne approximation as used by Landauer (Laibowitz, 1973). This gives in this limit

$$\rho = \frac{1}{3} f \rho_s = \frac{1}{3} f R_{ij} \frac{A}{t} \tag{10.3}$$

where ρ_s is the resistivity of the skin, A the area of contact between grains, t is the skin thickness and f is the volume fraction of thin film composed of insulator. If the grains are assumed to be cubical then

$$f = 3t\left[\sqrt{A} + 3t\right] \tag{10.4}$$

Which gives

$$R_{ij} = \rho\left(\sqrt{A} + 3t\right)/A \tag{10.5}$$

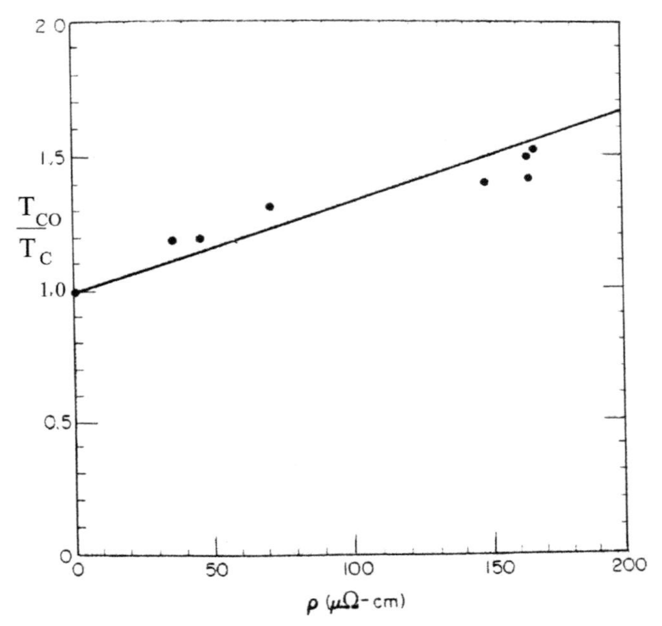

Fig. 10.7 Ratio of transition temperature in granular Nb film plotted against measured resistivity. Resistivity can be converted to intergrain resistance with the help of Eq. (10.5). Straight line corresponds to $Z_{eff} = 1.34$. as described in text. *(Reprinted from Laibowitz R.B., Broers, A.N., Stroud, D., Patton, B.R. 1980. Ratio of superconducting transition temperatures in granular Nb films. AIP Conf. Proc. 58, 278–281 with the permission of AIP Publishing).*

The transmission electron microscopy studies indicate the grain size varies from ~30-200 Å. We took $A = (30 \text{ Å})^2$ and the thickness of the tunneling barrier $t = 10$ Å. The resulting values of R_{ij} are given along with other sample properties in Table 10.1.

Fig. 10.7 shows the observed ratio of transition temperature T_{co}/T_c plotted against resistivity. The straight line is based on Eqs. (10.2) and (10.5) with $z_{eff} = 1.34$. While this value might seem low for a three-dimensional sample, we noted that in high-resistance materials near the percolation threshold (Shante and Kirkpatrick, 1971), the number of nearest neighbors is in the range 1–2. The agreement between theory and experiments as seen in Fig. 10.7 appears to be quite satisfactory and represents the first time this type of data has been systematically compared to a microscopical theory. Data on samples with larger ratios of T_{co}/T_c and from different materials would be desirable to further test the theoretical model.

In 1979 we were able to report on our measurements of the resistance and I-V properties of Nb nanobridges that were fabricated in a variable thickness bridge geometry and in which we were able to systematically vary the stripe length, keeping the overall device definition and quality at a high level (Laibowitz et al., 1979). The ac Josephson effect in the form of microwave-induced steps was observed in a series of samples in which the length was varied from about 1 to 0.1 μm. Studies of the phase slip centers in these samples enabled us to estimate the quasiparticle diffusion length to be 90 nm and the inelastic scattering time to be 12 ps as follows:

Additional steps were needed in the fabrication process to make the variable thickness bridges geometry and to provide electrical contact to the nanostripe. The contact areas were patterned using a vectorscan e-beam lithography system (Chang) and with PMMA resist. The pad separation (discussed below) determines the bridge length, in this way we have obtained bridge lengths as small as 120 nm. Fig. 10.8A–C are

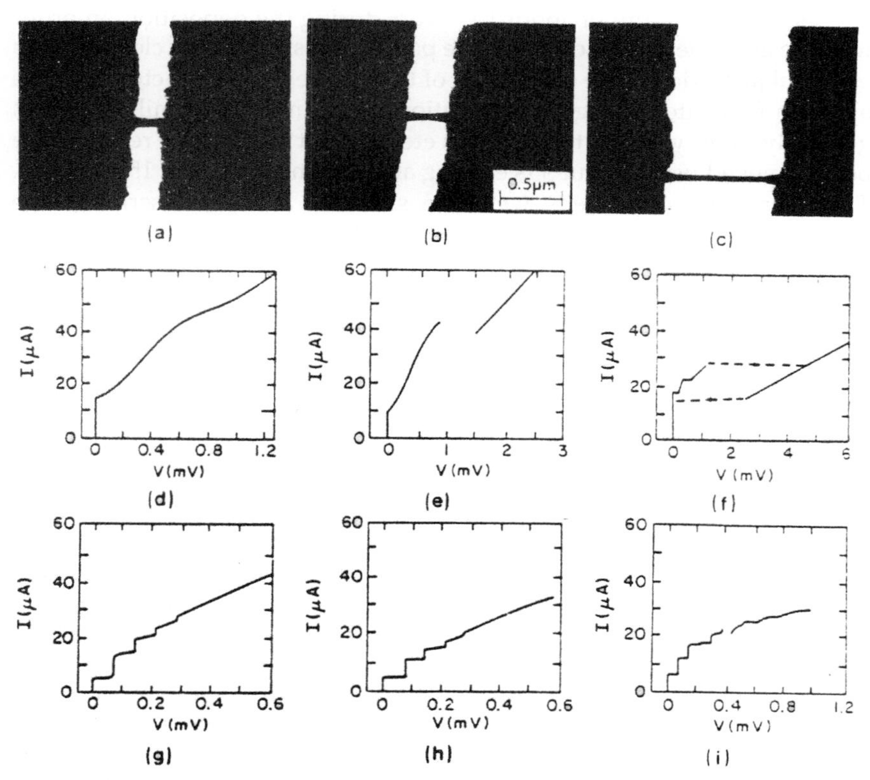

Fig. 10.8 (A) to (C) STEM micrographs of Nb bridges 30 nm thick. Bridge lengths (nm): (A) 120, (B) 28, (C) 900; widths (nm): (A) 7, (B) 60, (C) 55. Dimensions are accurate to 10%. (D)–(F); I-V characteristics at 4.2 K. (G)–(I): Effect of microwaves (36GHz) on the I-V characteristics at 4.2 K. *(Reprinted from Laibowitz R.B., Broers, A.N., Yeh, J.T., Viggiano, J.M. 1979. Josephson effect in Nb nanobridges. Appl. Phys. Lett. 35(11), 891–893 with the permission of AIP Publishing).*

scanning transmission micrographs of Nb bridges of three different lengths. From such micrographs the bridge length can be determined very accurately, an essential requirement in reconciling the experimental resistance measurements. It should also be observed that the Nb films are fine grained (grain size <10 nm), and as a result a high degree of bridge uniformity is achieved. The was particularly apparent from the edge definition. The films were deposited at room temperature.

The Nb pads were 80 nm thick and were fabricated using a lift-off process in which the niobium was deposited on top of the developed resist pattern. Niobium in the unwanted areas was then lifted off by dissolving away the remaining resist. After the pad were fabricated, the 30 nm thick film of Nb from which the nanostripes were former was deposited, The Nb was evaporated from an electron-beam heated source at a rate of about 30 Å/s and a pressure of about 10^{-7} Torr during the deposition. In order to make good electrical contact to the pads, an in situ plasma cleaning was essential just prior to the deposition of the bridge film. The actual devices are then fabricated using contamination resist and Ar ion milling. Nb is generally a low-yield material for ion etching, but satisfactory results have been achieved using about 35 s etching at a current density of 15 mA/cm^2. The later fabrication steps were the same as already described. The S_3N_4 membrane substrates were about 150 nm thick although we managed to us 20 nm membrane successfully.

Finally, the entire chip was then glued to a ceramic holder, and electrical contact was provided by aluminum wires, ultrasonically bonded to that part od the contact area which is not over the thin membrane.

After the fabrication and mounting the samples were slowly cooled to 4.2°K (2–4 h). In general, the windows were able to withstand several temperature cycles, although some windows did crack on the initial cool down. Four-terminal I-V measurements with and without microwaves (10 and 36 Ghz) and as a function of temperature were made. Temperature-dependent effects were also be studied with the samples either in the liquid helium or in a temperature-controlled exchange gas holder.

Although we gathered a huge volume of data on these bridges, I will concentrate here on the length dependence as shown in Fig. 10.8, where data taken at 4.2°K from bridges of three different lengths were found. The bridge transition temperature was generally about 7°K, while the contact pads could have a transition as high as 9.2°K. The I-V curves for the long bridges (900 nm in length) clearly showed the formation of phase slip centers (Skpcpol, n.d.; Skocpol et al., 1974) i.e., as the current was increased beyond the critical current, normal regions appeared along the stripes. Assuming that the spatial extent of such a phase slip center was about twice the quasiparticle diffusion length (Skpcpol, n.d.) Λ, then Λ could be calculated from the observed resistance values and the bridge length: $\Lambda = (R_n/R_{nn})l/2n$, where n was the number of phase slip centers, R_n is the resistances as given by the slope of the *I-V* curve (Fig. 10.8F) above the critical current, and R_{nn} was the normal resistance of the total bridge of length l. For Fig. 11.8F assuming $n=1$, we had $R_1 = 33\,\Omega$ and $R_{nn} = 165\,\Omega$, which gave $\Lambda = 90\,\text{nm}$. The quasiparticle inelastic relaxation time τ, could then be estimated from $\tau_r = \Lambda^2/D$, where $D = ⅓V_F l_{mfp}$. Taking a value (Mayadas et al., 1972) of V_F for Nb of $0.62 \times 10^8\,\text{cm/s}$ and $l_{mfp} \sim 3\,\text{nm}$, a τ_r of 13 ps was obtained. By using the next resistive part of the *I-V* curve of Fig. 10.8F, we estimated that $R_n = 91\,\Omega$, which gave a similar value for Λ assuming three phase slip centers were present. The 13 ps for Nb was, as expected, much shorter than the 800 ps obtained for Sn bridges. It had been shown that τ_r would be shorter in materials like Nb (Kaplan et al., 1976) where the electron–phonon coupling was large. In fact, we were able to estimate that τ_r for Nb was about 18 ps from the theoretical work of Kaplan et al (Kaplan et al., 1976), which compared quite well with our result ($\tau_r \approx \tau_0$ /8.4, where τ_0 is given in (Mayadas et al., 1972). Strictly speaking, this corresponds to the quasiparticle inelastic scattering time $T = T_c$, and at the fermi level it should be a good estimate for the quasiparticle inelastic relaxation time for the phase slip center because the most excess quasiparticles ae generated in the center where the gap is driven to zero periodically. The 250-nm-long bridge barely exhibited a continuous curve at 4.2°K. Thus, for the first time, Nb bridges (e.g., the 0.12 μm bridge) had been achieved which could successfully dissipate their energy (at 4.2°K) to the banks when portions of the bridge switched to the normal state. At lower temperatures, some hysteresis did appear in the short bridge.

Also shown in Fig. 10.8 are curves taken when the samples were irradiated with 36-GHz waves from an open-ended waveguide. Interestingly, all three samples showed the ac Josephson effect over the entire temperature range below their superconducting transition temperature. In general, this could not be achieved on larger Nb bridges (Laibowitz, 1973; Wang et al., 1977) (i.e., wide and thicker), nor was it possible on most bridges of any material. While such an effect was reasonable (Harris and Laibowitz, 1977) for the short bridge where hysteresis effects were minimized, the long bridge results needed further analysis. For such a bridge it appeared that at a particular level of microwave power, a step structure $(2eV) = nh\nu$ was developed across each phase slip center (Skocpol et al., 1974). The measured voltage was thus the sum of the two voltage drops. This is shown quite clearly in Fig. 10.8I, where a break in the *I-V* curve at higher voltages is observed. Above the break, part of the bridge is normal (as observed in Fig. 10.8F), and the step structure is modified. The measured voltage was now the sum of a series resistive voltage and a step structure. The effect became even more prominent at lower temperatures. The Nb nanobridges generally showed a complex resistive transition to

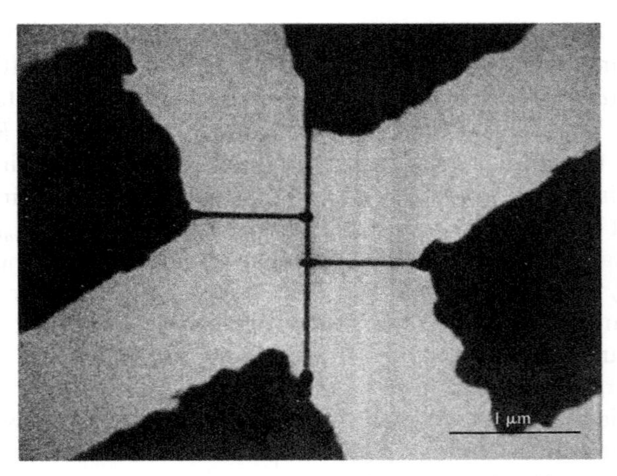

Fig. 10.9 STEM micrograph of Nb fine lines in a four terminal configuration. *(Reprinted from Laibowitz R.B., Broers, A.N., Yeh, J.T., and Viggiano, J.M. 1979. Josephson effect in Nb nanobridges. Appl. Phys. Lett. 35(11), 891–893. with the permission of AIP Publishing).*

the superconducting state. We then studied such behavior using the configuration shown in Fig. 10.9 in which fine-line voltages probes had been added along the stripe to eliminate pad and strips-to-pad effects. Our initial results indicated that the bridge transition occurred in two parts. Similar behavior had been seen in other systems (Deutscher and Rappaport, 1978a,b), and theoretical explanations (Deutscher et al., 1974; Patton et al., 1979) of this result in terms of granularity had been proposed. According to those explanations, the higher transition, generally above 8°K occurred when the Nb grains became superconducting while the lower transition, about 7°K, signaled the result of long-range superconducting (Josephson) coupling from between the grains, when the entire bridge became superconducting. Future measurements using nanostructures fabricated from either single crystal or amorphous films were needed to fully understand the resistive transition.

Fig. 10.9 also illustrates the power and versatility of electron beam lithography. Complex yet physically interesting structures could now be fabricated and studied even in a refractory material such as Nb. In addition, much shorter bridges (<120 nm) could be made by using the high resolution capability used to make the bridges to make the contact pads themselves.

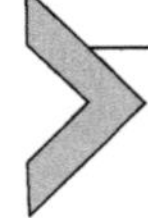

Quasiparticle charge-diffusion length in amorphous Re-W wires

In 1980, led by Praveen Chaudhari, we measured the quasiparticle charge-diffusion length in amorphous W-Re wires with a cross-sectional area of 2.5×10^{-12} cm^2 (250 Å wide by 100 Å thick). The normal electron diffusion length deduced from these data was in good agreement with values obtained from the application of localization theory on the same materials. This report provided the first quantitative support for the theory of localization in one-dimensional metallic wires. We also noted, parenthetically, that these W-Re wires had the smallest cross-section of any wires on which physical measurements had been made at that time. Most of the research reported in this section has been reprinted with permission from Chaudhari et al. (1980) Copyright 1980 by the American Physical Society.

The temperature dependence of the resistance of metallic wires had been measured by a number of investigators (Chaudhari and Habermeir, 1980a,b; Dolan and Osheroff, 1979, 1980; Garland et al., 1979, 1980; Giordano, 1980; Giordano et al., 1979; Overcash et al., 1980). In the case of polycrystalline and amorphous alloys, the temperature dependence changes as the cross-section of the wire approaches 10^{-11} cm^2. In contrast no evidence for localizations was obtained in experiments carried out on Bi whiskers (Garland et al., 1980) and on colloidal Ag particles (Garland et al., 1979) dispersed in KCL. All of these experimental investigations were prompted by Thouless's suggestions that in disordered metallic systems, there is a maximum metallic resistance of the order of 10 kΩ above which all such metal systems are unable to transport an electric current across the entire length of the specimen at 0°K (Thouless, 1977).

This behavior is associated with Anderson localization. In condensed matter physics, Anderson localization is the absence of diffusion of waves in a disordered medium. In this case, the transport of electricity by electrons can be considered the propagation of a wave, and the disordered medium is the W-Re amorphous material. Anderson localization finds its origin in the wave interference between multiple scattering paths. In the strong scattering limit, the severe interferences can completely halt the waves inside the disordered medium, and the film becomes an insulator.

At finite temperatures inelastic scattering events can delocalize the electrons and hence lead to finite resistance. The temperature dependence of the extra resistance associated with localizations is determined by the temperature dependence on inelastic scattering. Experimental results show that the extra resistance is inversely proportional to the square root of temperature leading to the conclusion that the inelastic scattering time constant is inversely proportional to temperature (Chaudhari and Habermeir, 1980a,b; Giordano et al., 1979). Based on Landauer (1970) ideas about the relation between the conductance and transmission coefficients across barriers in a linear chain, Anderson et al. (1980) have derived an expression for the temperature dependence of the one-dimensional wire. With use of this expression (Thouless, 1980) and the experimental data, the value of the inelastic-scattering time constant or the corresponding diffusion length can be extracted. In amorphous W-Re alloys the diffusion length is estimated to be approximately 250 Å at 4.2 K. A similar value is obtained for the polycrystalline Au-Pd alloys.

Although the temperature dependence of the inelastic scattering can be explained in terms of two-level scattering, the magnitude of the scattering is two orders of magnitude larger than the available theoretical estimates (Black et al., 1979). In fact, there is currently no published theory that can explain these short inelastic-scattering time constants. This situation has brought into question the validity of the interpretation of the experimental data in terms of localization theory. Lee and coworkers (Lee, 1980) have shown that a theory based on Coulomb interactions between electrons gives the same expressions as localization theory without running into the difficulty of explaining the values of inelastic-scattering times. Our data agree quantitatively with the Coulomb interaction theory. Given this situation it is desirable to measure the inelastic-scattering time by some independent technique that does not involve localization. We have carried out such a measurement on the same alloy on which we perform the localization experiments. Our method is based on the notion of phase-slip centers in superconducting wires (Laibowitz et al., 1979; Langer and Ambegaokar, 1967; Skocpol et al., 1974). We find that the value of the inelastic scattering time is comparable to that deduced using localization theory. These results lend support to the ideas of localization and pose a theoretical puzzle on the mechanism of such inelastic scattering times.

Amorphous W-Re films with approximately 60 at. % Re were prepared by electron-beam evaporation. The superconducting transition temperature (approximately 4 K) and normal-state resistivity of the evaporated films were very similar to those reported in the localization experiments. Using high-resolution electron beam lithography with the contamination resist, we fabricated wires with a width of approximately 250 Å and larger. The thickness if the wires was a nominal 100 Å. We were forced to use narrow wires as the zero-temperature coherence length of amorphous superconductors, as determined from critical field measurements, is usually <100 Å (Agyeman et al., n.d.). The zero-temperature coherence length of amorphous W-Re alloys was found in the same experiments to be approximately 60 Å. In order to apply the one-dimensional approximation our wires must be narrow and thin, and the data taken not far from the superconducting transition temperature where the temperature-dependent coherence length is comparable to the thickness or the width of our samples. We note, parenthetically, that these W-Re wires have the smallest cross-section of any wires on which any physical measurement has been made to date. Needless to say, it was pleasing that our work on building the high-resolution electron probe and discovering means to make such structures was justified in this way.

The length of the wires was varied and was typically between 0.5 and 1 μm. We have examined a number of samples prepared in different runs, i.e., different evaporation and subsequent fabrication series. The results are generally similar. We have found that the yield, i.e., the number of successful samples to the number of attempts, low. The samples burned out, presumably because of electrical discharges, even though we worked in a shielded room and with protective circuitry. We also found that the superconducting transition temperature broadened when the films were cut into wires. We have therefore used the variation of critical current with temperature to identify the transition temperature with zero critical current. For example, in one of the samples we found that the critical current measurements gave a transition temperature of approximately 3.5°K in contrast to the film value of approximately 4°K.

The quasiparticle-charge-diffusion length has to be corrected to obtain a normal electron-diffusion length. Calculations describing such corrections have been carried out for electron-photon inelastic processes (Kaplan et al., 1976; Chi and Clarke, 1980). One simple model (see (Black et al., 1979)) which gives the inelastic-scattering time of conduction electrons an inverse temperature dependence is a two-level system with a flat distribution of the excitation energy. Using this model, we obtained an expression for the quasiparticle-charge-relaxation rate, $\tau_{Q^*.T}^{-1}(E)$, for quasiparticles of energy E in the superconducting state as

$$\tau_{in,T_c}\tau_{Q^*,T}{}^{-1}(E) = 0.41(T/T_C)\int_{-\infty}^{\infty} d(E'/k_BT)\rho_\Delta(E-E')\left[\Delta^2E'/E(E-E')^2\right]$$
$$x\left\{1+\tanh(E'/2k_BT)\tanh[(E-E')/2k_BT]\right\}/2, \tag{10.6}$$

Where $\tau_{in,Tc}$ is the electron inelastic scattering time at T_c, Δ is the superconducting gap, and $\rho_\Delta(E-E')\equiv\ominus|E-E'|\ -\Delta)|E-E'|/[(E-E')^2-\Delta^2]^{1/2}$ is the BCS density of states. Because the normalized quasiparticle-charge-diffusion length, $\lambda^*(T)/\lambda(T-T_c)$, is equal to $[\tau(E)_{Q*,T}/\tau_{in,T}]^{1/2}$. In Fig. 10.10 the dashed and the dotted lines correspond to $\tau_{Q*,T}^{-1}(2\Delta_o)$ and $\tau_{Q*,T}^{-1}(3\Delta_o)$, respectively and the solid line corresponds to an averaged $\tau_{Q*,T}^{-}1$ (E) over a quasiparticle distribution which represents a small shift in the chemical potential (Pethick and Smith, 1979), i.e., $\delta f=(-\delta f_T/E)[E^2-\Delta^2)^{1/2}/E]\delta\mu/2$. The averaged value is independent of the amount of the chemical potential shift as long as the first-order expansion of the quasiparticle distribution with respect to δμ is adequate.

As we approach a temperature of $0.9T_c$, the calculated curves merge and yield similar values. This then is the temperature range over which we can reliably translate the quasiparticle-charge-diffusion length to the normal electron-diffusion length. Of interest to note here is that the correction factor at the temperature is also close to 2 (Pethick and Smith, 1979).

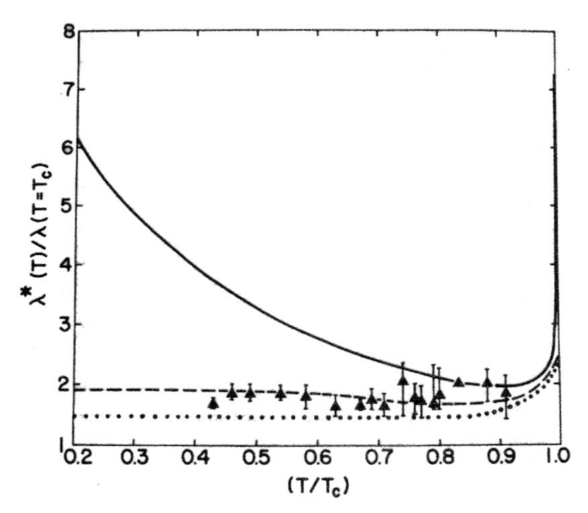

Fig. 10.10 Ratio of the quasiparticle-charge-diffusion length λ^* to the normal electron-diffusion length λ measured at T_c as a function of the reduced temperature. *(Reprinted with permission from Chaudhari P., Broers, A.N., Chi, C.C., Laibowitz, R.B., Spiller, E, Viggiano, J. 1980. Phase-slip and localization diffusion lengths in amorphous. W-Re Alloys Phys. Rev. Lett. 45, 930–932 copyright 1980 American Physical Society).*

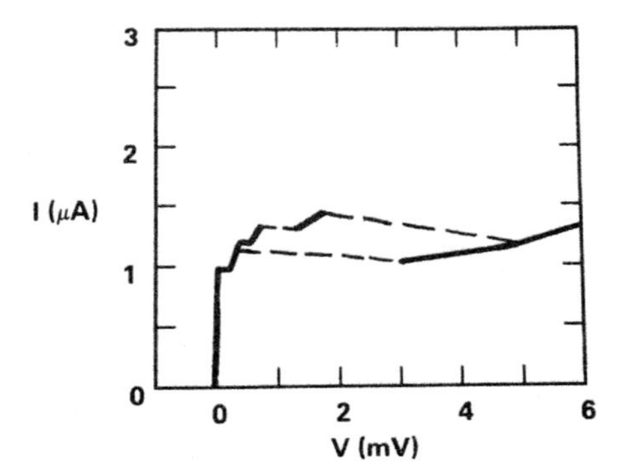

Fig. 10.11 An I-V curve of one of the samples showing three steps associated with two, three, and four phase-slip cavities. Other samples prepared from nominally the same material and fabrication processes showed one or two phase-slip cavities. *(Reprinted with permission from Chaudhari P., Broers, A.N., Chi, C.C., Laibowitz, R.B., Spiller, E, Viggiano, J. 1980. Phase-slip and localization diffusion lengths in amorphous. W-Re Alloys Phys. Rev. Lett. 45, 930–932 copyright 1980 American Physical Society).*

An I-V curve of one of the samples is shown in Fig. 10.11. The sequence of steps in this curve could be associated with two, three, and four phase-slip centers. We deduce these to be two three, and four rather than one, two and three phase-slip centers from the constancy of the extracted value of the quasiparticle-charge-diffusion length λ^*. If our assignment is wrong the value of the quasiparticle-charge-diffusion length averaged over the three steps is in error by approximately 33% on the low side as it is obtained from the relation $\lambda^* = LR_n/2nR_L$, where L is the length of the wire, R_L its normal-state resistance, n the number of phase-slip centers, and R_n the normal-state resistance associated with those centers (Skocpol et al., 1974).

The normal electron—diffusion length deduced from experiments carried out to verify one-dimensional localization theory was found for these alloys to be 260 ± 40 Å at 4.2°K. At 3.5°K this has an average extrapolated value of 285 Å. Using this number, we have normalized all of our measured values of the quasiparticle-charge-diffusion length. The results are shown in Fig. 11.10 for one of the samples. Very similar values were obtained for three other samples which have slightly different transition temperatures and were not measured over as great a temperature span as this sample. The agreement between theory and experiment is good near the transition temperature where we can reliably compare the two. At lower temperatures, theory and experiment cannot be compared in a meaningful way even though the agreement appears to be good because we do not know the average excitation energy of quasiparticles in the phase-slip center. We have also not included the effect of heating as the critical-current increases with decreasing temperature. However, the agreement near the transition temperature leads us to conclude that the inelastic-scattering times deduced from the localization model are comparable to those obtained from phase-slip measurements.

Nanobridge SQUIDS

In 1980, working with Richard Voss and Robert Laibowitz, I used the high-resolution electron probe and contamination resist to fabricate nanobridge dc SQUIDs (Superconduction QUantum interference Devices) with linewidths as small as 30 nm (Voss et al., 1980a,b). These SQUIDS demonstrated a minimum intrinsic energy resolution of about 3 h. This was the best value for any SQUID at that time and approached the limit of approximately Planck's constant h, set by the uncertainty principle. The dependence of SQUID voltage noise on bias parameters agreed with a general analysis of the SQUID noise made by Voss and Laibowitz. Most of the research reported here on nanobridge SQUID is reprinted from (Voss et al., 1980a,b) with the permission of AIP Publishing.

In the 1970s SQUIDs improved to the point that sensitivities limited by the uncertainty principle seemed possible. The accepted figure of merit for useable SQUID sensitivity (Clarke, 1977) is given by $\varepsilon = \phi_n^2/2\alpha^2 L$, where ϕ_n^2 is the effective mean square flux noise per Hz at the SQUID output, L is the SQUID loop inductance, and α is the coupling constant between the SQUID and an input coil. $\varepsilon_u B$ represents the minimum energy change in the input coil that the SQUID can resolve in a bandwidth B. At that time, however, a number of workers (Hu et al., 1979; Ketchen and Voss, 1979; Koch and Clarke, 1979; Voss et al., 1980a,b) had concentrated on optimizing the intrinsic energy resolution $\varepsilon = \phi_n^2/2$ independent of the problem of coupling an input coil to the SQUID. εB is the minimum change in internal energy of the SQUID that can be measured in a bandwidth B. Values of ε as low as 17h had been reported (Voss et al., 1980a,b) for an all-Nb tunnel junction dc SQUID at 1.6 K whose operation was limited by thermal noise. A value of ε of 5h independent of temperature was below 4.2 K has been reported (Ketchen and Voss, 1979) for a Pb-alloy tunnel junction ds SQUID.

Since both ε and ε_u represent the resolvable energy change per Hz of the SQUID and its input coil respectively, the uncertainty principle set a limit on their minimum values of order h. This limits has also been derived for a specific model of the DC SQUID under the assumption that the limiting noise was shot noise in the Josephson junctions (Tesche and Clarke, 1977). It has also been reaized (Gallop and Petley, 1976; Koch et al., 1980) that the zero point contributions to Johnson noise at the Josephson frequency of the oscillating supercurrent may limit ε. W showed that either zero point fluctuations or shot noise limit $\varepsilon > \sim h/2$.

To understand how physical noise sources limit ε, a model of the symmetric dc SQUID in which a superconducting loop of inductance L contains two Josephson elements was considered. Each element was characterized by a critical current I_o and an applied magnetic flux ϕ produced a modulation of the SQUID voltage $V(\phi)$ that was periodic in flux with period ϕ_0. The effective flux noise about any operating point was determined by the actual voltage noise V_n and the transfer function $|\delta V/\delta\phi|$ as $\phi_n^2 = V_n^2/|\delta V/\delta\phi|^2$.

ε was calculated assuming that the SQUID noise was due to the independent current noise sources i_1 and i_2 associated with each of the Josephson elements by using a small signal analysis about the operating point. These sources produced a circulating noise current $i_c = (i_1 - i_2)/2$ and a forward noise current $i_f = i_1 + i_2$ that were also independent contributions to V_n. i_f was coupled directly to V_n via the differential resistance $R_d = |\delta V/\delta I|$. i_c, on the other hand, produced a real flux noise $\phi_n = L i_c$ that coupled to V_n via $|\delta V/\delta\phi|$. Thus,

$$V_n^2 = 2\left(R_d^2 + L^2|\delta V/\delta\phi|^2/4\right)i_n^2. \tag{10.7}$$

Both terms of Eq. (10.7) were, in fact, found in the measured V_n^2 characteristics (see below). In general, R_d and $|\delta V/\delta\phi|$ were functions of L and the current and flux bias to minimize conditions. SQUID performance was optimized by varying these parameters to minimize

$$\varepsilon = \frac{R_d^2 i_n^2}{L|\delta V/\delta\phi|^2} + \frac{L i_n^2}{4}. \tag{10.8}$$

Tense to be worked on Typically, $|\delta V/\delta\phi| \approx R_d$ $|\delta V/\delta\phi| \approx 2I_c/\phi_0$, the usual SQUID design criterion (Clarke, 1977). In general, however, ε will be minimized by some specific choice for L ($\approx I_c/\phi_0$), I_b, and ϕ. We then expect contributions to ε from each term in Eq. (10.8) and

$$\varepsilon_{\min} = \gamma \frac{L i_n^2}{2}, \tag{10.9}$$

Where γ is a numerical constant of order 1. Even in cases where the two terms in Eq. (10.8) are not comparable $\gamma > ½$.

For a Johnson noise current source at a temperature T, $i_n^2 = 4k_BT/r$ and $\varepsilon_{min} \approx 2k_BTL/r$. This is a factor of 2 smaller than the prediction of the simple model of the SQUID (Clarke, 1977). If the limiting noise is the shot noise due to superconducting pairs tunneling at $I_b \approx I_c$, $i_n^2 \approx 4ei_0 = 2eI_c$ and $\varepsilon_{min} \approx eI_c/L \approx e\phi_0 = h/2$. Although this value is independent of temperature and consistent with the uncertainty principle, the existence of pair shot noise is controversial (Likharev, 1979). Moreover, the prediction is only valid when the charge carriers are statistically independent, as expected for tunneling. When the carriers are correlated as in a metal this contribution is dramatically reduced. Thus, we predicted that SQUIDs using Josephson elements of microbridges or SNS junctions could have a smaller ε_{min} due to the lack of shot noise. In such cases, the limiting noise source becomes the quantum corrections to Johnson noise (Callen and Welton, 1951):

$$i_n^2(f) = \frac{4}{r}\left(\frac{hf}{2} + \frac{hf}{e^{hf/k_BT} - 1}\right). \tag{10.10}$$

Where the first term represents the zero-point contributions to the current noise. It is important to remember that dc SQUID operation involves super-currents at the Josephson frequency $f_j = V/\phi_0$ and its harmonics nf_j. The low-frequency noise beats with the oscillating supercurrent at nf_j to provide low-frequency measurable fluctuations. For a limiting i_n^2 of 2hnfj/r, $\varepsilon_{min} \approx hnfjL/r = hnVL/\phi_0 r$. Typically, $V \approx rI_c/2$, $LIc \approx \phi_0$, and $\varepsilon_{min} \approx nh/2$. The actual value will depend on the contribution of each of the harmonics to the low-frequency noise. For SQUIDs having sharp characteristics operated close to Ic the higher harmonics can be extremely important. In any case, we expect $\varepsilon_{min} > h/2$.

In order to study the prediction of low ε_{min} for microbridge SQUIDs, all-Nb weak link dc SQUIDs were fabricated using these novel

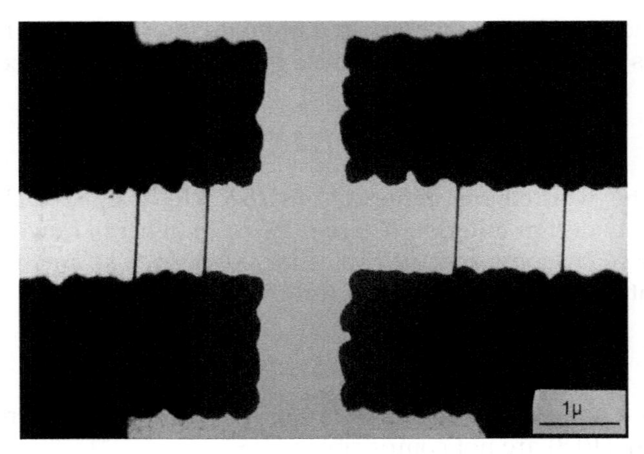

Fig. 10.12 STEM micrograph of two nanobridge SQUIDs. Each SQUID consists of two Nb nanobridges that are 30 nm thick and have a minimum width of about 30 nm and length of about 1 μm. *(Reprinted from Voss R.F., Laibowitz, R.B., Broers, A.N. 1980a. Niobium nanobridge SQUID Appl. Phys. Lett. 37, 656–658; Voss R.F., Laibowitz, R.B., Raider, S.I., Clarke, J. J. 1980b. Appl. Phys. March. 1980 with the permission of AIP Publishing).*

lithographic techniques capable of linewidths as small as 10 nm. Fig. 10.12 shows a STEM micrograph of two nanobridge SQUIDS. Each SQUID consisted of two narrow bridges 30 nm thick between pads about 80 nm thick. The minimum width is about 30 nm and the length of the bridges shown is about 1 μm. Bridges as short as 0.25 μm have been produced.

For electrical testing, the SQUID chips were bonded to headers. Al leads were ultrasonically bonded to the Nb pads., and the SQUIDs were mounted on a cryostat and immersed in liquid He inside a superconducting shield. Fig. 10.13 shows the modulation of the I-V characteristic of a SQUID at 4.2 K as the applied flux ϕ was varied by $\phi_0/2$. The sharp I-V characteristics are similar to those for the high I_c Pb-alloy tunnel junction SQUID (Ketchen and Voss, 1979), but without the resonance structure. L was estimated from $\Delta I_c/I_c \approx 0.26$ to be 150pH and appeared to be dominated by the kinetic inductance associated with the critical currents of the bridges.

For noise measurements, the SQUID was coupled directly to a room-temperature FET preamplifier. A constant current bias I_b was applied to the SQUID and $V(\phi)$ and V_n^2 were recorded digitally as a function of applied flux ϕ. Fig. 10.14A shows a typical $V(\phi)$ characteristic at 4.2K with $I_b = 32\,\mu A$. I_b was less than the maximum I_c of 34 μA and te SQUID was in the zero voltage state for a fraction of ϕ_0. Fig. 11.14B shows $|\delta V/\delta\phi|$ from a numerical differentiation of $V(\phi\}$. The maximum value is almost $6\,mV/\phi_0$. Fig. 11.14C shows the measured V_n across the SQUID at 100 kHz. The shorted preamplifier background of $3.2x10^{-18}V^2/Hz$ has been subtracted from the data in Fig. 11.14C. The peaks in the noise

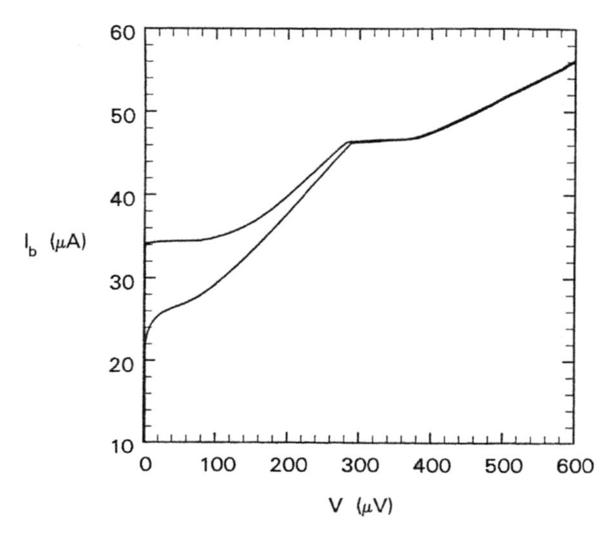

Fig. 10.13 Modulation of the I-V characteristic of a SQUID at 4.2 K as the applied flux was varied by $\phi_0/2$. *(Reprinted from Voss R.F., Laibowitz, R.B., Broers, A.N. 1980a. Niobium nanobridge SQUID Appl. Phys. Lett. 37, 656–658; Voss R.F., Laibowitz, R.B., Raider, S.I., Clarke, J. J. 1980b. Appl. Phys. March. 1980 with the permission of AIP Publishing).*

correspond to the peaks in $|\delta V/\delta\phi|$ as suggested by Eq. 10.7. Fig. 11.14D shows ε calculated from $V_n/2L|\delta V/\delta\phi|^2$. The minimum value of about 3h is reached at the points of maximum $|\delta V/\delta\phi|$.

Fig. 10.15 shows how $V(\phi)$ and V_n^2 vary with changing I_b for a similar SQUID ($I_c \approx 46\,\mu A$). Eq. 10.7 predicts a large V_n^2 at bias points of large $|\delta V/\delta\phi|$ or R_d. For $Ib < I_c$, there are two peaks in V_n^2 per period that correspond to the large $|\delta V/\delta\phi|$ associated with transitions from the zero voltage to the finite voltage states. For $I_b >_{\sim} I_c$, $|\delta V/\delta\phi|$ is reduced and the single V_n^2 peak per second period corresponds to the maximum R_d near $\phi_0/2$.

Several different SQUIDs gave about the same ε_{min} at 4.2 K. I_c increased rapidly as the temperature was lowered and the I-V characteristics became hysteretic below 4.2 K. The fact that.

ε_{min} is close to the quantum limit of order h/2 suggests that i_n^2 may be partially due to zero point fluctuations. The highest frequency harmonics in SQUID operation corresponds to the voltage at which flux modulation is observed. As shown in Fig. 10.13, flux modulation occurs out to 270 μV or a Josephson frequency such that $hf_j/k_BT \approx 1$ where zero-point fluctuations are comparable to the thermal contribution.

M. Ketchen, J. Harper, J. Cuomo, C.C. Chi, S. Raider, and W. Grobman contributed helpfully to our discussions on this work, and J. Viggiano, R. Drake, T. Donohue, J. Powers, J. Kuran, and J. Speidell provided technical expertise.

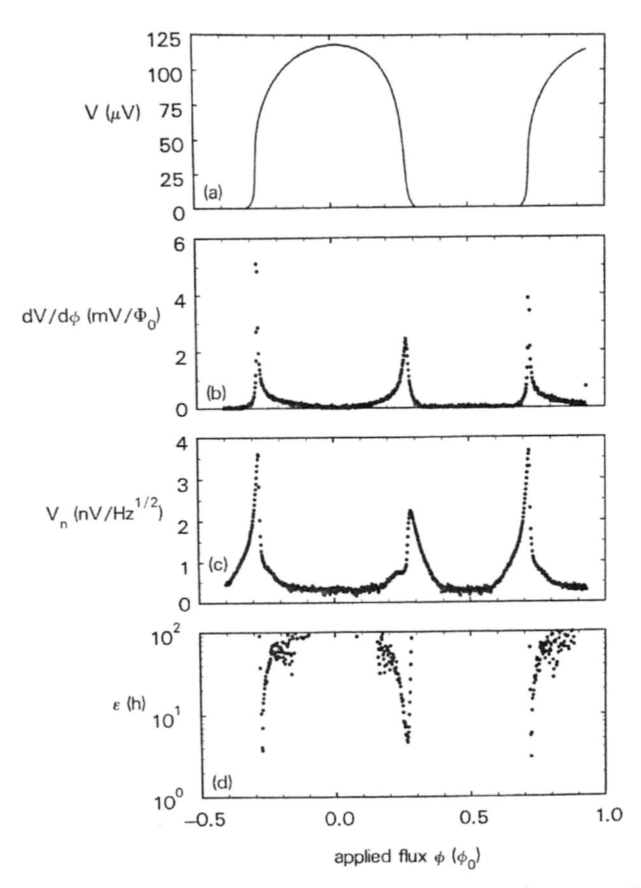

Fig. 10.14 (A) Shows a typical $V(\phi)$ characteristic at 4.2 K with $I_b = 32\,\mu A$. (B) shows $|\delta V/\delta\phi|$ from a numerical differentiation of $V(\phi)$. (C) Shows the measured V_n across the SQUID at 100 kHz. (D) shows ϵ calculated from $V_n^2/2L|\delta V/\delta\phi|^2$. *(Reprinted from Voss R.F., Laibowitz, R.B., Broers, A.N. 1980a. Niobium nanobridge SQUID Appl. Phys. Lett. 37, 656–658; Voss R.F., Laibowitz, R.B., Raider, S.I., Clarke, J. J. 1980b. Appl. Phys. March. 1980 with the permission of AIP Publishing).*

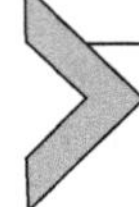

Observation of h/e Aharonov-Bohm Interference effects in submicron diameter, normal metal rings

Ever since Aharonov and Bohm published their original article in 1959 (Aharonov and Bohm, 1959), there has been considerable interest in demonstrating experimentally that charged particles are affected by potentials even in the region where all the fields, and therefore the forces on the particles, vanish. Electron interference effects in small metallic cylinders and rings became of great theoretical and experimental interest in

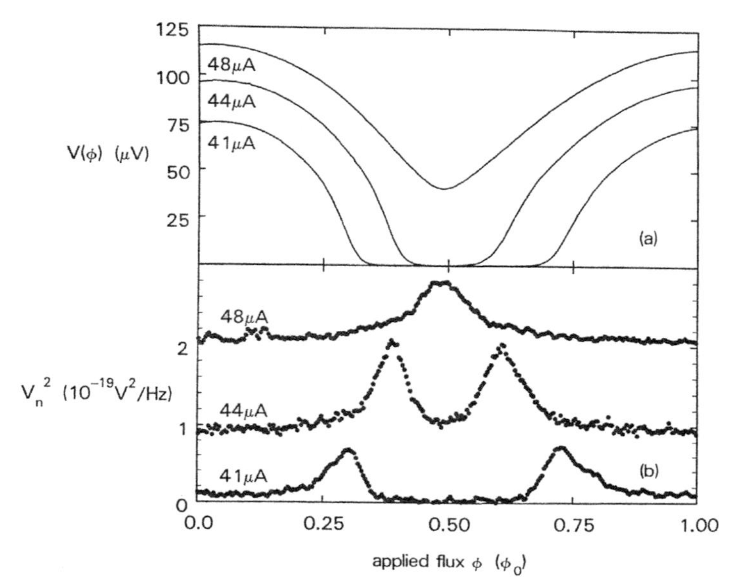

Fig. 10.15 shows how V(ϕ) and V_n^2 vary with changing I_b for a similar SQUID (Ic≈46 μA). *(Reprinted from Voss R.F., Laibowitz, R.B., Broers, A.N. 1980a. Niobium nanobridge SQUID Appl. Phys. Lett. 37, 656–658; Voss R.F., Laibowitz, R.B., Raider, S.I., Clarke, J. J. 1980b. Appl. Phys. March. 1980 with the permission of AIP Publishing).*

the mid-1980s. I was busy moving back to Cambridge at that time but Cory Umbach, who had taken over the high-resolution electron probe in the Thomas J Watson Research Laboratory, used it to fabricate some sub-micron diameter gold rings that showed clear evidence of Aharonov Bohm oscillations with respect to the flux quantum $\phi_0=h/e$. He and S. Washburn, R.A. Webb, R. Koch, M. Bucci, R.B. Laibowitz, and I showed that these magnetoresistance oscillations developed below T≈1 K and amazingly persisted without attenuation for >1000 periods (Umbach et al., 1986). Most of the research reported in this section is reprinted with permission from (Umbach et al., 1986). Copyright 1986, American Vacuum Society.

The Aharonov-Bohm effect had been observed with electrons traveling in a vacuum (Chambers, 1960; Tonomura, 1982) and also in solid metallic cylinders in which the electron scattering length was greater than the cylinder diameter (Brandt et al., 1976, 1982). In cylinders and rings with strong elastic scattering, two theories had been developed which also predict that interference effects would occur, provided that the inelastic and magnetic scattering were negligible The first theory (Al'tshuler, 1981a,b) averaged the coherent backscattering effects of electrons in all possible conduction paths around a cylinder or ring and predicted periodic oscillations in the magnetoresistance with a fundamental period corresponding to $\phi = h/2e$. The phase of these oscillations is fixed so that the resistance at zero field is either a minimum or maximum, depending on the spin-orbit scattering in the ring. Such periodic structure was observed in both cylinders (Al'tshuler et al., 1982; Dolan, 1985; Gijs, 1984a,b; Gordon, 1984; Ladan and Maurer, 1983; Yu, 1981a,b; Yu and Sharvin, 1984) and arrays of rings (Pannetier et al., 1984; Pannetier et al., 1985; Bishop et al., 1985). The second theory (Buettiger et al., 1985; Buettiker et al., 1983, 1984; Gefen, 1984a,b), however, considered the transport of electrons around a ring containing only a single conduction path or a small number of phase-coherent paths, and predicted that the fundamental period of the magnetoresistance oscillations should be with respect to the flux quantum $\phi_0 = h/e$. The phase of these oscillations depends on the details of the scattering in each of the conduction channels and is not fixed with respect to zero field. Past experiments (Umbach et al., 1984; Webb et al., 1985) with normal rings had not shown clearly periodic magnetoresistance oscillations due to superimposed aperiodic structure resulting from magnetic flux passing through the lines forming the rings. With these gold rings, however, we did show clearly h/e interference effects.

Ring fabrication

The window substrates used for the gold ring samples were fabricated with the process described in Section "Thin window substrate" in Chapter 6. They had Si_3N_4 windows which were 200 nm thick. The first step was to evaporate a 38 nm thick gold layer onto the samples which were then spin-coated with terpolymer resist. Standard electron beam lithography was used to expose the terpolymer resist with the patterns for the contact pads. The high-resolution electron probe was then used to expose the patterns for the gold rings and the current and voltage leads

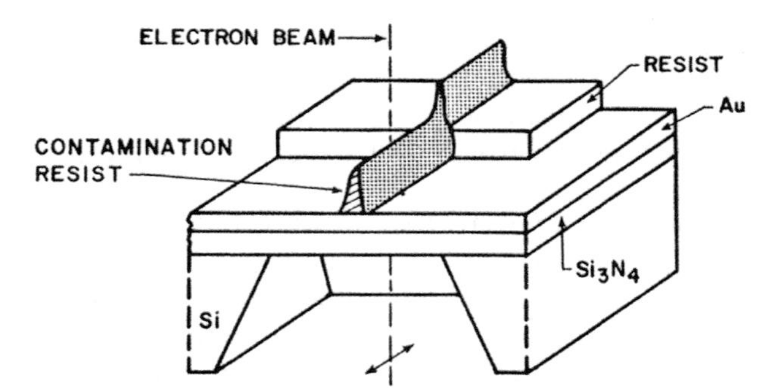

Fig. 10.16 Schematic of sample during fabrication procedure. *(Reprinted with permission from Umbach C.P., Washburn, S., Webb, R., Koch, R., Bucci, M., Broers, A.N., and Laibowitz, R. 1986. Observation of h/e Aharanov-Bohm interference effects in sub-micron diameter, normal metal rings. J. Vac. Sci. Technol. A B4(1), 383–385. Copyright 1986, American Vacuum Society).*

using contamination resist. Fig. 10.16 is a diagrammatic view of the sample at this point in the fabrication procedure. Ion milling was then used to remove the gold that was not protected by the contamination resist or the terpolymer resist. Finally, the terpolymer resist was removed by ashing in a radio-frequency oxygen plasma, and aluminum wires were ultrasonically bonded to the gold contact pads.

The samples were measured using standard low-frequency techniques in a ^{3}He-^{4}He dilution refrigerator which could reach temperatures lower than 10mK. The magnification of the STEM was calibrated using 2.2 μm diameter latex spheres supplied by the E.F. Fulham company.

Results and discussion

A STEM micrograph of a gold ring 825 nm in diameter with lines 40 nm wide is shown in Fig. 10.17. Magnetoresistance oscillations developed in this ring when it was cooled to below $T \approx 1$. A plot of these oscillations at $T = 50$ mK is shown in Fig. 10.18. Field periodic resistance oscillations corresponding to a $\phi_0 = h/e$ ($\Delta B = 77$ G) are clearly visible. The Fourier power spectrum of these data shown in Fig. 10.19 shows a strong peak at a field corresponding to h/e and a weaker peak at a field of h/2e. The h/2e peak may result from the coherent backscattering of electrons around the ring (Al'tshuler, 1981a,b). Alternatively, it may be an artifact of the deconvolution on a nonsinusoidal h/e fundamental period in the magnetoresistance.

The magnetoresistance is plotted from 0 to 8 T in Fig. 10.20. The irregular background structure results from magnetic flux passing through the

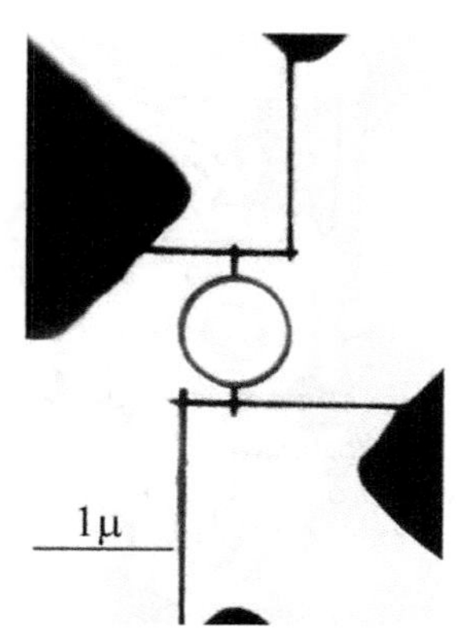

Fig. 10.17 STEM micrograph of gold ring measured in this experiment. Ring diameter 0.85 μm. *(Reprinted with permission from Umbach C.P., Washburn, S., Webb, R., Koch, R., Bucci, M., Broers, A.N., and Laibowitz, R. 1986. Observation of h/e Aharanov-Bohm interference effects in sub-micron diameter, normal metal rings. J. Vac. Sci. Technol. A B4(1), 383–385. Copyright 1986, American Vacuum Society).*

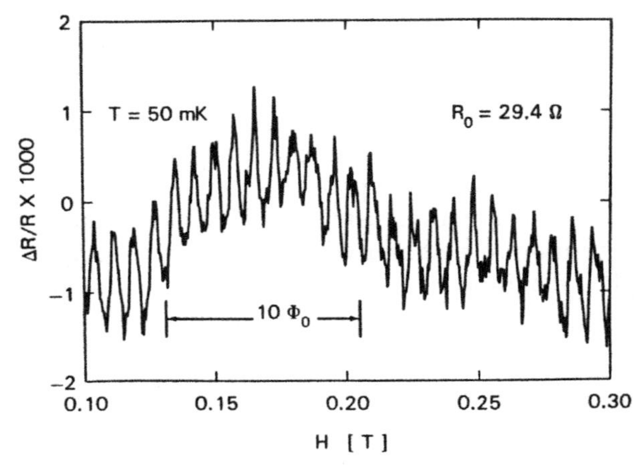

Fig. 10.18 Plot of the Magnetoresistance of the ring shown in Fig. 10.17 at T=50 mK from 0.1 to 0.3T. *(Reprinted with permission from Umbach C.P., Washburn, S., Webb, R., Koch, R., Bucci, M., Broers, A.N., and Laibowitz, R. 1986. Observation of h/e Aharanov-Bohm interference effects in sub-micron diameter, normal metal rings. J. Vac. Sci. Technol. A B4(1), 383–385. Copyright 1986, American Vacuum Society).*

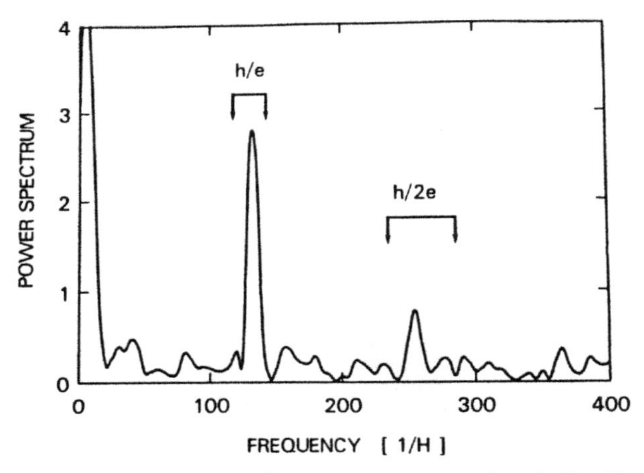

Fig. 10.19 Fourier power spectrum of magnetoresistance data in Fig. 10.16. *(Reprinted with permission from Umbach C.P., Washburn, S., Webb, R., Koch, R., Bucci, M., Broers, A.N., and Laibowitz, R. 1986. Observation of h/e Aharanov-Bohm interference effects in sub-micron diameter, normal metal rings. J. Vac. Sci. Technol. A B4(1), 383–385. Copyright 1986, American Vacuum Society).*

lines forming the ring. The physical, microscopic origin of this structure is currently under active theoretical investigation (Lee and Stone, 1985; Stone, 1985). Expanding the field scale on Fig. 10.20 reveals the h/e periodic oscillations. Surprisingly, these oscillations show no sign of attenuation out to 8T. If the oscillations resulted from the backscattering of electrons

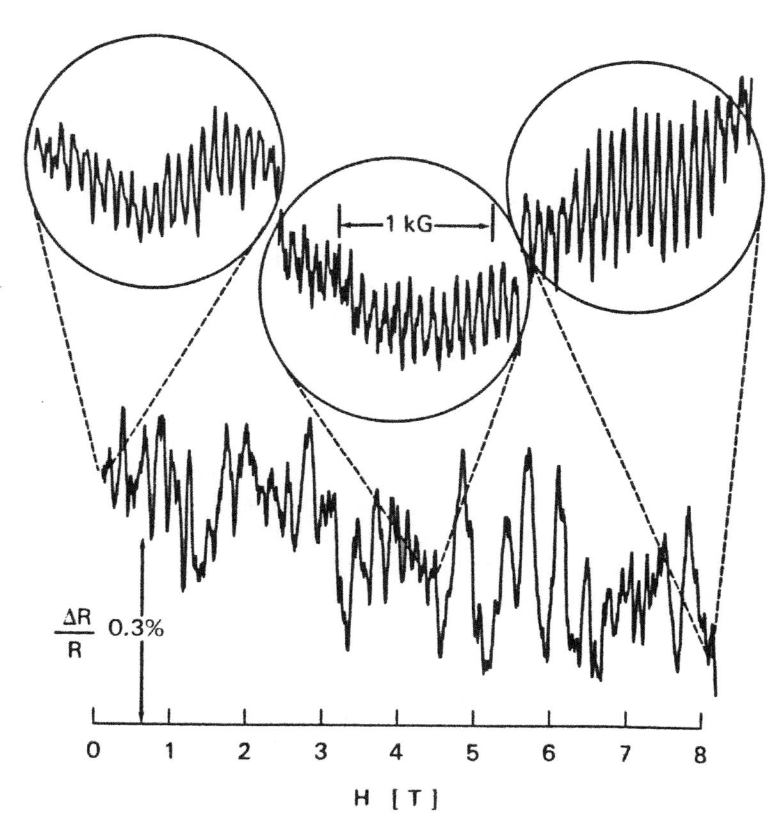

Fig. 10.20 Magnetoresistance of ring in Fig. 10.15 at T = 50 mK from 0 to 8 T. The insets show portions of the data where the field axis has been expanded. *(Reprinted with permission from Umbach C.P., Washburn, S., Webb, R., Koch, R., Bucci, M., Broers, A.N., and Laibowitz, R. 1986. Observation of h/e Aharanov-Bohm interference effects in sub-micron diameter, normal metal rings. J. Vac. Sci. Technol. A B4(1), 383–385. Copyright 1986, American Vacuum Society).*

(Al'tshuler, 1981a,b), they should die away at high fields. Alternatively, since oscillations produced by electrons traveling around the inner and outer circumferences of the ring will be 180° out of phase after roughly five periods, one might expect the experimentally observed oscillations produced by all of the concentric electron paths in the ring to die away at this point. In contrast, the oscillations in the magnetoresistance of the Au ring persist for over 1000 periods. One possible explanation for this behavior is that the electron paths around the ring are sufficiently erratic that all electrons sample the entire ring (Browne et al., 1984; Carina et al., 1984). In this case all electron paths enclose the same area and only one fundamental h/e periodicity will be observed.

All theories (Imry and Gefen, n.d.; Al'tshuler, 1981a,b), which calculate conductance by averaging over a large number of possible electron scattering configurations uniformly predict oscillations in the magnetoresistance with a fundamental period corresponding to h/2e. In contrast, the fundamental period of the oscillations in our samples was with respect to the flux quantum $\phi_0=h/e$. This is clear evidence that conventional ensemble impurity averaging theoretical techniques are not valid in describing some of the transport properties of small, quasi-one-dimensional samples, such as the Au ring.

The number of ring segments separated by a phase coherence length in doubly connected samples play a significant role in determining the transport properties. In samples containing a large number of these phase incoherent ring segments, such as cylinders and arrays of rings, only h/2e oscillations are predicted and observed. It has been conjectured (Imry and Gefen, n.d.) that the h/e oscillations, for which the phase is not fixed, from each of the constituent ring segments, add incoherently in a multiring sample and wash out. The phase of h/2e oscillations in the constituent rings, however, is fixed, so h/2e oscillations persist in cylinders and arrays of rings. This interpretation while appealing, remains to be proven.

Conclusions

Periodic oscillations have been observed in the low-temperature magnetoresistance of single, sub-micron diameter, normal metal rings. The fundamental period of the oscillations is that corresponding to the flux quantum $\phi_0=h/e$. The oscillations persisted without attenuations for over 1000 periods. These results are evidence that at low temperatures short, quasi-one-dimensional samples enter a new transport regime where conduction effects specific to individual electron scattering configurations dominate the transport properties.

References

Agyeman K., Tsuei, C.C., Chaudhari, P. Personal communication.

Aharonov, Y., Bohm, D., 1959. Significance of electromagnetic potentials in the quantum theory. Phys. Rev. 115, 485.

Al'tshuler, B.L., Aronov, A.G., Spivak, Z., Yu, D., Sharvin, V., 1982. Pis'ma Zh. Teor. Fiz 35, 585.

Al'tshuler, B.L., Aronov, A.G., Spivak, B.Z., 1981a. The Aaronov-Bohm effect in disordered conductors. JETP Lett 33, 94.

Al'tshuler, B.L., Aronov, A.G., Spivak, B.Z., 1981b. Pis'ma Zh. Eksp. Teor. Fiz., 101.

Ambegaokar, V., Baratoff, A., 1963. Tunneling between superconductors. Phys. Rev. Lett. 10, 486.

Anderson P.W., Thouless, D.J., Abrahams, E.. Fisher, D.S. 1980. Personal communication.

Bishop, D.J., Licini, J.C., Dolan, G.J., 1985. Lithium quench-condensed microstructures and the Aharonov–Bohm effect. Appl. Phys. Lett. 46, 1000.

Black, J.L., Gyorffy, B.L., Jackle, J., 1979. Philos. Mag. 49, 331.
Brandt, N.B., Gitsu, D.V., Nikolaeva, A.A., Ya, G., 1976. Ponomarev, Fiz. Nizk. Temp 24, 304.
Brandt, N.B., Gitsu, D.V., Nikolaeva, A.A., Ya, G., 1982. Sov. J. Low Temp. Phys. 8, 358.
Broers, A.N., Laibowitz, R.B., 1978. High resolution electron beam lithography and applications to superconducting devices. In: Deavor Jr., B.S., et al. (Eds.), Future Trends in Superconducting Electronics. Am. Instit. Phys, N.Y, pp. 289–297.
Browne, D.A., Carina, J.F., Muttalib, K.A., Nagel, S.R., 1984. Phys. Rev. Lett. B 30, 6798.
Buettiger, M., Imry, Y., Landauer, R., Pinhas, S., 1985. Generalized many-channel conductance formula with application to small rings. Phys. Rev. B 31, 6207.
Buettiker, B., Imry, Y., Landauer, R., 1983. Phys. Rev. Lett., A 96, 365.
Buettiker, M., Imry, Y., Ya. Azbel, M., 1984. Phys. Rev. A 30, 1982.
Callen, H.B., Welton, T.A., 1951. Phys. Rev. 83, 34.
Carina, J.P., Muttalib, K.A., Nagel, S.R., 1984. Phys. Rev. Lett. 53, 102.
Chambers, R.G., 1960. Phys. Rev. Lett. 5, 3.
Chaudhari, P., Habermeir, H.-U., 1980a. One-dimensional quantum localization in amorphous W—Re films. Solid State Commun. 34, 687.
Chaudhari, P., Habermeir, H.-U., 1980b. Quantum localization in amorphous W-Re alloys. Phys. Rev. Lett. 44, 40.
Chaudhari, P., Broers, A.N., Chi, C.C., Laibowitz, R.B., Spiller, E., Viggiano, J., 1980. Phase-slip and localization diffusion lengths in amorphous W-Re Alloys. Phys. Rev. Lett. 45, 930–932.
Chi, C.C., Clarke, J., 1980. Phys. Rev. 21, 333.
Clarke, J., 1977. In: Schwartz, B.S., Foner, S.H. (Eds.), Superconductor Applications: SQUIDs and Machines. Plenum, NY.
Deutscher, G., Rappaport, M.L., 1978a. J. Physiol. Paris 39. C6–581.
Deutscher, G., Rappaport, M.L., 1978b. J. de Physique C6, 581.
Deutscher, G., Imry, Y., Gunther, L., 1974. Phys. Rev. B. 10. 4598.
Dolan, G.J., 1985. Bull. Am. Phys. Soc 30, 395.
Dolan, G.J., Osheroff, D.D., 1979. Phys. Rev. Lett. 43, 721.
Dolan, G.J., Osheroff, D.D., 1980. Bull. Am. Phys. Soc. 25, 355.
Gallop, J.C., Petley, B.W., 1976. J. Phys. E 9, 417.
Garland, J.C., Gully, W.J., Tanner, D.B., 1979. Bull. Am. Phys. Soc 24, 280.
Garland, J.C., Gully, W.J., Tanner, D.B., 1980. Search for maximum metallic resistance in random metal-particle composites. Phys. Rev. B 22, 507.
Gefen, Y., Imry, Y., Ya. Azbel, M., 1984a. Quantum oscillations in small rings at low temperatures. Surf. Sci. 142, 203.
Gefen, Y., Imry, Y., Ya. Azbel, M., 1984b. Quantum oscillations and the Aharonov-Bohm effect for parallel resistors. Phys. Rev. Lett. 52, 129.
Gijs, M., van Haesendonct, C., Bruynseraede, Y., 1984a. Phys. Rev. Lett. 52, 2069.
Gijs, M., van Haesendonct, C., Bruynseraede, Y., 1984b. Phys. Res. B30, 2964.
Giordano, N., 1980. Bull. Am. Phys. Soc. 25, 355.
Giordano, N., Gibson, W., Prober, D.E., 1979. Phys. Rev. Lett. 43, 725.
Gordon, J.M., 1984. Phys. Rev. B30, 6770.
Harris, E.P., Laibowitz, R.B., 1977. IEEE Trans. Mag. MAG 13, 724.
Hu, E.L., Jackel, L.D., Epworth, R.P., Fetter, L.A., 1979. IEEE Trans. Magn. MAG-15, 585.
Imry Y., Gefen, Y. Private communication.
Kaplan, S.B., Chi, C.C., Langenberg, D.N., Chang, J.J., Jafarey, S., Scalapino, D.J., 1976. Phys. Rev. B 14, 4854.

Ketchen, M.B., Voss, R.F., 1979. Appl. Phys. Lett. 35, 812.
Koch, R., Clarke, J., 1979. Bull. Am. Phys. Soc. 24, 264.
Koch, R.H., van Harlingen, D.J., Clarke, J., Tesche, C.D., 1980. Bull. Am. Phys. SAoc. 25, 198.
Ladan, F.R., Maurer, J.C.R., 1983. Acad. Sci. 297, 227.
Laibowitz, R.B., 1973. Appl. Phys. Lett. 23, 407.
Laibowitz, R.B., Broers, A.N., Viggiano, J.M., Cuomo, J.J., Molzen, W.W., 1978. Bull. Am. Phys. Soc. 23, 357.
Laibowitz, R.B., Broers, A.N., Yeh, J.T., Viggiano, J.M., 1979. Josephson effect in Nb nanobridges. Appl. Phys. Lett. 35 (11), 891–893.
Laibowitz, R.B., Broers, A.N., Stroud, D., Patton, B.R., 1980. Ratio of superconducting transition temperatures in granular Nb films. AIP Conf. Proc. 58, 278–281.
Landauer, R., 1970. Philos. Mag. 21, 863.
Langer, J.S., Ambegaokar, V.A., 1967. Phys. Rev. Lett. 164, 498.
Lee, P.A., 1980. Bull. Am. Phys. Soc. 25, 355.
Lee, P.A., Stone, A.D., 1985. Phys. Rev. Lett. 55. 1622.
Likharev, K.K., 1979. Rev. Mod. Phys. 51, 101.
Mayadas, A.F., Laibowitz, R.B., Cuomo, J., 1972. Appl. Phys. Lett. 43, 1287.
Overcash, D.B., Ratman, R.A., Skove, M.J., Stillwell, E.P., 1980. Phys. Rev. Lett. 44, 1348.
Pannetier, B., Chaussy, J., Rammal, R., Gandit, 1984. Phys. Rev. Lett. 53, 718.
Pannetier, B., Chaussy, J., Rammal, R., Gandit, 1985. Phys. Rev. B31, 3209.
Patton, B.R., Stroud, D., Lamb, W., 1979. Bull. Am. Phys. Soc. 24, 356.
Pethick, C.J., Smith, H., 1979. Ann. Phys. (N.Y.) 119, 133.
Shante, V.K.S., Kirkpatrick, S., 1971. Advan. Phys. 20, 325.
Skocpol, W.J., Beasley, M.R., Tinkham, M., 1974. J. Low. Temp. Phys. 16, 145.
Skpcpol W.J., Private communication.
Stone, A.D., 1985. Phys. Rev. Lett. 54, 2692.
Tesche, C.D., Clarke, J., 1977. J. Low Temp. Phys 29, 301.
Thouless, D.J., 1977. Phys. Rev. Lett. 39, 1167.
Thouless, D.J., 1980. Solid State Commun. 34, 683.
Tonomura, A., 1982. Phys. Rev. Lett. 48, 1443.
Umbach, C.P., Washburn, S., Laibowitz, Webb, R.A., 1984. Phys. Rev. Lett., B 30, 4048.
Umbach, C.P., Washburn, S., Webb, R., Koch, R., Bucci, M., Broers, A.N., Laibowitz, R., 1986. Observation of h/e Aharanov-Bohm interference effects in sub-micron diameter, normal metal rings. J. Vac. Sci. Technol. A B4 (1), 383–385.
Voss, R.F., Laibowitz, R.B., Broers, A.N., 1980a. Niobium nanobridge SQUID. Appl. Phys. Lett. 37, 656–658.
Voss, R.F., Laibowitz, R.B., Raider, S.I., Clarke, J., 1980b. J. Appl. Phys. March, 1980.
Wang, L.-K., Callegari, A., Deavor Jr., B.C., Barr, D.W., Mattauch, R.J., 1977. Appl. Phys. Lett. 31, 306.
Webb, R.A., Washburn, C.P., Umbach, C.P., Laibowitz, R.B., 1985. In: Bergmann, G., Bruynsersede, Y., Kramer, B. (Eds.), Localization, Interaction and Transport Phenomena in Impure Metals. Springer, Heidelberg, p. 121.
Yu, D., Sharvin, V., 1981a. Pos'ma Zh. Eksp. Teor. Fiz. 34, 285.
Yu, D., Sharvin, V., 1981b. JETP Lett. 47, 272.
Yu, Sharvin, V., 1984. Physica B 126, 288.

CHAPTER ELEVEN

Semiconductor lithography and processing in the 1980s and 1990s

Contents

When I started working in IBM's Thomas J Watson Research Center in Yorktown Heights, New York, in March 1965, I became a Research Staff Member (RSM). All members of the Research Staff, including the managers, were RSMs. Salaries were adjusted for performance and for holding management responsibility, but all the research staff were nominally RSMs. The technicians who supported the RSMs had various grades. They were exceptionally talented, and many held higher education degrees. Overall, this was a flat and effective organization.

My first manager, Alan Brown, left about a year after I arrived to work on a range of other technologies, including magnetic core memories, displays, and the superconducting Josephson devices used in the company's pioneering superconducting computer project. Alan was an outstanding team leader who brought out the best in everyone in the electron beam group.

Advances in Imaging and Electron Physics, Volume 231
ISSN 1076-5670
https://doi.org/10.1016/B978-0-443-31462-9.00011-3

Richard Thornley, another one of Professor Charles Oatley's students from Cambridge, took over as manager of the group and remained for 3 years before moving to the IBM laboratory in Boulder, Colorado, to work on magnetic memory projects, including the detection of magnetic fields in the SEM. He was a world leader in the design of electron optical systems and brought great expertise to IBM. I became the manager in 1968 and continued to manage the group until 1977 when I became an IBM Fellow. The group, by then, had expanded to about 25 people and included optical and X-ray lithography.

The IBM Fellow program was founded in 1962 by the company's Chairman and CEO, Thomas Watson Jr., to promote creativity among a small group of technical professionals. It continues today in 2024. The title IBM Fellow is granted in recognition of technical achievements and leadership in engineering, programming, services, science, design, and technology. Recipients of Fellowships are encouraged to use the freedoms associated with the position to further their own research. They are no longer part of the standard management structure and are free to determine for themselves what they do. The only responsibility I had at the time was to write a short report each year to the CEO describing what I had done over the previous year. I stepped down from management for several years and concentrated on my research. As already described, this included the scanning electron microscopy of biological samples, the fabrication of devices for scientific study, and the measurement of the resolution of scanning electron microscopy of bulk samples and of electron beam lithography.

In 1982, however, I accepted a challenge from Holly Caswell, who managed IBM's development laboratory in East Fishkill, to join him and work on the strategy for developing IBM's high-speed bipolar chips. These were to be used in the company's 3090 family of computers that were to be announced in 1985. The 3090 family succeeded IBM's System 370 family that had followed IBM's famous System 360 family launched in 1960. Holly Caswell had earlier led a group that worked on early attempts to build a superconducting computer using cryotrons.

The working environment in the East Fishkill laboratory was entirely different from that in the Research Division. Everything was driven by strategic aims and schedules rather than discovery and science, although innovation was key to achieving the aims and meeting the schedules. The laboratory had about 2000 people working on all aspects of the electronics used in IBM's processors. This included the advanced emitter coupled logic (ECL) chips that were to be used in the 3090 family of computers. Interestingly, Hannon York, the inventor of ECL, became manager of the Fishkill electron beam lithography group.

FET (Field Effect Transistors) Memory chips were developed and manufactured in Burlington Vermont. In Fishkill, we only worked on the bipolar transistors for IBM's high-speed logic chips. At the time, they were much faster than FETs. In fact, BM's mainframe computers used bipolar transistors in their processors for the next 15 years. It was only in 1997 that CMOS FET chips finally outperformed bipolar chips.

Bipolar transistors are power hungry, and to dissipate the power, IBM developed in E Fishkill what was called the Thermal Conduction Module (TCM) (Blodgett and Barbour, 1982; Blodgett, 1983). It contained 133 5 mm chips, each containing 704 circuits. The chips were mounted on an incredible 33-layer ceramic substrate. Aluminum pistons were pressed against the chips to keep them cool, and the chips and rods were immersed in Helium to increase thermal conduction between the rods and the chips. Each TCM weighed two kilograms and consumed 300 W but delivered the same computing power as an entire medium-sized System 370 computer of only a decade before. Each 3090 computer had about 24 TCMs. Looking forward a decade or so, however, a single Pentium II Xeon microprocessor chip had comparable performance to these huge mainframes, and today we have moved orders of magnitude beyond the Pentium.

Holly Caswell asked me to spend my first few weeks reviewing what was going on in the laboratory and feeding back to him my observations. It was all new to me, and much of it was in technical areas where I had little knowledge. For example, there was a large group of computer programmers writing the programs used to design the chips and the processors. I was astounded to learn that these programs contained millions of lines of code. I was favorably impressed with the quality and volume of the technical work and was interested to observe the competitive environment in which it was carried out. IBM had over 400,000 employees then, so there was no shortage of people. At times, the company would give similar objectives to competing groups in the belief that people would work more effectively against competition than the clock. This led to considerable stress, especially if targets were not being met, but it produced remarkable progress.

The company used its size in developing semiconductor chips by having separate teams in the company's Research, Advanced development, Development, and Manufacturing divisions, each with its own fabrication facility. To enhance technology transfer, the groups in Research and Advanced Development were gathered together under one director to ensure efficient technology transfer. The first director of this group was Robert A Henle who had been appointed an IBM Fellow in 1964 and was awarded the IEEE Edison medal in 1987. This combined organization made the first integrated circuits with fully operating 1 μm transistors.

The ideas and advancements made by Henle's organization were passed on to the product development groups in East Fishkill for logic devices and in Burlington for memory devices. Both the East Fishkill and Burlington laboratories had their own fabrication facilities. The technologies emerging from the product development lines were transferred to pre-production lines that scaled up the processes so that they were finally ready for transfer to manufacturing.

IBM, therefore, used four fabrication lines, whereas other companies would only have two: one for research and development and one for manufacturing. Others would often evaluate new processes and materials in separate sectors on the manufacturing line.

I became manager of Technology Tools, later called Semiconductor Lithography and Process Development. This was about a quarter of the laboratory, with around 500 people working on lithography exposure systems, photoresists, dry and wet etching systems, and the software used to control processing. There was even a small chemical manufacturing factory with 70 chemists and technicians producing novel resists, etchants, and solvents. We had the development fabrication line in which new generations of integrated circuits were developed until they met criteria that ensured that they were ready to be transferred to the pre-production line. We had a budget of about $500 million a year.

There had been an ambitious project in East Fishkill to build a completely automated semiconductor fabrication line in which human involvement was minimized to eliminate contaminating particles. This was known as AMS (Automated Manufacturing System). Wafers were handled by robots and inspected with computer-controlled measurement tools as they proceeded through the thousands of processing steps. It was not a success. It became clear that human intervention was needed throughout the manufacturing process, especially in the lithography steps. The application of photoresist, its exposure, development, and subsequent use as a mask for etching or deposition required close inspection by experts at every stage. This remains the case today, although progress in artificial intelligence might make the dream of a fully automated processing facility feasible.

Semiconductor technology in 1980

This was an important time in the evolution of integrated circuits, which, by 1980, were universally called silicon chips. Things had gone well for the first 20 years, and some engineers suggested that progress would accelerate in the next decade. Others, including myself, thought that we would be lucky to maintain the remarkable pace achieved in the sixties and seventies. It was already getting difficult to find low-hanging solutions to the problems arising as transistor circuits became smaller and smaller.

Gordon Moore, a co-founder of Fairchild Semiconductor and chairman and CEO of Intel, observed in 1965 that the number of transistors on a chip was doubling every year. His observation became famous and known as Moore's Law. It set the pace of development for the industry for the years ahead. He originally predicted that it would continue for a decade, but in 1975 he extended his prediction but increased the time for doubling to 18 months. Progress at this rate was nonetheless unprecedented in the history of technology and amazingly continues today in 2024, although since about 1980, many have based their plans on doubling every 3 years. Up until about 15 years ago, progress was sustained mainly by reducing the size of the transistors and the width and spacing of the interconnecting wires, but recently, this has slowed considerably. For the last decade, the increase in the number of transistors on a chip has relied on advances in transistor design, especially those that use the vertical dimension. Field effect transistors called FinFETs are an example of this. FinFETs use a vertical fin as the channel, rather than a horizontal channel, and take less area on the wafer.

In 1980, there were significant strategic issues in all aspects of the manufacture of semiconductor integrated circuits. I will limit myself mainly to discussing lithography, but important advances were being made in all stages of device fabrication, and decisions had to be made about which to use and when to use them. I will just mention four areas of interest.

First, the diameter of the cylindrical ingots of silicon from which silicon wafers were cut had grown from 15 mm in the early 1960s to about 150 mm by 1980. Larger wafers could contain larger chips and/or more chips, and serendipitously the cost of processing a wafer remained roughly the same. Today, the largest wafers are 450 mm in diameter, much larger than a dinner plate. Every time it was decided to adopt larger wafers, most of the fabrication tools had to be replaced, and they were expensive. Fabrication facilities in 1980 cost millions of dollars. Today, they cost tens of billions of dollars.

Serious problems with electro-migration arose as the aluminum interconnecting wires became smaller. Electro-migration is an atomic transport process. The momentum of the conduction electrons is transferred to the metal atoms moving them, and as they move, there is a depletion of material "upstream" and an accumulation "downstream," leading to voids and eventually to wires burning out. The process was contained by doping the aluminum with silicon. The silicon migrated to grain boundaries in the aluminum and slowed the migration. Eventually, copper was to replace aluminum completely, but that was more than a decade away. Copper also presented difficulties because it readily diffused into silicon and copper wires had to be encapsulated with a diffusion barrier to prevent corrosion and electrical leakage between adjacent conductors.

Ion implantation increasingly replaced thermal diffusion in creating the different regions of transistors. However, it was more expensive and introduced crystallographic damage. Fortunately, the damage could usually be annealed out, but ion implantation was more complex and expensive.

The growth of SiO_2 also gained a lot of attention as layers a few nanometers thick were needed to get the gate of field effect transistors close enough to the conducting channel to control the current between source and drain. Steady progress over the decades produced atomic-level perfection in these critical dielectric layers.

In summary, progress was made throughout the chip fabrication process, but in almost all cases, it increased complexity and cost. I have mentioned four of the areas where progress was made and where IBM and others were devoting resources, but the area that gained the most attention was lithography. It was the ability to make smaller devices that set the pace for the miniaturization needed to keep up with Moore's Law. There must have been more than 300 people in IBM working on chip lithography.

Lithography in 1980

Up until 1980, almost all lithography systems used short-wavelength light. They were either simple contact or proximity printers, or projection cameras. However, it was thought that it would not be possible to extend these forms of UV light lithography to dimensions below about 0.5 μm. Scanning electron beam lithography was being used to make micron and sub-micron transistors in the laboratory, confirming the predictions that smaller transistors were faster, lower power, and cheaper, but the serial way in which it wrote patterns made it too slow and expensive for use in manufacturing except for the personalization of gate-array chips. Its main uses were in research and for writing masks for UV cameras. The leading contender for manufacturing devices smaller than 0.5 μm was proximity printing with X-rays. It was referred to simply as X-ray lithography. X-ray masks were fragile, and conventional X-ray sources were too weak to provide adequate throughput, but 0.5 μm seemed a long way off in 1980, and there seemed to be plenty of time to solve these difficulties. In any case, more than a dozen alternatives using electrons, ions, and photons were being pursued should X-ray lithography fail.

Optical lithography

Resolution and contrast for light optical systems have already been discussed in Chapter 9.

Both reflecting and refracting lenses have been used in optical microcircuit cameras. In the early 1970s, the cameras that replaced contact and proximity printers used 1:1 mirror lenses. They had a resolution of about 2 μm. There was then a period when demagnifying refracting lenses were used to reduce dimensions from about 2 to 0.5 μm. To further increase resolution, however, it was necessary to reduce the exposing wavelength to a level where the materials available for refractive elements were no longer transparent, and demagnifying optical systems using mirrors were introduced. Today, the highest-resolution cameras use extreme ultra-violet wavelengths and demagnifying mirrors with multi-layer coatings. These mirrors have to be fabricated with mechanical tolerances that approach atomic dimensions.

Scanning mirror cameras

By 1980, most semiconductor companies were using Perkin Elmer Micralign scanning optical aligners (Greed and Markle, 1982; Markel, 1974; Offner, 1979). These 1:1 projection systems used the same masks as the contact and proximity printers they replaced but could use the master masks and avoid the need to make working masks. The master mask could be used 100,000 times compared with 10 times for a working mask in a proximity printer. This significantly reduced cost, but more importantly, the projection printers separated mask and wafer and hugely improved yield by eliminating the contact problems with dirt and sticking emulsion encountered with proximity printing.

Micraligns used a 1:1 magnification mirror imaging system to project the image of a crescent on the mask onto the wafer (Fig. 11.1). Exposure of the full wafer was completed by scanning the mask and wafer past the imaging system. A roof prism was used so that the parity of the image allowed the mask and wafer to be scanned in the same direction and, therefore, be mounted on a single holder. An entire 7.5 cm diameter wafer was exposed in less than a minute.

The numerical aperture (NA) of the mirror imaging system was 0.16, and the first systems produced a resolution of about 2 μm and could expose 3-in. wafers. The first systems allowed the entire output of the mercury illumination lamp to be utilized for photoresist exposure. Full advantage was taken of the 365, 404.7, and 435.8 nm peaks, resulting in exposure times that were measured in seconds rather than minutes and making it possible to reduce deleterious standing wave effects in photoresist exposure.

Fig. 11.2 shows the M.T.F.s for the 0.16 N.A. Micralign cameras for three exposure frequencies. Subsequent models in the Micralign series (100, 200, 300, and 500) were designed to expose larger wafers. The illumination system was improved and made more flexible, allowing exposure in three wavelength regimes (UV $\lambda = 400$ nm), mid-UV ($\lambda = 300$ nm), and deep UV ($\lambda = 250$ nm). The linewidths at which the image contrast fell to 60% are shown to be 1.5 μm at 400 nm, 1.125 μm at 300 nm, and 0.9 μm at

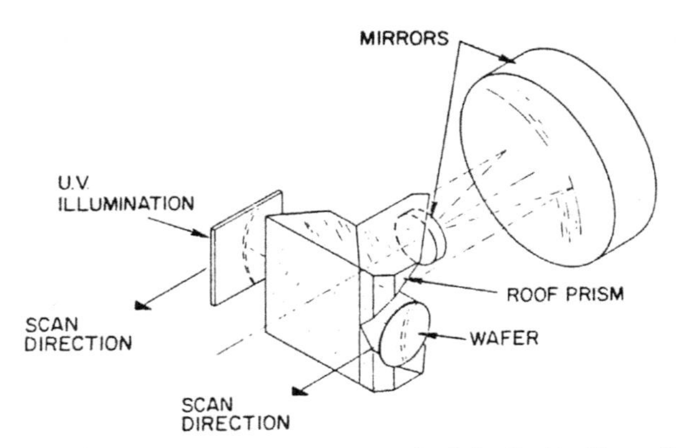

Fig. 11.1 1:1 optical scanning projection camera built by Perkin Elmer (Markel, 1974). The camera was widely used in the 1970s and can reproduce linewidths down to about 2 μm.

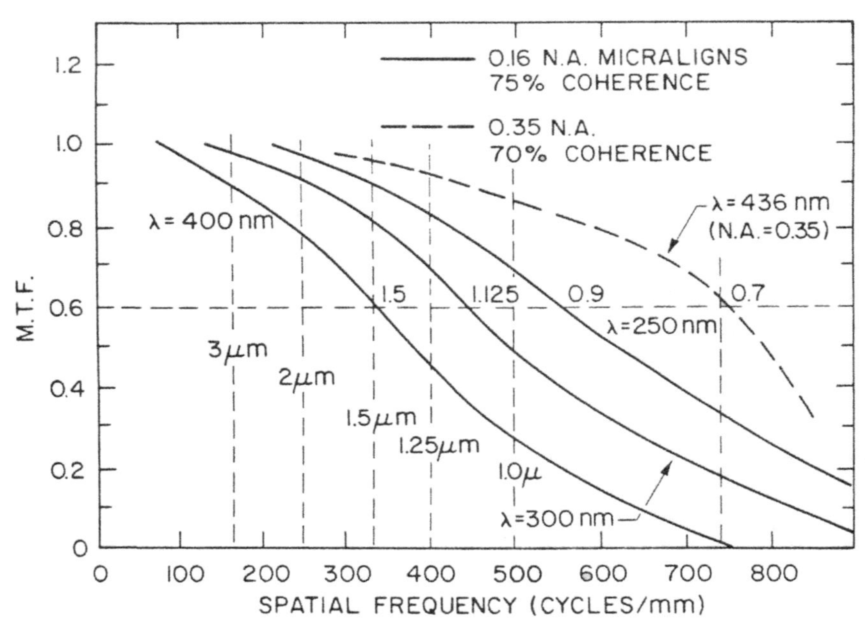

Fig. 11.2 Modulation transfer function for Perkin Elmer Micralign cameras. *(Reprinted with permission from Broers, A.N., 1984. Practical and fundamental aspects of lithography, Materials for Microlithography. American Chemical Society, Chapter 2. Copyright 1984 American Chemical Society).*

250 nm. The case for a 0.35 N.A. optical system operating at 436 nm is also shown where 60% contrast is reached at a linewidth of 0.7 μm.

Major advances with the last in the series, the Micralign 500 (Fig. 11.3), were the ability to expose 150 mm diameter wafers, compared to 100 mm for the earlier cameras, and to adjust magnification. The latter made it possible to compensate for the inevitable microscopic changes in wafer size that arose during device processing. Image size in the direction of the slit was changed by small axial motions of the "stronger" pair of refractive elements. Corrections in the opposite direction were made by micro-scanning the mask during exposure to slightly increase or decrease the scan length for the mask compared to that for the wafer.

If distortions of the sample were truly isotropic, the only errors that remained after magnification correction were residual distortions in the optics and mask errors. The easiest way to reduce mask errors was to

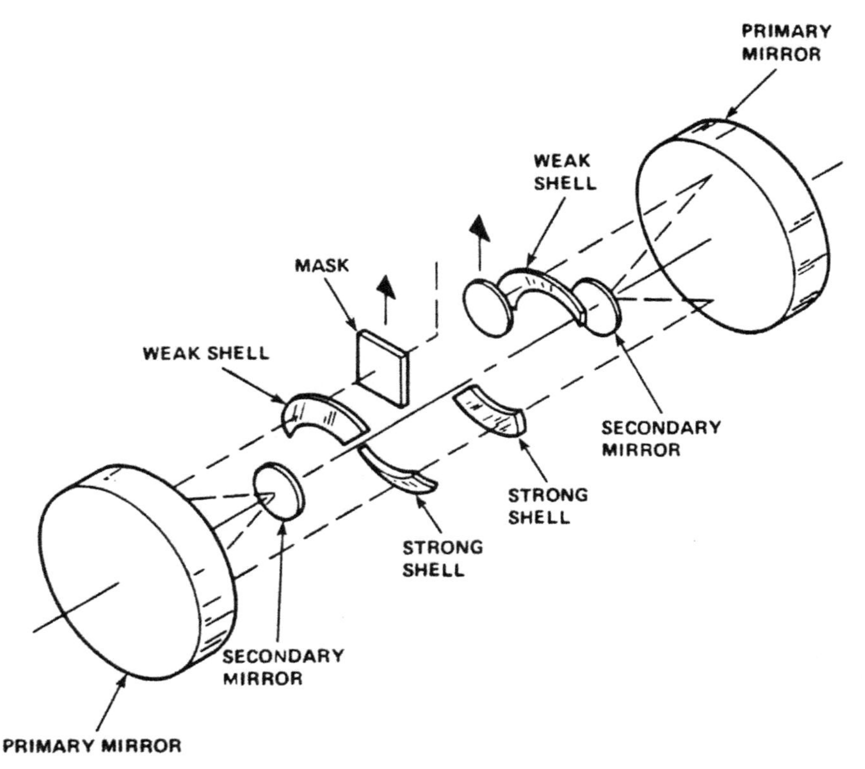

Fig. 11.3 Perkin Elmer 500 advanced full field camera. *(Used with permission of The Royal Society from Broers, A.N., 1988. Limits of thin film microfabrication, the Clifford Paterson lecture. Proc. Roy. Soc. Lond. A 416, 1–42, permission conveyed through Copyright Clearance Center, Inc.).*

go to 5× or 10× S/R systems, but eventually, it should have been possible with electron beams to reduce mask errors to insignificant levels. Distortion in the optics of the model 500 was already below 0.1 μm and $\pm 3\sigma(\alpha)$ overlays of less than 0.5 μ had been obtained.

Micraligns revolutionized the industry, improving quality and dramatically reducing cost, so much so that companies like Intel initially kept their use secret.

As already mentioned, resolution with mirror lenses can be improved by operating at shorter wavelengths. The difficulty was to find resist/lamp combinations that gave short enough exposure times. Several resists had already been found that were satisfactory for 300 nm exposure, and there were others with potential for operating at 250 nm, but it took a lot of searching to find resins that were transparent enough to allow uniform exposure to the bottom of the resist at the shorter wavelengths.

By the mid-1980s, half-micron device technology had been proven in the laboratory. What was needed were manufacturing processes, especially lithography cameras and resist processes, that could manufacture these devices at a cost that took advantage of the immense gains in speed, power, and cost available when dimensions were reduced. It looked impossible to extend the resolution of a full field scanning system to 0.5 μm. But by moving to a scanning system that only exposed a portion of the wafer, it should have been possible to increase the NA to 0.35. Eventually, it might have been possible to build a scanning optical system with an NA of 0.4, which, at a wavelength of 250 nm, would have produced 60% contrast for linewidths of 0.5 μm. The depth of field would have been small (±0.8 μm for the incoherent case), but capacitive sensing could have been used to monitor the wafer surface to ±0.1 μm, and piezo-electric manipulators could have been used to correct focus. I am sure that Perkin Elmer considered these advances, but I do not think a system that incorporated them was ever built.

The more available route to higher resolution was the S/R cameras, which projected a much smaller, demagnified image onto the wafer, typically that of a single chip, and stepped and repeated it over the wafer. Fig. 11.4 is a diagrammatic view of a step-and-repeat camera.

Step and repeat cameras

Step and repeat cameras in the early 1980s used refractive lenses with N.A.s of 0.2 to 0.4. The best lenses exposed a field of about 2 cm × 2 cm at a numerical aperture of 0.3. Higher NA was available for smaller field sizes. The field size and NA for S/R camera lenses and the Perkin Elmer Micraligns are shown in Fig. 11.5. Most were designed to operate at a single wavelength that corresponded to a strong line in the mercury

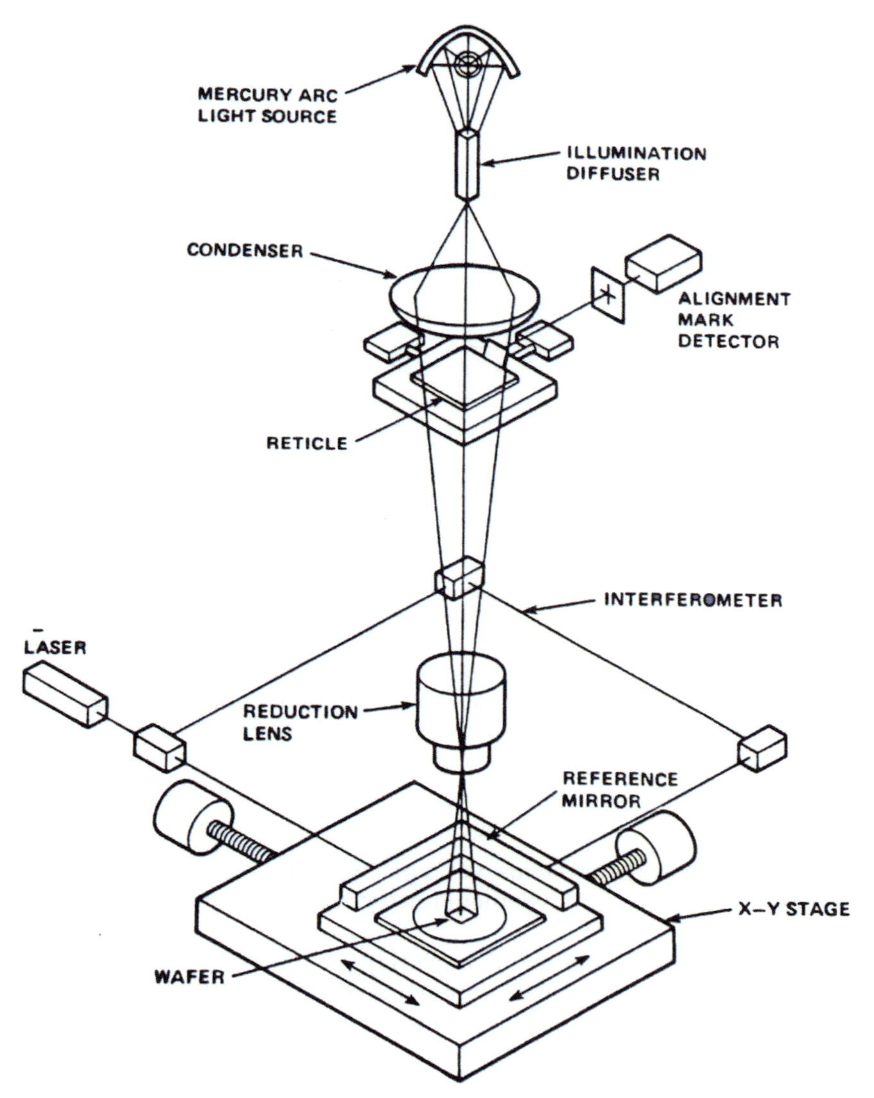

Fig. 11.4 Diagrammatic view of optical step and repeat camera. *(Used with permission of The Royal Society from Broers, A.N., 1988. Limits of thin film microfabrication, the Clifford Paterson lecture. Proc. Roy. Soc. Lond. A 416, 1–42, permission conveyed through Copyright Clearance Center, Inc.).*

spectrum (365, 405, or 436 nm), but lenses had also been corrected for two wavelengths (405 and 436 nm) to reduce the effects of standing waves interference in the resist, With two-layer resists, double wavelength exposure was not really necessary. Fig. 11.3 shows the optical systems for the most advanced of the Perkin Elmer Micralign cameras, the 500, where the

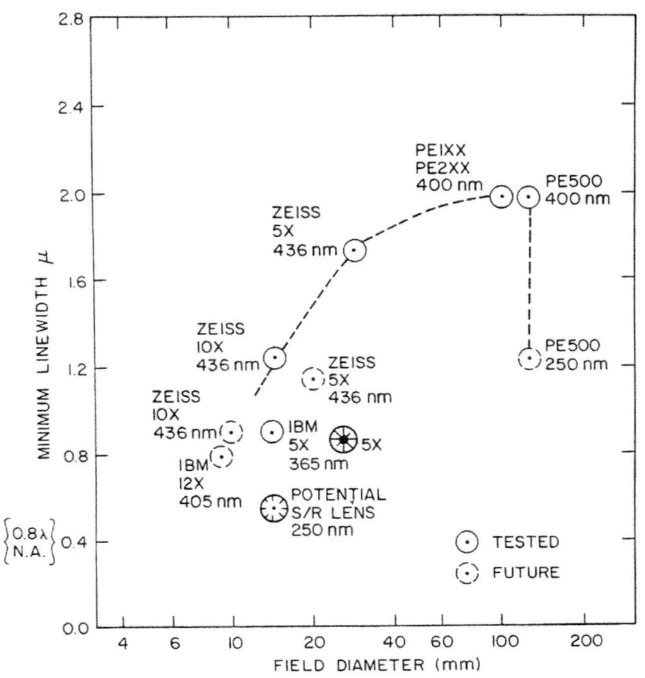

Fig. 11.5 Linewidth vs field size of optical lithography systems. *(Reprinted with permission from Broers, A.N., 1984. Practical and fundamental aspects of lithography, Materials for Microlithography. American Chemical Society, Chapter 2. Copyright 1984 American Chemical Society).*

roof prism of the earlier systems was replaced with a second spherical mirror. Fig. 11.6 shows a multi-element microcircuit camera lens designed at IBM Research.

As dimensions became smaller, overlay accuracy became increasingly important, placing stringent requirements on image distortion. It proved difficult to reduce this below about 0.1 μm, making it a significant

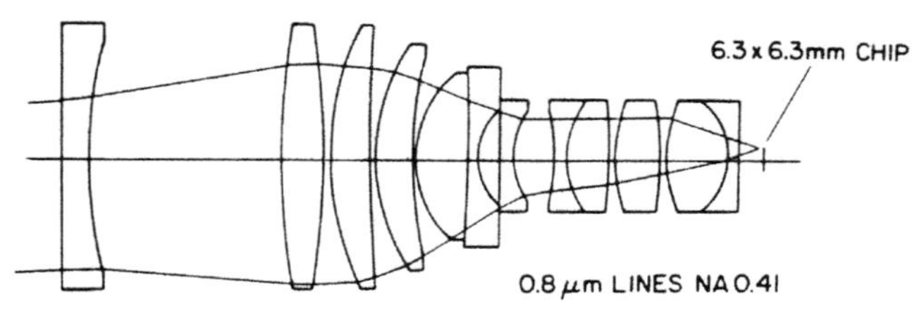

Fig. 11.6 Microcircuit lens designed by Wilczynski and Tibbets (Tibbets and Wilczynski, 1980) from Lens is capable of imaging 0.8 μm features over a field of 6.3 mm × 6.3 mm.

contributor to the final overlay when different cameras were used for different layers.

The sample position in the S/R cameras was tracked by a laser interferometer after an initial mask-to-wafer alignment, or the mask and sample were aligned with respect to each other at every chip site (Dey et al., 1980; Dubreoueq et al., 1980; Lauria et al., 1980; Mayer et al., 1980; Offner, 1979; Wilczynski, 1979; Wittekoek, 1980). Alignment at every chip site avoided errors due to drift between the wafer and the interferometer reference point, but it was difficult to implement. For certain resist thicknesses, the light was reflected equally from the reference mark and the background surface, and the alignment marks "disappeared." To avoid this, it was necessary to carefully control resist thickness (<10 nm control was needed in some cases) or to remove the resist from the marks. Both solutions complicated the process. The alignment marks could be examined with dark-field or bright-field illumination, and it was best to have both available.

In principle, the resolution of optical lithography can be similar to that of optical microscopy. A lens with a numerical aperture of 0.95 would produce linewidths down to less than 0.4 μm over an area of about 0.2 mm × 0.2 mm. The small field would not be an inherent limitation because the image could be stepped or scanned, but the depth of field would only be about ± 0.2 μm. Improvement could be obtained by operating with mid-UV or deep-UV light, but refractive lenses were difficult to build for deep-UV wavelengths because of the lack of a range of glasses that were transparent to the short wavelength light and had high refractive indices. If such lenses could have been made, the excimer lasers would have produced more than enough illuminating power at these frequencies.

Despite the problems with alignment and overlay, it was recognized in the mid-1980s that optical lithography would prevail until dimensions reached 0.5 μm, perhaps even 0.35 μm. Below this, it was still believed that new technologies would be needed. Exposure systems were being developed around the world that used X-rays, electrons, and ions, but none had been shown at that time, even in R&D laboratories, that they could compete with optical lithography for the fabrication of chips. The two that were gaining the most attention were scanning electron beam lithography and x-ray lithography.

Electron beam lithography

JEOL and the Cambridge Instrument company made the first electron beam lithography systems for widespread application in the late 1960s. Both companies had been making scanning electron microscopes for many years and had the skills and technologies to do this. After early research in Tubingen and Cambridge universities, several companies including IBM, Karl Zeiss, Westinghouse, SRI, GE, AEI, and Mullard became involved, and in the late 1960s, interest in applying electron beam lithography to the manufacture of integrated circuits built rapidly at Texas Instruments, Western Electric, Hughes Aircraft, Thomson CSF, Hitachi, Western Electric and Leica. In 1974, AT&T Bell Labs announced their development of the EBES mask writer described later. By the middle of the 1970s, there were e-beam systems in most laboratories where integrated circuits were being developed.

The Cambridge Instrument and Leica systems used round electron beams, which were electronically scanned to write patterns under computer control. The scanned area was about 5 mm × 5 mm, and the pattern was written in a vector-scan mode where the beam was only deflected to positions on the sample where the resist was to be exposed. Raster scanning, as used in television imaging, could also be used but was slower because of the time wasted while the beam traversed areas that did not require exposure. A fast, highly accurate deflection system was needed for vector-scanning. After writing each pattern, the sample was mechanically stepped to the next location, and the pattern was repeated. The general mode of operation for a vector scan system is shown in Fig. 11.7.

Philip Chang, who had with Garry Stewart designed the Cambridge Instrument Company's Vectorscan system, moved to IBM Research in the 1970s and, with Alan Wilson and others, built the IBM VS1 vector scan

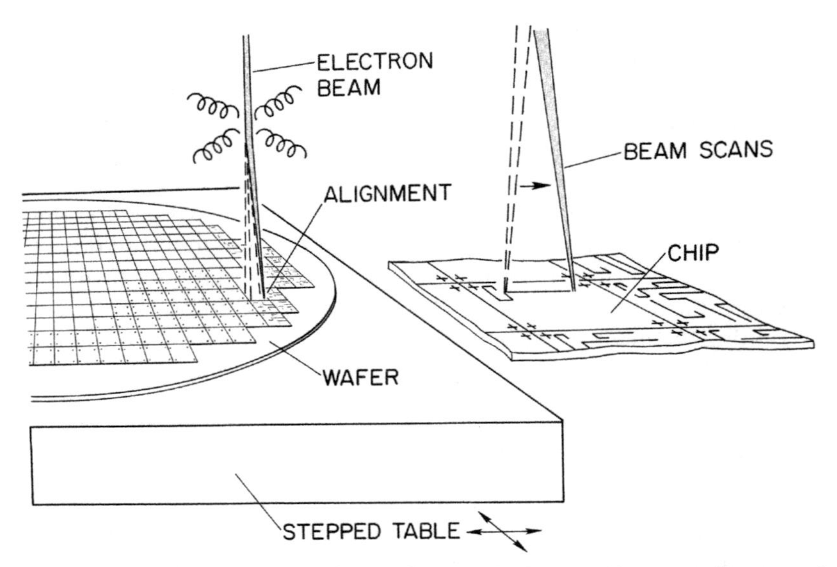

Fig. 11.7 Vectorscan writing with electron beams. The beam is electronically scanned to write each chip. The sample is then mechanically moved to the next chip site.

system (Chang et al., 1974). This was used to build the earliest sub-micron FET chips that were vital in establishing the Dennard scaling of FETs. Round beam vector scan systems were also built by Pearce-Percy et al. (1979) and Eidson et al. (1981).

The first electron beam lithography tool to be used in the manufacturing of integrated circuits was in the early 1970s. It was a raster scanning system with a 2.5 μm square beam built by a large group in IBM East Fishkill under the management of Hannon Yourke, A Oberai and Dick Moore and the technical leadership of Hans Pfeiffer, Werner Stickel, and others (Pfeiffer, 1975). The square beam increased the exposure rate by more than an order of magnitude compared to the round beam systems.

The throughput was sufficient to write the patterns for the metal interconnection wires for IBM's advanced gate-array chips. These chips could be manufactured to order with a delivery time of 2 weeks because there was no need to make masks for the custom metal layers. This was the first success in manufacturing for direct-write electron beam exposure of wafers. The square beam system was too slow and expensive to be used for all of the lithography steps, but its ability to provide a quick turn-around of custom chips proved extremely valuable.

At about the same time, AT&T Bell labs built the EBES (Electron Beam Exposure System) (Herriott et al., 1975) mask writer that combined electronic scanning with mechanical scanning. The pattern was written with a round electron beam that was scanned over a small area while the mask substrate was mechanically scanned beneath it. Errors in the position of the continuously moving mechanical table were detected by laser interferometers and corrections fed back to the beam deflection coils. Mechanical scanning allowed much larger areas to be exposed than electronic scanning. A 0.5 μm diameter round electron beam was used to write mask patterns with minimum linewidths of about 2.5 μm. Fig. 11.8 shows the general mode of operation of the system. EBES mask writers were able to make the masks for the Perkin Elmer Micralign optical projection cameras mentioned above. It was a success for making masks, but like the vectorscan systems, it was too slow and expensive to use for general lithography.

A series of EBES systems followed EBES I to meet the need for sub-micron dimensions and higher throughput; EBES II (Alles et al., 1975) and EBES4 (Alles et al., 1987) (Fig. 11.8).

In the mid-1970s, the IBM group that successfully built the square beam system was expanded to almost 100 people and given the challenge of developing a high throughput sub-micron lithography system for manufacturing. They decided not to use the continuous mechanical movement of the wafer used in the EBES systems but to retain the step-and-repeat approach, so the chip pattern was written using electronic scanning

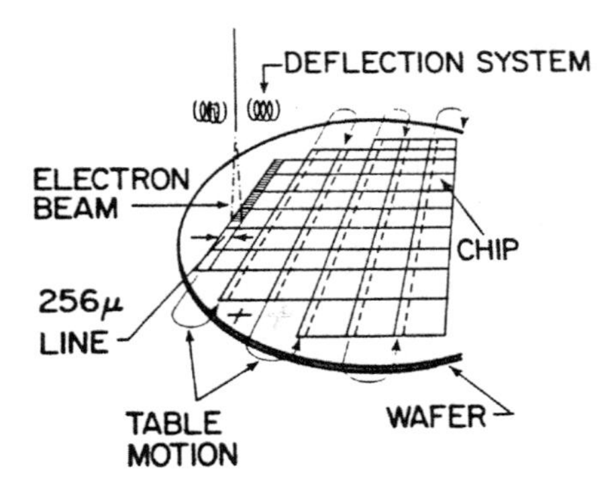

Fig. 11.8 Operating mode for the EBES electron mask maker (Herriott et al., 1975), The beam is electronically scanned over a line 256 μm long and the sample moved in a serpentine manner under the beam to complete the full wafer mask. A laser interferometer tracks the exact position of the mask.

with the mechanical stage stationary. The wafer was moved mechanically to the next chip site, and the pattern repeated. The electron beam position on the wafer was checked before each exposure by scanning alignment marks on the wafer. Laser interferometers were used to detect any residual mechanical movement after each step and feed corrections to the beam deflection system.

Throughput of this new IBM manufacturing system was to be increased by using a variably shaped beam (VSB). This further increased the average number of pixels simultaneously exposed by the beam. A range of rectangular shapes was produced by varying the position of a shaped illuminating beam on a fixed square aperture. Fig. 11.9 shows how the shaped beam was formed. Fig. 11.10 shows the reduction in beam addresses gained with fixed and variably shaped electron beam systems compared to round beam systems.

A higher brightness electron gun and a faster and more accurate deflection system were also used to increase writing speed. Several laboratories started to work on variable-shaped beams at the same time (Goto et al., 1978; Moore et al., 1981; Pfeiffer, 1978; Thompson et al., 1978).

It was hoped that together, these advances would make the system fast enough for widespread use in chip manufacture. But this was not to be the case. A significant challenge was to prevent the Boersch effect from blurring the beam. Electron-electron interactions increase the energy spread in the beam to a level where the chromatic aberrations of the focusing lens and the deflection system blur the beam. The beam would be sharply defined at low current density and small shapes, but as the number of pixels and the current density increased, pattern definition degraded. Hans Pfeiffer and Werner Stickel managed to contain the Boersch effect but only at a throughput that was too low to compete with optical

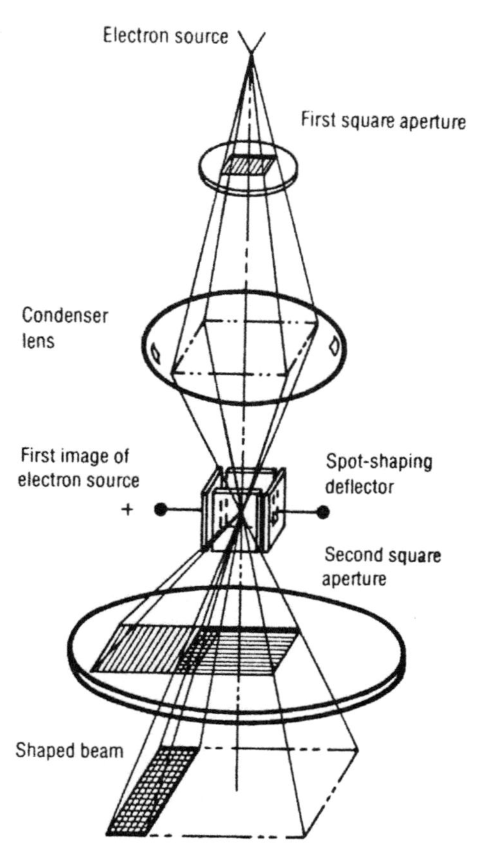

Fig. 11.9 Spot shaping upper column used by Hans Pfeiffer (Pfeiffer, 1978) to vary the shape of the electron beam. *(Used with permission of The Royal Society, from Broers, A.N., 1988. Limits of thin film microfabrication, the Clifford Paterson lecture. Proc. Roy. Soc. Lond. A 416, 1–42, permission conveyed through Copyright Clearance Center, Inc.).*

lithography. Their advances were nonetheless significant. For an image size of 1 μm, a field size of 5 mm × 5 mm, and an overlay accuracy of 0.4 μm, 10–20 3-in. wafers could exposed per hour at 10 μC/cm^2 with a beam current density of 50 A/cm^2 For an image size of 2 μm, a field size of 10 mm × 10 mm, and overlay accuracy of 0.7 μm, a throughput of 20–45 wafers/h was predicted. In the end, this did not prove great enough to offset the cost of the systems, and they were not used for high-volume exposure of wafers. The Perkin Elmer 500 cost less, and could expose 50,125 mm wafers an hour. However, the variable-shaped beam electron beam systems were successfully applied to the customization of bipolar chips and mask-making. By the 1980s, IBM had installed more than 30 of these variable-shaped beam systems in their factories worldwide (Moore et al., 1981).

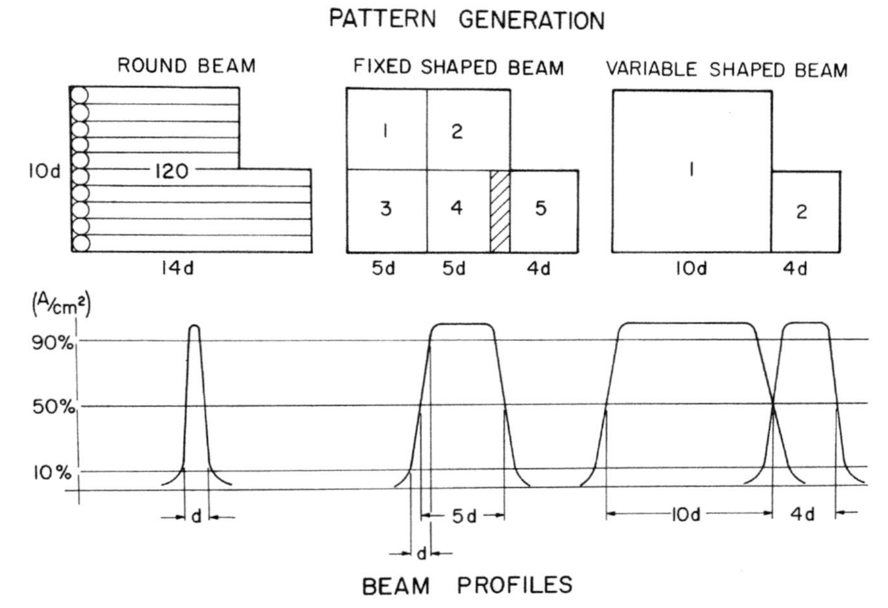

Fig. 11.10 Methods in which pattern elements are filled with round beams, fixed square beams, and varuiable shaped beams. Note the reduction in beam addresses with the shaped beams.

More sensitive resists would have allowed smaller beam currents for a given writing speed, potentially avoiding the Boersch effect, but this approach is unacceptable when the number of electrons needed to expose a single pixel falls below about 100 and the image becomes noisy. This noise limit is more serious as dimensions are reduced. For a pixel size of 0.05 μm, which is needed for 0.25 μm dimensions, the noise limit would be reached at an exposure level of 10^{-6} C/cm^2. This is largely an academic limit as there were few proven resists with a sensitivity below 10^{-6} C/cm^{2,} but even if they had existed, they could not be used because of the noise limit.

For less sensitive resists, throughput depends on beam brightness, the aberrations of the focusing and deflection system, and the Boersch effect. We had no idea at that time that 40 years later, in 2024, device dimensions would be about 25 nm, and the electron optical constraints and Boersch effect would have made it impossible to reach electron arrival rates of 100 per pixel. I will return to these limits in the Epilogue.

Throughput limits for electron beam lithography

Early electron beam systems operated at pixel exposure rates of about 250 kHz. By 1990, this had grown to the equivalent of 1 GHz, but this only balanced the increase in chip capacity from 1 kbit to 4 Mbit over the same period. Wafer areas had also increased by almost two orders of magnitude, and wafer throughput had to be maintained to gain essential cost-per-chip reductions. As a result, the throughput of electron beam systems measured in wafers per hour had, on average, decreased while optical projection systems had maintained constant throughput despite resolution improvements.

There were prospects for a further increase in pixel exposure rates for scanning systems, but again, they were scarcely sufficient to keep up with the predicted increase in the number of pixels per wafer. Fig. 11.11 shows the throughput for three types of electron beam systems that were being considered in the mid-1990s for 0.25 μm devices as examples of what might eventually be achieved. Table 11.1 shows the parameters that were used in calculating the throughput of the three systems. Only the variable-shaped beam short focal length lens case comes close to producing competitive throughput. The VSS continuous table system uses a conventional long focal length final lens.

It was unlikely that the short focal length case would ever be built and I don't think it was. I include it as an illustration of the limiting case. Short focal length lenses were being used in microscopes and in the high-resolution probe described in Chapter 8. The scan field would inevitably be small but would be used with a refined version of the continuously moving table concept used in the EBES systems. The short focal length lens would produce the highest current and smallest beams.

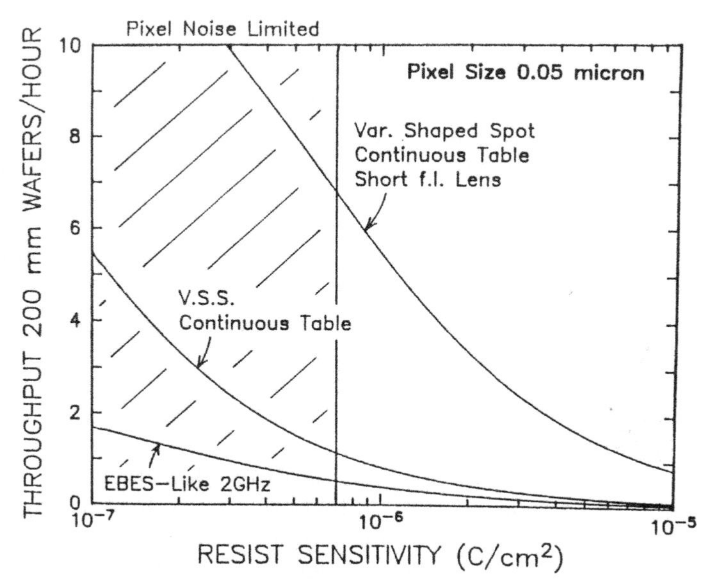

Fig. 11.11 Throughput for potential future electron beam systems. The case for the Variable Shaped Spot (VSS) short focal lengthlens is an artifical case as discussed in the text.

Table 11.1 Parameters used in calculating the throughput of electron beam systems shown in Fig. 11.11.

	V.S.S. cont. mvmt. table	**V.S.S. cont. mvmt. table short f.1. lens**	**EBES-like vector scanning**
Pixel size	0.05 μm	0.05 μm	0.05 μm
Spot current density	$20\,A/cm^2$	$200\,A/cm^2$	$300\,A/cm^2$
Resist sensitivity (100 electrons/pixel)	$7 \times 10^{-7} C/cm^2$	$7 \times 10^{-7} C/cm^2$	$7 \times 10^{-7} C/cm^2$
Pixels per shape	30	30	1
Pixels per sub-field	10^6	10^6	–
Chip size	20 mm	20 mm	20 mm
Wafer size	200 mm	200 mm	200 mm
Fraction exposed	25%	25%	25%
Sub-field to sub-field time	5 μs	5 μs	–
Shape to shape time	2 ns	2 ns	0.5 ns
Chip to chip time	0	0	0
Throughput (200 mm wafers/h)	1.2	7	0.5

A significant advantage of such lenses is that the electron-electron interactions would be minimized. The problems of implementing the short focal length lens are due to eddy current and deflection aberrations.

Much longer focal length lenses were preferred because they allowed the sample to be kept outside the lens magnetic field and the beam deflection angle to be minimized for a given field size. With the short focal length lens, the sample would have to be placed inside the lens and consequently be immersed in the magnetic field, where it would be difficult to avoid eddy current interference. Moving parts, even the samples themselves would have current induced in them, and these currents produce magnetic fields that result in spurious deflections of the beam. Optimum deflection systems also require that the final deflection takes place in the middle of the final lens. The coil that produces this deflection is placed inside the lens and, therefore, induces eddy currents in the lens pole pieces if they are electrically conducting. This problem has traditionally been solved by using ferrites instead of soft iron for the pole pieces. With the optimum short focal length lens, the ferrites would saturate. The maximum field in a long focal length lens is about 0.4 Tesla whereas about 2 Tesla is needed in the short focal length lens. These problems are not fundamental; for example, the deflection could be accomplished before the lens, and the beam only deflected when the sample was stationary, but the practical constraints would be severe.

Fig. 11.11 also shows the performance of two potentially realizable systems, but their throughput is small. The VSS continuously moving table case, which uses a conventional long focal length lens, only reaches a little over one 200mm wafer per hour for 0.25 μm dimensions, and the speeded-up EBES only 0.5 wafers per hour. The VSS system will have sufficient throughput for custom devices but not for volume production. The advanced EBES will be adequate for masks but not for direct-write. In both realistic cases, the throughput is mainly limited by the time taken to expose the resist and not by the overhead times associated with the electronic and mechanical systems. This is clear from the variation of throughput with resist sensitivity. Throughput would be increased with higher beam current density.

Brighter electron guns can be used to increase the current density in electron probes, but as just discussed, the shaped beam systems that provided the highest effective pixel exposure rates were limited in current density by the Boersch effect and not by beam brightness. Resist heating also becomes troublesome at higher exposure rates. Brighter electron guns can play a more important role at smaller dimensions (<0.1 μm) where the total beam currents are smaller but would not help much for integrated circuit applications until dimensions became much smaller.

Eventually, chips would become so complex and wafers so large that it would take tens of hours to write a wafer-scale mask with a single electron beam, and multi-beam systems would be needed. But this was decades ahead, and I will return to this in the Epilogue.

In the late 1980s, electron beams were more than adequate for mask-making, but it was becoming clear that they would never be cost-competitive for mainstream lithography. It was also thought that optical lithography would never have the resolution for sub-tenth-micron line widths. To fill the gap, innumerable alternatives were being worked on that used electrons, ions, and X-rays. Examples included proximity printers, scanning systems, and projection printers. They included full projection electron beam systems, an approach that Marc Heritage had pioneered in the IBM Research lithography group. I will return to some of these in Section "Other lithography exposure methods being worked on in 1980" toward the end of this chapter.

X-ray lithography

Proximity printing with soft X-rays (wavelength between 0.4 and 10nm) was pioneered by Spears and Smith (1972a) and Smith et al. (1973) at Lincoln Labs in the early 1970s, and by 1980 was the most promising of the parallel exposure methods that competed with optical projection for volume production. It appeared to be the easiest route to dimensions below 0.5 μm and had extendibility to below 0.25 μm. Soft X-rays are not scattered in the resist, as are electrons, and their wavelength is so short that diffraction effects can be neglected, at least down to about 0.2 μm, even for a mask-to-wafer spacing of 50 μm. Vertical resist profiles are produced even in thick, single-layer resists. These single-layer resists were thought to be cheaper to use than the surface-acting resists needed for optical lithography below 0.5 μm.

The mask could not be made on the quartz plates used for optical lithography because the plate would absorb the X-rays, so it was formed on a thin (<10 μm) membrane. The membrane was generally made from a combination of two materials. One is chosen for dimensional stability (silicon, boron nitride, etc.) and the other for ruggedness (e.g., polyimide). The heavy metal absorber pattern was generally patterned with electron beam lithography. Despite the attempts to produce a rugged and stable mask, the mask remained the major difficulty with X-ray lithography.

Initial experiments used conventional electron bombardment X-ray sources. These were point sources and their large divergence gave rise to image distortions if the mask and wafer were not perfectly flat (see Fig. 11.12). To keep distortion below 0.1 μm for a 125 mm wafer and a source-to-wafer distance of 50 cm, the mask and wafer had to be held flat to better than 1 μm. Wafer buckling that arose during many standard

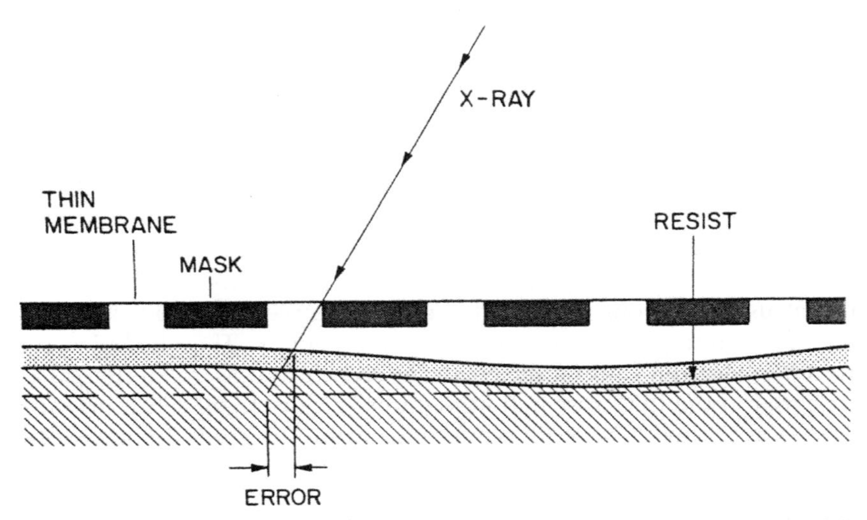

Fig. 11.12 With a conventional electron bombardment X-ray source the position of the shadow image varies with the mask and wafer spacing. Wafers frequently become buckled due to processing. *(Reprinted with permission from Broers, A.N., 1984. Practical and fundamental aspects of lithography, Materials for Microlithography. American Chemical Society, Chapter 2. Copyright 1984 American Chemical Society).*

integrated circuit processes typically exceeds 10 μm, making it unlikely that full wafer exposure could be accomplished for wafers larger than about 7.5 cm.

The run-out error could have been avoided if resist sensitivity was improved to the point that exposure could be made at a point further removed from the source. This required a sensitivity approaching 1 mJ/cm^2 and while there were resists with this sensitivity (Taylor and Wolf, 1981), they were yet to produce high aspect ratios and therefore lacked the major advantage of X-ray lithography over optical lithography. High aspect ratio could have been obtained using a tri-layer resist process, but this option was also open to optical lithography.

The run-out error could have been avoided by going to a S/R approach and limiting mask size to a few centimeters. This would also have alleviated problems due to mask and/or wafer distortions, but exposure times with the resists and available sources were too long for this to be economical. Exposure times of a few seconds would have been needed for a source to wafer distance of about 50 cm, but this would have required the same sensitivity as the full wafer case.

Electron storage ring

One source that did seem to produce X-ray beams with the properties needed for S/R exposure was the electron storage ring (Grobman, 1980; Spiller et al., 1976). The X-ray beams emitted from a storage ring are so intense and have such small divergence, that the problems of run-out error and long exposure times would be eliminated.

A storage ring consists of a circular stainless-steel vacuum tube in which electrons with an energy of 500 MeV to 1 GeV circulate at speeds close to the speed of light, as shown diagrammatically in Fig. 11.13. The electrons are injected into the ring from an accelerator, and their energy is increased and subsequently maintained by passing them through microwave cavities each time they pass around the ring. The electrons are held in orbit by a series of magnets. Each time they are deflected by one of the magnets, they emit radiation with a spectrum that extends from visible light to hard x-rays. By choosing the appropriate electron energy (~1 GeV), copious soft X-rays suitable for X-ray lithography are emitted.

An alternative method for generating the X-rays is to "wiggle" the electrons while they pass down the straights in the ring between the magnets. Wiggling is accomplished with multi-pole magnets that produce a sequence of alternating magnetic fields. The bending radius can be smaller, and the X-ray energy at each deflection adds up to yield increased total X-ray output for a given electron energy. Lower energy reduces the cost of the ring.

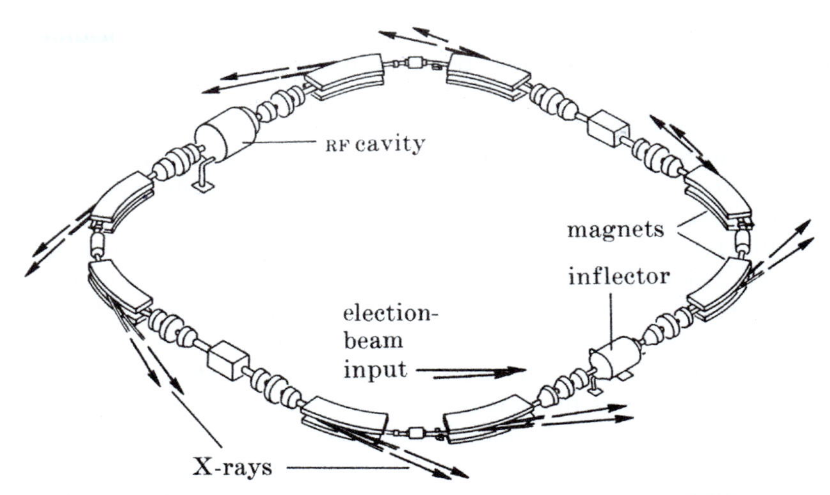

Fig. 11.13 Electron synchrotron storage ring. *(Used with permission of The Royal Society from Broers, A.N., 1988. Limits of thin film microfabrication, the Clifford Paterson lecture. Proc. Roy. Soc. Lond. A 416, 1–42, permission conveyed through Copyright Clearance Center, Inc.).*

The radiation from a storage ring is emitted in a broad sweep with a very narrow vertical spread. The beam is typically 1 cm high and 30 cm wide at a distance of about 10 m from the ring. An oscillating mirror is used to broaden it to the desired height. It was predicted that enough radiation would be emitted at each of the bending magnets, or wigglers, to expose 40–50,125 mm wafers per hour (Grobman, 1983), a rate similar to that of optical S/R cameras available in 1980, so a fully utilized ring would have produced the equivalent of 6–12 optical S/R cameras. The optical cameras cost between $500 K and $1 M, so fully loaded a storage ring costing $5–10 million would have been competitive with optics and would have provided better resolution and higher aspect ratio resist patterns. With hindsight, this all looks optimistic, but it assumed that the electron beam current in the ring would be about 1 A, which proved very difficult to achieve.

Full-scale synchrotrons were far too expensive. For example, Diamond, the UK's national synchrotron facility, cost about $1 billion in 2007. However, several laboratories were said to be working on designs for "compact" storage rings suitable for lithography that would be much lower cost and have the potential to achieve the specifications needed for production environments (Gronig, 1981; LeDuff, 1981). In the end, only Oxford Instruments in the UK built a compact storage ring that was available for commercial lithography. Other companies, such as NTT, were working on X-ray lithography and may have had their own rings. The

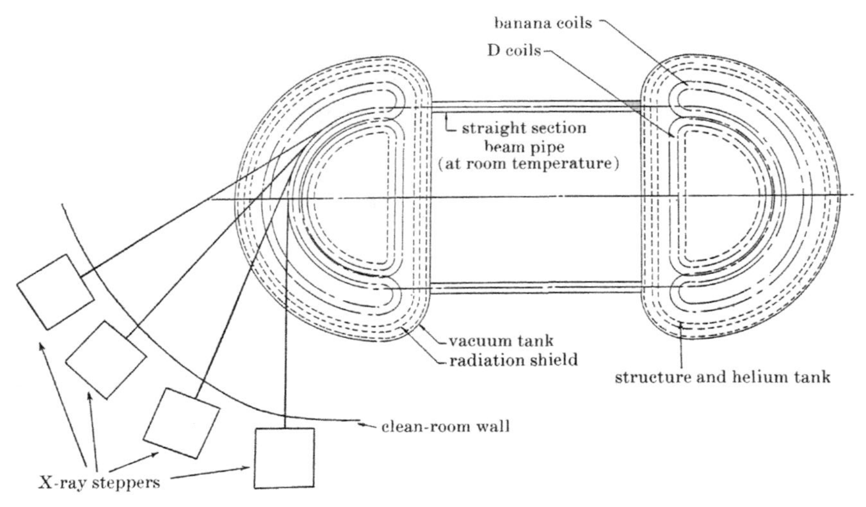

Fig. 11.14 Configuration of compact electron synchrotron storage ring for lithography. Race-track rather than circular storage ring were first proposed by Oxford Instruments. *(Used with permission of The Royal Society from Broers, A.N., 1988. Limits of thin film microfabrication, the Clifford Paterson lecture. Proc. Roy. Soc. Lond. A 416, 1–42, permission conveyed through Copyright Clearance Center, Inc.).*

Oxford Instruments ring was called Helios, and Helios 1 became the centerpiece of what was called the Advanced Lithography Facility (ALF) at IBM East Fishkill. ALF was used to explore the application of X-ray lithography for large-volume sub-half micron devices at IBM East Fishkill in the early 1990s. Fig. 11.14 shows a simplified view of Helios. The beam energy was 700 MeV, and the ring was designed to illuminate 20 exposure stations. (I will return to this later when I discuss lithography in the 1990s in the Epilogue.)

X-ray lithography with a storage ring offered high throughput and a step-and-repeat approach that was relatively tolerant to distortions of mask and wafer, but alignment accuracy would have been no better than that with optics because the alignment methods were to be optical. This was not necessarily a problem because, as already described, high accuracy could be obtained with optics. Two basic methods were tried for X-ray lithography. The first used simultaneous imaging of mask and wafer with an optical system operating either with two different wavelengths or with diffracted beams from gratings. The second used zone plates placed on the mask and wafer. Both methods were theoretically capable of achieving accuracies of less than 0.1 μm, and this was demonstrated in the laboratory but was difficult to achieve in a device process.

In addition to alignment errors, overlay accuracy depends on mask errors and mask-to-wafer distortions. Errors and distortions were relatively serious because the mask was fragile and had the same dimensions as the wafer. With the conventional sources, this was somewhat alleviated because isotropic changes in wafer and mask size could be corrected by changing the mask-wafer gap, but an adequate overlay for sub-micron devices was not demonstrated until the 1990s.

Resolution of X-ray lithography

Two factors set the resolution for x-ray proximity printing: (1) diffraction between mask and wafer, and (2) the range of the photoelectrons excited when the x-ray photons are absorbed in the resist. The same criterion used for optical proximity printing can be applied to X-rays to give the relationships shown in Fig. 11.15.

For dimensions larger than the photoelectron range, the resolution of X-ray lithography, as with UV proximity printing, is set by Fresnel diffraction between the mask and the bottom of the resist. The minimum linewidth that can be reproduced with adequate fidelity for lithography

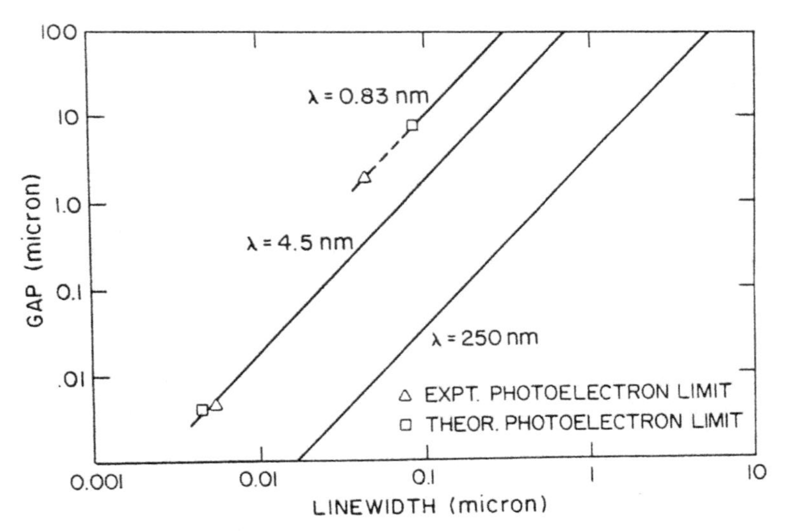

Fig. 11.15 Linewidth vs gap for deep UV and X-ray printing. Theoretical points correspond to the Gruen range for the maximum energy photoelectrons. Experimental points were measured by Feder and Spiller (Spiller and Feder, 1977). *(Reprinted with permission from Broers, A.N., 1984. Practical and fundamental aspects of lithography, Materials for Microlithography. American Chemical Society, Chapter 2. Copyright 1984 American Chemical Society).*

is said, rather optimistically, to be given by the square root of the product of the wavelength (λ) and the gap (g) between the mask and the wafer. At this dimension, the intensity at the wafer first reaches the background intensity in the absence of the mask. When the openings in the mask are about $(\lambda g)^{0.5}$, however, only a few optical "zones" of constructive and destructive interference pass through the mask, and the intensity at the center of a line varies rapidly with linewidth. This is illustrated in Fig. 11.16 which shows the intensity distribution above and below $(\lambda g)^{0.5}$ for monochromatic illumination. The situation is better with the broadband illumination produced by an electron storage ring, but processing will still be difficult. In practice, it is possible, at least down to dimensions of 0.25 μm, to keep the gap small enough (20 μm) to allow minimum linewidth to be at least $2 \times (\lambda g)^{0.5}$.

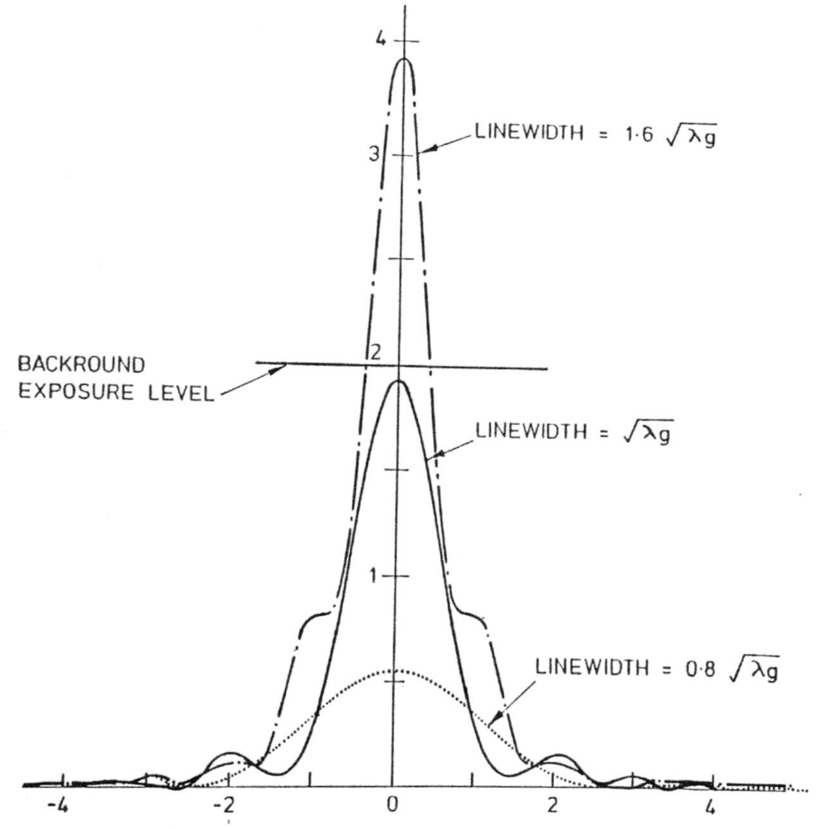

Fig. 11.16 Variation in intensity distribution for proximity printing through a slit with dimensions close to $(\lambda g)^{0.5}$, where λ is the wavelength of the radiation and g is the gap between the slit and the plane of measurement.

To reach 0.1 μm and maintain the minimum linewidth at $2 \times (\lambda g)^{0.5}$, the wavelength has to be reduced to 0.5 nm and the gap to 5 μm. These were worrying numbers. 0.5 nm X-rays are relatively hard, and the mask absorber would have to be at least 1 μm thick leading to extraordinary aspect ratios approaching 10:1. A gap of 5 μm would have seriously increased the risk of mask wafer contact with consequent yield loss. The alternative was to give up the factor of two and operate at the limit of $(\lambda g)^{0.5}$ where much of the large process window of X-ray exposure would have been lost.

Ultimately, assuming that the mask and wafer gap is so small that diffraction effects are negligible, the resolution will be determined by the range of photoelectrons. Spiller and Feder (1977) have measured this experimentally. The Gruen depth dose relationship has also been used to estimate the photoelectron range, although Gruen only confirmed that relationship over the relatively high energy range of 5–54 keV. It has been suggested (Gruen, 1957) that the minimum linewidth, as set by the photoelectron range, will be equal to the Gruen range (R_G)

$$R_G(m) = \frac{2.57 \times 10^{-11} E(\mathrm{eV})^{1.75}}{(\mathrm{g/cm^3})} \simeq 10^{-23} \lambda^{1.75}(m),$$

where E (eV) is the minimum photoelectron energy associated with $\lambda(m)$, the wavelength of the exposing X-rays, and (g/cm^3) is the density of the resist (1.2 g/cm^3 for PMMA). The Gruen range and Feder and Spiller's experimental results are shown in Fig. 11.15 As can be seen, the highest resolution predicted is about 5 nm using carbon characteristic radiation (4.5 nm). In support of this prediction, 4.5 nm X-rays have been used to shadow-print biological samples into.

PMMA with a spatial resolution of about 10 nm (Feder, 1977). In this case, only a relief profile was obtained in the resist rather than a fully developed image. The resist was metalized, and the sample was examined in a surface SEM, as shown in Fig. 11.17. Lines of about 20 nm have also been replicated with X-ray lithography using a mask that was fabricated by alternatively shadowing tungsten and carbon onto the side of a surface step (Flanders, 1979), see Fig. 11.18.

The aspect ratio of resist thickness to minimum linewidth produced by X-ray lithography can be much higher than it is with optical and electron beam methods. Fig. 11.19 shows 1 μm wide resist lines in several microns of resist.

The advantage is very large at dimensions greater than ~0.1 μm but disappears compared to electron beam lithography for dimensions below 0.1 μm (see Fig. 11.20), where degradation due to diffraction with x-rays exceeds that due to scattering of high-energy electrons (~100 keV). Aspect ratios are nonetheless satisfactory with x-rays down to linewidths of a few tens of nanometers.

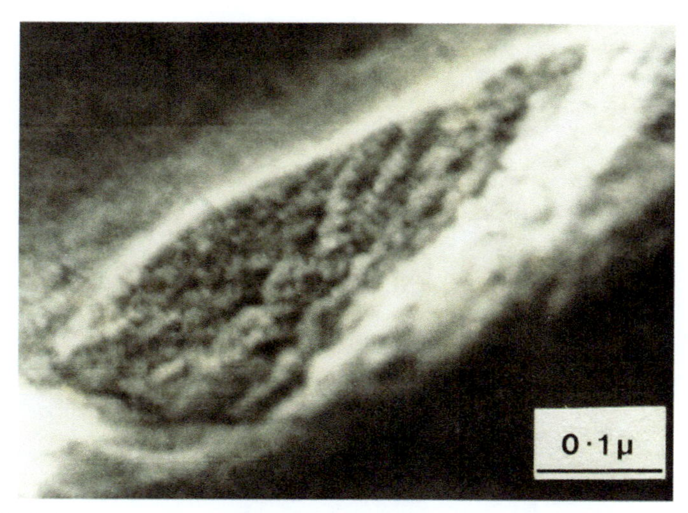

Fig. 11.17 Soft x-ray contact print of a biological sample in PMMA. After exposure the PMMA layer was developed and coated with a thin layer of gold. Details smaller than 10 nm are visible. *(Used with permission of American Association for the Advancement of Science from Feder et al. (1977) permission conveyed through Copyright Clearance Center, Inc.).*

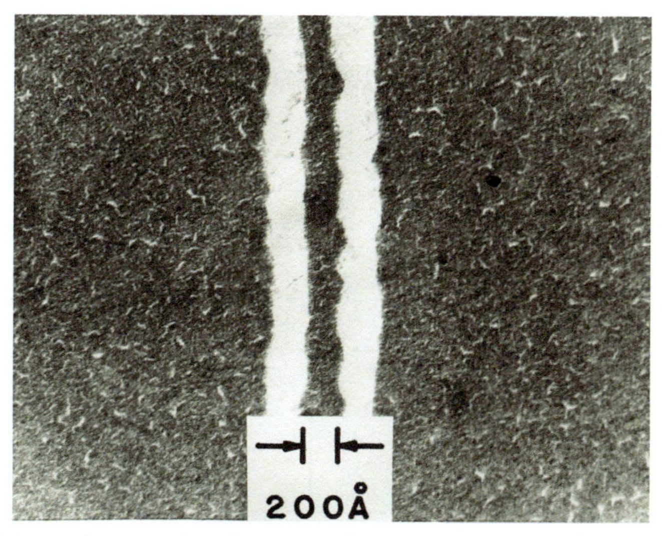

Fig. 11.18 20 nm lines produced in PMMA by replication with soft X-rays (Flanders, 1979). *(Reprinted with permission from Broers, A.N., 1984. Practical and fundamental aspects of lithography, Materials for Microlithography. American Chemical Society, Chapter 2. Copyright 1984 American Chemical Society).*

Fig. 11.19 High aspect resist structure created by X-ray lithography. Spiller et al. (Spiller and Feder, 1977). *(Reprinted with permission from Broers, A.N., 1984. Practical and fundamental aspects of lithography, Materials for Microlithography. American Chemical Society, Chapter 2. Copyright 1984 American Chemical Society).*

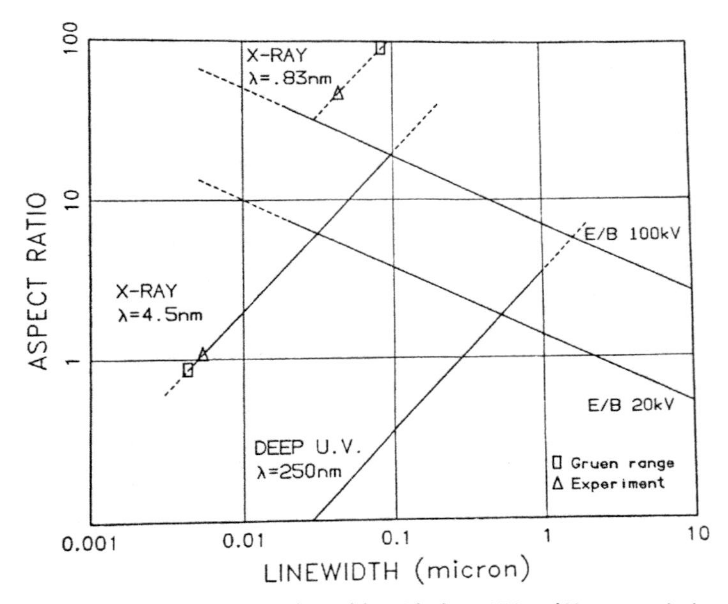

Fig. 11.20 Maximum aspect ratio achievable with deep UV and X-ray proximity printing and with electron beam exposure. With proximity printing the resist thickness is determined by Fresnel diffraction. With electron beam exposure, lateral scattering in the resist determines resolution. *(Reprinted with permission from Broers, A.N., 1984. Practical and fundamental aspects of lithography, Materials for Microlithography. American Chemical Society, Chapter 2. Copyright 1984 American Chemical Society).*

For integrated circuit applications, the significant factor to be observed from Fig. 11.15 is that a gap over 100 μm can be tolerated for 1 μm resolution. With conventional sources, penumbral blurring due to the finite source size is more likely to limit resolution.

In the 1980s, many thought that proximity printing with soft X-rays was the most promising method for large-scale production of sub-0.75 μm devices because it would produce wider process margins than optics and lower costs than scanning electron beams. But problems with mask stability and alignment and with source cost and reliability prevented its successful application. For dimension below 0.1 μm, the gap between the mask and the wafer becomes so small that the process is no longer practicable. I will return to the resolution limits of the method in the Epilogue.

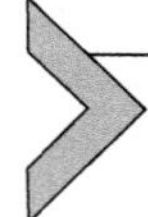

Other lithography exposure methods being worked on in 1980

More than 10 types of lithography were investigated in the early 1980s many to discuss in detail in this volume. Several are listed below with a brief comment on their state of development in 1980s, too and 1990s. They are placed in an order that reflects my views about their relative importance in 1980.

(1) Laser pattern generators were used to write the reticles for reduction S/R cameras. Their cost was slightly lower than electron beam mask writers because they did not need a vacuum system, but this was not very significant. The cost of mask generators is dominated by the costs of the precision measurement system required to meet the extreme accuracy requirements, the digital system for handling and supplying the data needed for multiple complex patterns, and the mechanisms for handling the reticles. The advantages were that the throughput could be higher than for electron beams and that they could use standard optical resists. Resolution was more than adequate for 5× masks for optical step and repeat cameras but inadequate for 1:1 X-ray masks.

(2) Ion probes were available with beam sizes of a few tens of nanometers. They could be used for exploratory device fabrication, but most importantly, to repair masks. Their writing rate for resist exposure was slower than scanning electron beam systems because the brightness of ion sources was relatively low, and the aberrations of the electrostatic lenses that had to be used were relatively high. It was also not possible to overlap deflection and focusing fields as is possible with magnetic lenses and deflection coils, so the size of the scan field was smaller. Compared to electron exposure, ions produce little or no proximity effect, but this was not a particularly significant advantage because the proximity effect with electrons can be brought within acceptable limits by operating at high enough accelerating voltages.

(3) Both 1:1 and reduction electron and ion projection systems were built in the 1970s and 1980s because they potentially offered higher throughput than the scanning systems. Some had stencil masks, and some electron beam projectors used photocathodes instead of masks. The masks had stability problems, and it proved very difficult to reduce pattern distortion distortions to an acceptable level for sub-micron dimensions. Most of the projects were abandoned by 1990.

(4) New ideas for electron beam projection systems, however, appeared again in the late 1990s and culminated in two large projects that addressed the need for sub-100 nm devices. The first at Lucent Technologies' Bell Labs was called SCALPEL (Herriott, 1999), and the second at IBM was called PREVAIL (Dhaliwal et al., 2001). Both operated at 4× demagnification and projected a full chip pattern onto the wafer.

SCALPEL is an acronym for SCattering with Angular Limitation Projection Electron Lithography which relates to the key invention that was thought to make this new form of e-beam projection feasible. The reticle was not an open stencil that required two masks if all shapes were to be allowed, but a two-layer mask with a contiguous thin layer of a low molecular weight material and a patterned layer of a high molecular weight metal. The metal layer scattered the electrons more strongly into a wider cone than the low molecular weight material, and a contrast aperture the size of the unscattered beam intercepted the scattered electrons and provided the intensity contrast in the projected image. PREVAIL also used as a mask with a thin substrate but had supporting struts to improve its dimensional stability and to dissipate the heat produced by the illuminating electron beam.

The 4× mask in PREVAIL contained a full chip image that was divided into 1 mm squares. These squares were illuminated sequentially by a 20–50 μA square electron beam. The uniform beam was created by a special large-area cathode electron gun and directed to each 1 mm area by a variable axis lens/deflection system. Below the mask was a 4× reduction projection system that projected the mask image onto the wafer. This consisted of an amazingly complex series of variable axis lenses combined with beam deflectors and dynamic field correctors that canceled the off-axis distortions in the focusing field. The diameter of the electron optical column was much larger than that of previous electron probe systems.

Despite the excellence of the variable axis lens systems, which were a tour de force in electron optical design similar in magnitude to the correction systems for spherical aberration in electron microscopes, the resolution and accuracy of the electron beam projectors were marginal for 0.1 μm dimensions. More importantly the throughput was lower than that of the optical projectors that operated at deep UV and extreme UV wavelengths that I now discuss in (5).

(5) By the 1980s, considerable progress had been made with multi-layer X-ray mirror lenses in many laboratories around the world. These were used in microscopes and telescopes, and up to 60% reflection efficiency was achieved. It became clear that it was possible to build a reduction imaging system using these mirrors with adequate quality for lithography. and that projection at soft X-ray wavelengths might eventually become a contender for volume lithography of 100 nm devices. The challenge of building a large field, diffraction-limited X-ray lens was daunting, however, as was the development of a high-output soft X-ray source. There was also the problem of finding photoresists that were both sensitive enough at these short wavelengths and had sub-tenth micron resolution.

I put soft-X-ray projection systems at the end of my list in 1984. This was a mistake, but at least I included the technology that was to be the winner in the race to find a lithography for the manufacture of devices with sub-tenth-micron dimensions. It is perhaps fortunate that 13.5 nm is the wavelength of choice for the radiation as it falls just outside the commonly accepted range of X-ray wavelengths of 10 picometers to 10 nm, so this form of lithography is called extreme UV lithography and will not be confused with X-ray proximity printing that is frequently referred as X-ray lithography.

The problem of finding resists for these wavelengths was solved by Grant Willson, Jean Frechet, and Hiroshi Ito, who used a catalyst to amplify the sensitivity of the photoresist (Willson, 1983). It was not obvious that amplified resists would have high resolution, but they did, and this advance proved to be of immense value. We measured the resolution of one of these resists using the high-resolution electron probe and found that in the negative mode, it had the same resolution as PMMA (Umbach et al., 1988).

The future

I will discuss the further evolution of lithography systems in the Epilogue. Dimensions have now converged upon the resolution limits set by the molecular nature of conventional resist processes that we discovered 60 years ago, and attention is turning to resistless processes and even to the elimination of masks. However, it seems unlikely that masks will be eliminated as they are the most efficient way to store the trillions of bytes of data needed to describe modern chips, and the data can be delivered in a fraction of a second. These data volumes are now so high that the high throughput variable-shaped electron beam systems developed in an attempt to meet mainstream lithography throughputs can barely expose the pattern for a reticle in an acceptable time. It takes more than 10 h to expose a single reticle. Systems using hundreds of thousands of electron beams have now been developed to try to reduce this time.

References

Alles, D.S., Ashley, F.R., Collier, R.J., Gere, E.A., Herriott, D.R., Johnson, A.M., Thomson, 1975. Supplement to International Electron Devices Meeting. IEEE, New York, p. 1.

Alles, D.S., et al., 1987. EBES4: a new electron-beam exposure system. J. Vac. Sci. Technol. B 5 (1), 47–50.

Blodgett, A.J., 1983. Microelectronic packaging. Sci. Am. 27, 86.

Blodgett, A.J., Barbour, D.R., 1982. IBM J. Res. Dev. 26, 30.

Chang, T.H.P., et al., 1974. In: Bakinh, L. (Ed.), Proc. of the 7th Internat. Conf. on Electron & Ion Beam Technology. Electrochemical Society Inc., Princeton, NJ, p. 97.

Dey, J., et al., 1980. In: Kramer, R.P. (Ed.), Microcircuit Engineering 80. Delft University Press, Holland, p. 211.

Dhaliwal, R.S., Enichen, W.A., Golladay, S.D., Gordon, M.S., Kendall, R.A., Lieberman, J.E., Pfeiffer, H.C., Pinckney, D.J., Robinson, C.F., Rochrohn, J.D., Stickel, W., Tressler, E.V., 2001. PREVAIL—electron projection technology approach for next-generation lithography. IBM J. Res. Dev. 45 (5), 615–638.

Dubreoueq, G., et al., 1980. In: Kramer, R.P. (Ed.), Microcircuit Engineering 80. Delft University Press, Holland, p. 181.

Eidson, J.C., et al., 1981. J. Vac. Sci. Technol. 19, 932.

Feder, R., Spiller, E., Topalian, J., Broers, A.N., Gudat, W., Panessa, B.J., Zadunaiski, J.Z., Seedat, J., 1977. High resolution soft X-ray microscopy. Science 197, 259.

Flanders, D.C., 1979. X-ray lithography at ∠100 Å linewidths using x-ray masks fabricated by shadowing techniques. J. Vac. Sci. Technol. 16, 1615.

Goto, E., et al., 1978. J. Vac. Sci. Technol. 15, 883.

Greed, J.J., Markle, D.A., 1982. SPIE Conference, Santa Clara, California.

Grobman, W.D., 1980. IEDM DIgest. IEEE NY, p. 415.

Grobman, W.D., 1983. J. Vac. Sci. Technol. B 1, 1300.

Gronig, E., 1981. Microcircuit Engineering 81. Swiss Federal Instit. Technol., Lausanne, Switzerland, p. 122.

Gruen, A.E., 1957. Naturforsch 12a, 89–95.

Herriott, L.R., 1999. SCALPEL–projection electron beam technology. In: Proceedings of the 1999 Particle Accelerator Conference, New York, NY, pp. 595–599.

Herriott, D.R., Collier, R.J., Alles, D.S., Stafford, J.W., 1975. EBES: a practical electron beam exposure system. IEEE Trans. Electron Dev. ED-22 (7), 385–392.

Lauria, J., et al., 1980. In: Kramer, R.P. (Ed.), Microcircuit Engineering 80. Delft University Press, Holland, p. 171.

LeDuff, J., 1981. Microcircuit Engineering 81. Swiss Federal Instit. Technol., Lausanne, Switzerland, p. 130.

Markel, D.A., 1974. Solid State Technol. 17, 1977.

Mayer, H.E., et al., 1980. In: Kramer, R.P. (Ed.), Microcircuit Engineering 80. Delft University Press, Holland, p. 191.
Moore, R.D., Caccoma, G.A., Pfeiffer, H.C., Weber, E.V., Woodard, O.C., 1981. J. Vac. Sci. Technol. 19, 950.
Offner, A., 1979. Photogr. Sci. Eng. 23, 50.
Pearce-Percy, H.T., Spicer, D.F., Abbot, M., Winborn, C., Varnell, G.L., 1979. High speed electron optics for direct slice writing. J. Vac. Sci. Technol. 16, 1794–1799.
Pfeiffer, H.C., 1975. New imaging and deflection concept for probe-forming microfabrication systems. J. Vac. Sci. Technol. 12, 1170.
Pfeiffer, H.C., 1978. Variable spot shaping for electron beam lithography. J. Vac. Sci. Technol. 15, 887–890.
Smith, H.I., Spears, D.L., Bernacki, S.E., 1973. X-ray lithography: a complementary technique to electron beam lithography. J. Vac. Sci. Technol. 10 (6), 913–917.
Spears, D.L., Smith, H.I., 1972a. High-resolution pattern replication using soft x-rays. Electron. Lett. 8, 102–104.
Spiller, E., Eastman, D.E., Ffeder, R., Grobman, W.D., Gudat, W., Toopalian, 1976. Application of synchrotron radiation to x-ray lithography. J. Appl. Phys. 47, 5450.
Spiller, E., Feder, 1977. In: Queisser, H.J. (Ed.), X-ray Lithography, X-ray Optics. Springer Verlag, Berlin.
Taylor, G.N., Wolf, T.N., 1981. Microcircuit Engineering 81. Swiss Federal Instit. Technol., Lausanne, p. 381.
Thompson, M.G.R., et al., 1978. J. Vac. Sci. Technol. 15, 891.
Tibbets, R.E., Wilczynski, J.S., 1980. In: Fischer, R.E. (Ed.), Proc. 1980 International Lens Design Conference, SPIE. vol. 237. Society of Photo-Optical Instrumentation Engineers, Bellingham, Washington DC, p. 321.
Umbach, C.P., Broers, A.N., Willson, C.G., Koch, R., Laibowitz, R.B., 1988. Nanolithography with an acid catalyzed resist. J. Vac. Sci. Technol. B6 (1), 319–322.
Wilczynski, J.S., 1979. J. Vac. Sci. Technol. 16, 1929.
Willson, C.G., 1983. Introduction to microlithography. In: Thompson, L.F., Willson, C.G., Bowden, B.J. (Eds.), Am. Chem. Soc. Symp. Ser. vol. 219. American Chemical Society, Washington DC, p. 87.
Wittekoek, S., 1980. In: Kramer, R.P. (Ed.), Microcircuit Engineering 80. Delft University Press, Holland, p. 155.

Further reading

Spears, D.L., Smith, H.I., 1972b. Solid State Technol. 15, 21.

Nanolithography at 350 kV

Contents

By 1984 I had become responsible for what was called "Advanced Development" in IBM's East Fishkill laboratory and pressure was building for a major drive for leadership in the technologies needed for sub-micron devices. If I stayed in E Fishkill, I would have had to abandon my research, which by then was already restricted to weekends and was placing a strain on my family life, and I decided I was not prepared to do this. I reasoned that it would not be possible to return to experimental research if I stepped away for too long, whereas I would probably be able to return to management. I also decided that it would be better to move to academia because my research on the limits of microfabrication was not of immediate interest to industry. So, I decided to leave IBM and find an academic position.

When I informed IBM of this decision, they offered me a position on the Corporate Technical Committee in the company's headquarters in

Advances in Imaging and Electron Physics, Volume 231
ISSN 1076-5670
https://doi.org/10.1016/B978-0-443-31462-9.00012-5

Armonk. They thought that I needed time to think about whether I really wanted to do this, and if I did, to give me time to find a suitable academic position. I had in fact been considering such a move for about a year since Professor Jan Le Poole had visited me just before he retired in 1962 to suggest I apply for the chair that he occupied with such distinction at Delft University. I was more than honored that he considered me a worthy candidate, but pointed out that, while my ancestors were from the Netherlands, I did not speak Dutch, and this would make it difficult for me. He said it would not matter as most people in Holland spoke English. I still thought it would be difficult, especially with the technical staff. Nonetheless, Delft University approached me, and I was tempted as Le Poole's group included several world-leading researchers, and the experimental facilities were outstanding. Finally, I decided it would be too much of a change, although financially, it would have been easier than a move to Cambridge because professors in Delft were paid twice as much as Cambridge professors in the 1980s.

All of this came to a head early in 1984 when, by coincidence, the Professorship of Electrical Engineering in Cambridge became vacant. This was the chair that Sir Charles Oatley had occupied, and members of the University's selection committee suggested that I apply. I did and was short-listed. The next step was to go to Cambridge for an interview. I booked my flight to London but was still so uncertain about the move that I hesitated on entering the terminal at Kennedy Airport and returned to the car just as Mary was about to drive back to our home in Westchester. I told her I had changed my mind and wasn't going. She told me not to be so stupid and get on the plane. I had the interview, and the committee offered me the professorship.

IBM made the decision easier because, as well as asking me to be a member of the Company's Science Advisory Committee, they generously allowed me to return whenever I wished to my laboratory in the Thomas J Watson Research Laboratory, where I continued to work with Cory Umbach. They even said I could return to my position at IBM any time in the next 5 years if I wished. As an IBM Fellow, I could go to Cambridge for a couple of years without leaving IBM. I did not accept this offer remembering my father's advice that if I went to another country to work, I should assume that I was going forever so that I and others would see that I was committed to the move. This was wise as it proved more difficult than I had predicted to find the resources to build my laboratory in Cambridge. If it had been easy to return, I might just have given up and returned to IBM. In the end, it worked out very well in Cambridge and I was able to acquire a new JEOL 400 kV electron microscope and build a Class 100 clean room to house the equipment needed to continue my exploration of nanofabrication.

Being a member of IBM's science advisory committee had the huge benefit that I was also able to keep in touch with IBM and with the semiconductor industry and participate in the discussions in the 1980s that led to the decision in 1990 to build what was known as the Advanced Semiconductor Technology Center (ASTC) in East Fishkill. The ASTC included the Advanced Lithography Facility (ALF), where the HELIOS electron synchrotron X-ray source was to be used for X-ray lithography. My close links with IBM continued until 1993, when IBM's SAC was dissolved after the company suffered a severe financial downturn. The company lost almost US$16 billion between 1991 and 1993, which was said to be the largest ever for an American company. By this time, I was deeply embedded in Cambridge, mainly with my research, although my time as a researcher was to end in another 3 years.

Teaching and research at Cambridge 1984–1996

I returned to Cambridge in September of 1984 to take up the professorship of Electrical Engineering and to be a Fellow of Trinity College. There was a temporary difficulty with my Fellowship of Trinity College because there was an agreement that professors should be fairly distributed between the 30 colleges of the University regardless of their wealth or size, a bit like the Electoral College in the USA, and Trinity was over its quota. The rules for this agreement are buried in the University's statutes where it is also stated that this statute "... shall not debar ... the Regius Professor of Greek from becoming a Professorial Fellow of Trinity College".

When the Master of Trinity College, Nobel Laureate Sir Alan Hodgkin, wrote to me in 1983 offering me a Fellowship, Trinity College was within its quota; however, before I arrived, Professor Eric Handley was elected to the Regius Professorship of Greek and took up his right to become a Fellow. This put the college over its quota preventing them from offering me a fellowship. The College was very generous and allowed me to have the privileges of a fellow until the quota cleared, so this had no impact on my return. The college also allowed Mary and me to use the upper floor of a house they owned in Cambridge while we bought our own house. The ground floor of this house was occupied by Anne Barton, the renowned American-English scholar and Shakespearian critic, who was also waiting for the quota to clear so that she could become a Fellow. All of this brought home to me that I was now back in the rich cultural world of Cambridge, which I thoroughly enjoyed. Ann was a charming and friendly person who grew up in Westchester County, near where we lived for 20 years. She drove a splendid antique Alvis car that she drove furiously with all the windows open.

On arrival in Cambridge, I immediately became deeply involved in teaching and research in the Engineering Department. Teaching was very challenging for me, as I had never done anything like it before. I had given many lectures at technical conferences and taught short courses to industrialists on semiconductor lithography and scanning electron microscopy, but the latter were informal lectures to small groups who could ask you questions if they did not follow what you were trying to teach them. Most importantly, they did not have to sit formal examinations to see if you had actually taught them anything.

To ensure that I learned about examinations, I was asked to be chairman of first-year examinations and was given the task of setting the questions for the electrical systems and devices examination. I thought that setting exam questions would be an intriguing and rewarding experience that would take me a few days and that I would probably enjoy it. After several days of struggling to think of suitable questions, only to find that they were too similar to questions already asked in previous years, I changed my mind. It was one of the most difficult things I had ever had to do. There was a rule that those who lectured to the first-year students were not allowed to set the exam questions relating to their lectures, but they did have the chance to comment on the suitability of the questions that others had set. Fortunately, my colleagues were amazingly helpful and tolerant in their comments about my questions, although at times, I sensed that they found it difficult to contain their amusement as they watched this industrialist struggling to complete what were regarded as elementary academic tasks.

It took 4 or 5 years before I settled into the tasks of lecturing and examining. I taught the first-year lectures in electronic devices and circuits and courses to the final year on Digital Circuits and Devices. The Engineering Faculty had led the University in introducing student assessment of lecturers long before I arrived, and I was interested to see that the popularity of my lectures declined as the years passed, but the exam performance of the students, if anything, improved. I think in the early days; the students were more entertained by watching me learn to be an academic than they were in learning themselves.

In pulling together a research group, I was on more familiar ground, and this was relatively easy in Cambridge because the students and staff were outstandingly talented. I did not want a large group and wished to remain in the laboratory myself, where I could continue to explore and apply the highest-resolution electron beam nano-fabrication techniques. The Head of the Department, Professor Jacques Hyman, was very helpful in finding space for the group to build a state-of-the-art clean room in the basement of the main building, and there was strong technician support within the Electrical Division and a well-equipped mechanical workshop.

JEOL 4000 EX electron microscope

There was little money to purchase the equipment we needed, but within a year, we managed to gain a grant of £400k (equivalent to about $1.5 million in 2024 US$)) from the Research Councils that was sufficient to pay for the clean-room and purchase a 4000 EX high-resolution transmission electron microscope from the leading Japanese company JEOL. This state-of-the-art microscope became the centerpiece of the laboratory (Broers, 1989). JEOL significantly discounted its price, which was an immense help. It was a new experience for me using equipment that I had not designed myself and I was concerned that keeping it operating at peak performance might be difficult, but this was not the case. The microscope was reliable and proved easy to modify to be used for nano-fabrication studies (Figs. 12.1–12.4). We operated the microscope in three modes: the conventional projection imaging mode, a scanning transmission imaging mode, a secondary electron surface scanning mode, and a pattern writing mode.

Roger Koch, who I had worked with in the IBM research laboratory, came to Cambridge and installed the same type of pattern generator we had used in the Yorktown laboratory. This was an IBM PC pattern generator that produced patterns containing rectangles, circles and triangles that he had designed. The pattern data were entered in a simple format, and there was no need for post-processing. The beam was deflected with double deflection coils located between the third condenser lens and the objective lens. The 4000 EX is shown with the pattern generator in Fig. 12.5.

Fig. 12.1 Mrs. Timbs, wife of Arthur Timbs, chief design engineer at the Cambridge University Engineering department who designed the vibration isolation system for the JEOL 4000 EX microscope.

Fig. 12.2 Mounting frame and air isolation units for the 4000EX vibration isolation system.

Fig. 12.3 Finished vibration isolation system for the 4000EX microscope.

The microscope was not equipped with a blanking system, so one was designed and built by Andrew Hoole, one of my PhD students. This was a complicated and difficult task because of the limited space within the column, the multiple operating modes of the microscope, and the high energy of the electrons. To ensure adequate speed, electrostatic deflection was chosen. This required long and closely spaced electrodes and a relatively high blanking voltage. After debating where to locate the plates, Andrew decided that the only position large enough was below the deflection coils. The final unit had 29 mm long plates spaced by 2 mm and a blanking voltage of 400 V. Blanking was so fast that the movement of the beam during blanking was not a problem. The unit operated very well.

The theoretical resolution of the 4000 EX operating in TEM mode was 0.2 nm but this could only be obtained if no mechanical vibration or

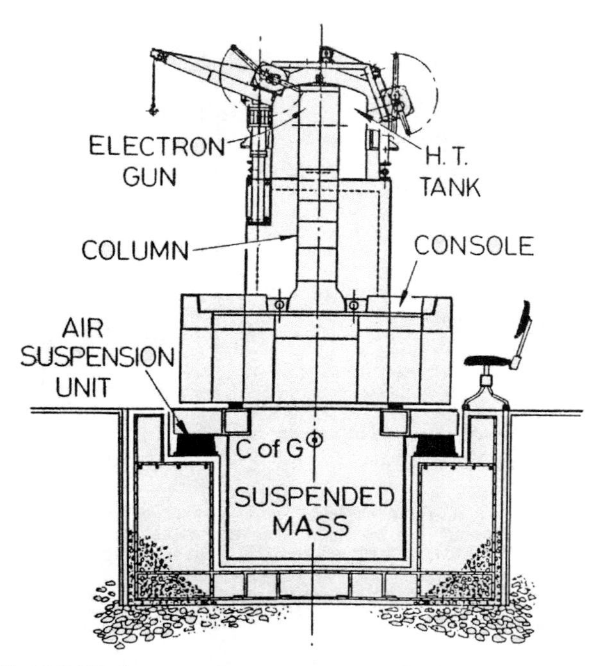

Fig. 12.4 JEOL 4000EX electron microscope on its vibration isolation system.

electromagnetic noise interfered with its operation. To eliminate mechanical vibration, Arthur Timbs, the Engineering Department's Chief Design Engineer, designed the servo-controlled air suspension system shown in Figs. 12.1–12.4. As the microscope was located in the basement of the building, it was possible to excavate a hole beneath it that was large enough to contain a mass equal to that of the microscope itself so that the center of gravity of the total system was level with the top of the suspended frame. This reduced any tendency of the microscope to rock when perturbed. The total suspended mass was 3500Kg, and the suspension system reduced the sensitivity to external vibrations by more than a factor of 10. Without it, beam deflections of 1 to 1.5 nm were observed when vehicles passed up and down the road that was 30m from the building. With the system active, no deflections were observed, provided the room was quiet. Someone talking loudly near the microscope, however, would produce deflections approaching a nanometer. Beam deflection was measured by observing a highly magnified (1–2 million times) image of the beam on the TEM screen at the bottom of the microscope column. Electromagnetic interference was reduced by fitting one of Spicer Consulting's Magnetic Field Canceling Systems. I knew Dennis Spicer well. He had worked for Texas Instrument on electron beam lithography and had immense experience in creating electro-magnetically quiet

Fig. 12.5 JEOL 4000EX electron microscope and IBM PC pattern generator.

spaces. The canceling system consisted of orthogonal Helmholtz coils placed on the walls, ceiling, and floor surrounding the microscope and a sophisticated digital control unit that adjusted the magnitude and phase of the a.c. and d.c. currents flowing in the coils. As with mechanical vibration, the effect of electromagnetic interference was monitored by observing the highly magnified image of the beam on the TEM screen. Figs 12.6 and 12.7 show the two highly magnified images of the beam, one before and one after correction. By turning on a "gun" soldering iron near the column, it was possible to generate Lissajous figures as the stray magnetic field and the field from the soldering iron as the fields were superimposed. This made it easy to verify that the frequency of the correcting field was correct (Fig. 12.8).

The probe system consisted of the three TEM condenser lenses and the upper half of the objective lens. For beam writing, these lenses were adjusted so that the diameter of the Gaussian image of the gun cross-over

Fig. 12.6 Magnified image of electron beam on final screen of the 4000 EX TEM with a.c. magnetic field interference (Broers et al., 1989). *(Reprinted with permission from Elsevier).*

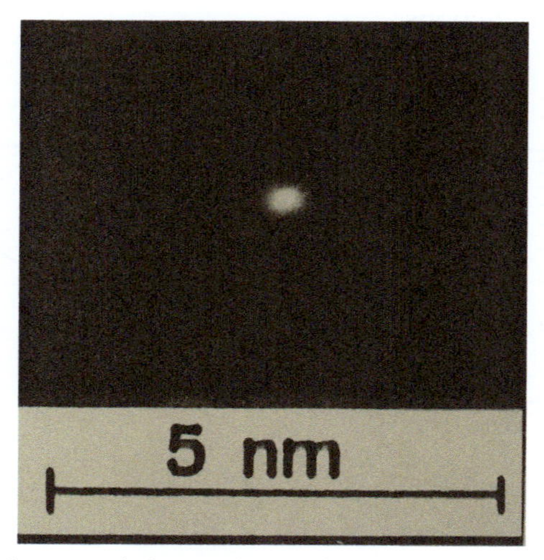

Fig. 12.7 Magnified image of electron beam after cancellation of the magnetic field with the Spicer correction coils (Broers et al., 1989). *(Reprinted with permission from Elsevier).*

at the sample was about 0.2 nm. The diameter of a physical aperture in the gap of the third condenser lens and the focal length of the lens were chosen so that the final lens was operating at the optimum aperture (α_{OPT}) for a given beam current. Adjustments for different beam currents were conveniently made by changing the excitation of the third condenser lens.

The upper half of the objective lens acts as the final probe-forming lens. The pole-piece dimensions were not ideal for this purpose because the lens was designed for transmission microscopy, with the lower half of the lens field acting as the objective lens. Nonetheless, the aberration coefficients (C_s=2.6 mm, C_c=2.8 mm), as modeled from the pole piece geometry using Eric Munro's computer programs, were still low enough to produce a beam diameter of 0.5 nm for a current of 10^{-12} Amp (Fig. 12.9).

To calculate the beam current versus beam diameter (defined here as the diameter containing 80% of the beam current), it was also necessary to know the brightness of the beam and the energy spread. The brightness was measured at the sample at about 2.10^7 Amp/cm^2.ster. at 350kV, and the energy spread was assumed to be less than 3 eV based on previous measurements of the energy spread of LaB_6 cathode electron guns operating at similar total currents. The relationship between the beam diameter and the beam current calculated using these data is shown in Fig. 12.8. Experimental measurements for small beam currents are also shown in Fig. 12.8. They were obtained from images similar to those in Fig. 12.7. The beam diameter was close to the theoretical predictions.

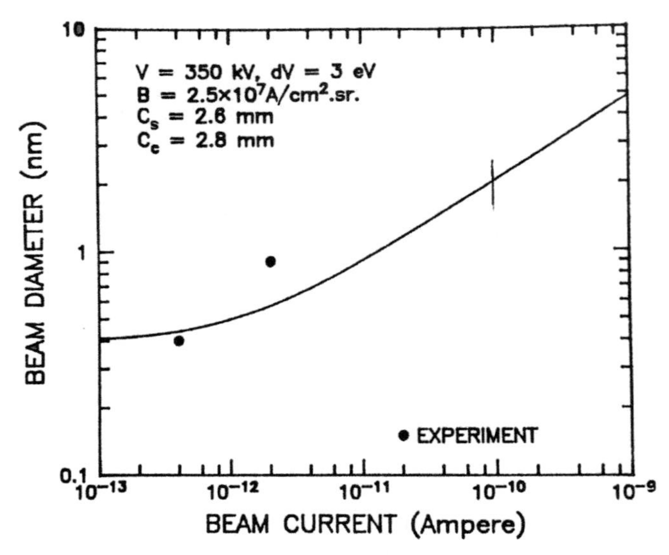

Fig. 12.8 Beam Current vs beam diameter for 4000 EX (Broers, 1989). *(Reprinted with permission from Elsevier.).*

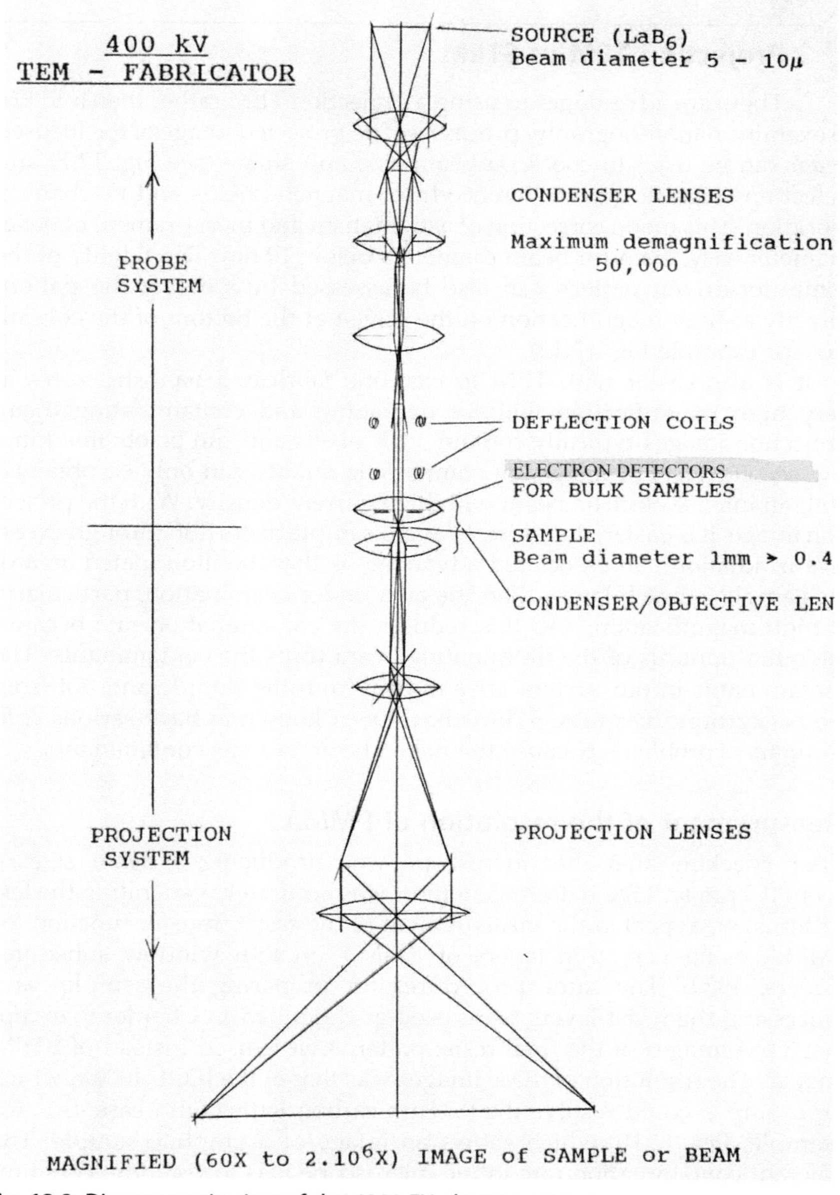

Fig. 12.9 Diagrammatic view of the 4000 EX electron microscope.

Projection TEM vs STEM

There are advantages to using a projection TEM rather than a STEM to examine nanolithography processes. The projected image of the focused beam can be used to check on beam size and shape (see Fig. 12.7) and detect any significant interference from magnetic fields and mechanical vibration. This made correction of astigmatism and measurement of beam diameter easy, even for beam diameters below 10 nm. The fidelity of the computer-driven pattern can also be assessed by viewing the pattern directly at high magnification on the screen at the bottom of the column (see for example Fig. 12.12).

It is also easier with TEM to examine fabricated nanostructures at very high magnification without damaging and contaminating them. Projection images typically contain $>10^7$ pixels and can be obtained in a few seconds. STEM images of comparable quality can only be obtained with an intense electron beam with high current density. With the projection image, it is easier, therefore, to avoid sample distortion through excessive irradiation. An associated advantage is that the illuminated area of the sample is much larger than the area under examination, particularly at high magnification, and this reduces the contamination rate because the outer annulus of the illuminating beam traps the contaminants. The contaminants in our system arise mostly from the sample and not from the background vacuum. STEMs have been known to have serious contamination problems because the naked beam attracts contaminants.

Measurement of the resolution of PMMA

After checking that the microscope was producing a beam size of about 0.5 nm and the pattern generator was accurately generating the test patterns, we repeated the measurement of the resist transfer function for PMMA using very thin layers of PMMA on thin window substrates (Broers, 1989). The same procedures for preparing the samples and processing the resist layers were used as described in Chapter 9, except that TEM images of the final resist patterns were used instead of STEM images. The resolution of these images was that of the JEOL 4000X, which for example, could resolve the 0.34 nm carbon lattice with ease. See, for example, Fig. 12.10, which shows an image of a graphite sample. The inherent contamination rate in the microscope was low and no contamination was visible when operating in the TEM mode.

Figures 12.11 and 12.12A show part of the test pattern used to evaluate the resolution of PMMA. Fig. 12.12B was obtained with no sample present but with the beam focused exactly where the sample's surface would be placed. The highly magnified image of the pattern was recorded on the TEM screen at the bottom of the column. After checking that this pattern

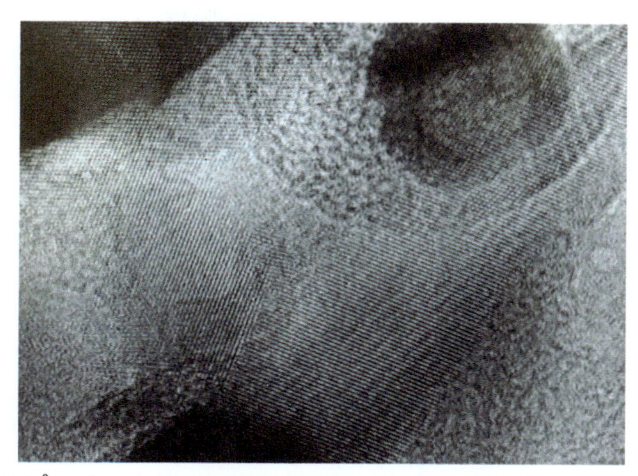

Fig. 12.10 3.4 Å carbon lattice image obtained with the 4000 EX.

was accurately focused by examining the image at high magnification and checking that the 3.2 nm line was precisely and clearly defined, the sample was introduced, and the pattern was written in the PMMA resist layer. The 350kV electron beam had a diameter below 1 nm.

Figures 12.11 and 12.12A are TEM images of a developed and shadowed PMMA resist layer supported on a 100nm thick silicon nitride membrane. The resist was shadowed at 40° with 3 nm of PtPd to enhance contrast and to indicate the resist thickness. The detail in this film is clearly visible in the micrographs. Fig. 12.11 enlarges a portion of Fig. 12.12A. The size of the pattern was chosen so that the minimum 'written' linewidth (3.2 nm) was about four times smaller than the half-width of the fundamental exposure distribution, and the largest linewidth was about 10 times greater than the half-width. For each experiment, the pattern was repeated at 7 different exposure doses. The lightest dose was below what was needed to open up the largest shapes. The heaviest dose was high enough to ensure that the resist developed through to the substrate in the site of the narrowest lines. The resist patterns shown in Figs 12.11 and 12.12A were exposed at a dose slightly greater than that needed to open up the largest shapes.

As mentioned before, this method for measuring the resist contrast function avoids the need to measure the actual width of the resist lines. Only the written linewidths and the exposure dose needed for each line to develop through to the substrate are needed. This was important when the developed lines were examined using STEM operation because the high instantaneous intensity of the focused beam could damage the resist pattern. This does not occur in TEM mode, and it is, therefore, possible to measure the width of the developed lines precisely. The increase in width is because there is, inevitably, some lateral development of the unexposed resist during development.

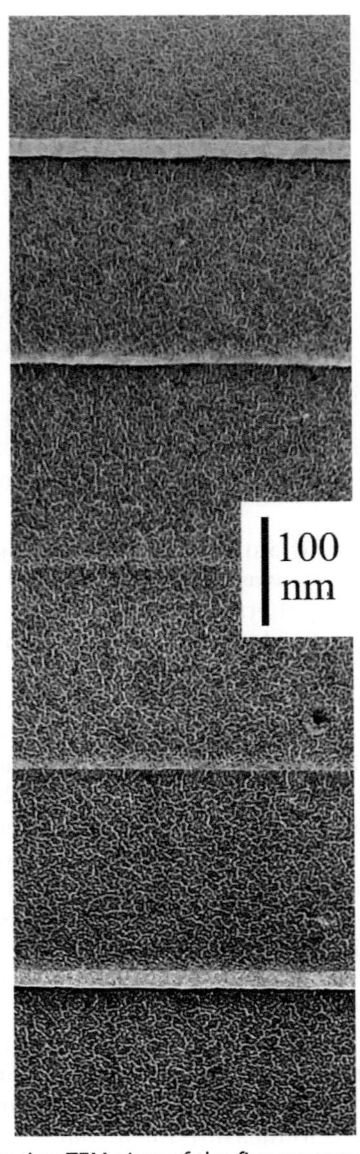

Fig. 12.11 High magnification TEM view of the five narrowest lines in Fig. 12.12

Fig. 12.13 shows the fraction of the large area dose required to expose the resist for linewidths below the width of the exposure distribution for 350 kV exposure These data suggest an exposure distribution with a width of about 15 nm and, therefore a minimum useful linewidth for isolated lines of about 100 Å, and a minimum pitch of about 250 Å nm for closely spaced lines.

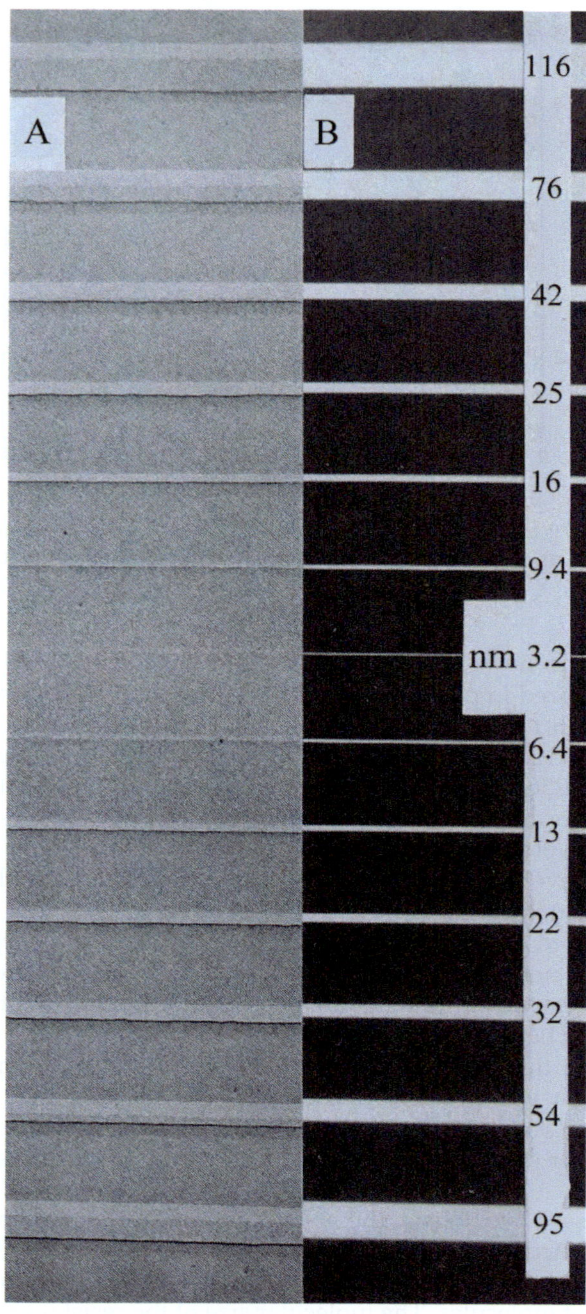

Fig. 12.12 Test pattern used for measuring the resist contrast function for PMMA. (A) Developed resist pattern. (B) Test pattern as recorded at high magnification on the screen at the bottom of the microscope column.

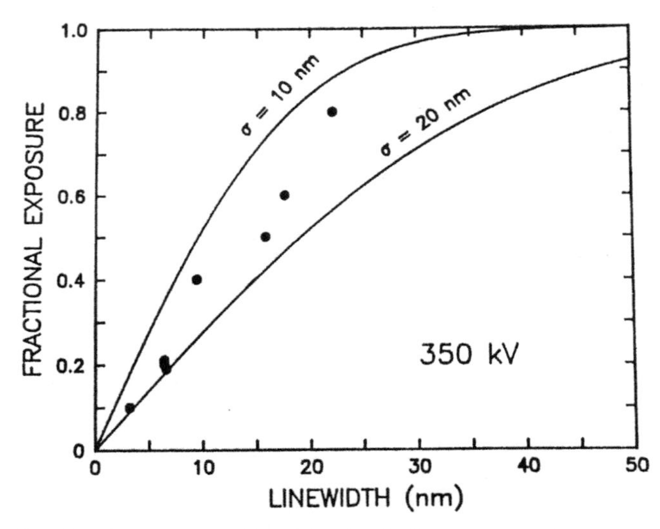

Fig. 12.13 Fractional exposure vs linewidth for PMMA exposed at an energy of 350keV (Broers, 1989). Dots are experimentsl points measured using the test pattern shown in Fig. 12.12.

To examine how well the resist contrast function predicted what would be realized in practice, we simulated the case of two parallel lines of equal width for spacings of 200Å, 250Å, and 300Å. The results are shown in Fig. 12.14. The simulation predicts that the minimum spacing will be about 250Å, which agrees with what is found in practice.

A simple examination of Figs 12.11 and 12.12A suggests that these predictions were correct. At an exposure slightly more than that required to develop through to the substrate in large areas, the nominal 32Å line had not developed through to the substrate. It was just a shallow v-groove in the resist. The nominal 64Å line was a deeper groove but had still not developed through to the substrate. The 94Å line had just developed through to the substrate but had a width that was about 100Å. The 130Å line was a cleanly developed line with parallel sides but had developed out to a width of about 150Å. These results indicate that it was possible to produce isolated lines with a width of about 100Å in 300Å thick resist, but the exposure dose had to be exactly correct. Higher doses produced wider lines, and lower doses discontinuous lines.

As discussed previously, the resolution limit does not depend on the size of the electron beam, which in this case was 20 times smaller than the exposure width, but because of the interaction range of the electrons with the resist molecules. The molecular size of the resist does not appear to play a role either, although these measurements were made with

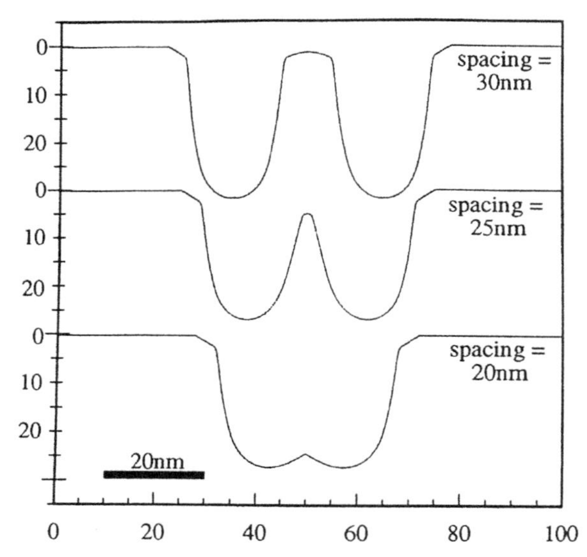

Fig. 12.14 Simulation of PMMA profiles after development of two parallel lines with spacings of 20 n, 25 nm, and 3 nm using the Gaussian sigma of 15 nm for the resist contrast function. Simulation suggests that the minimum spacing will be about 25 nm, which agrees with experimental evidence. *(Used with permission of Elsevier Science and Technology Journals, from Broers A.N., Hoole, A.C.F., Ryan, J.M. 1996. Electron beam lithography—Resolution limits, Microelectron. Eng. 32, 131–142 permission conveyed through Copyright Clearance Center, Inc.).*

PMMA with molecular weights of 20,000 to 500,00, and it has been suggested that the resolution might improve with much lower molecular weight, although I am not aware that this had ever been demonstrated. The following section throws some light limitations set by the molecular weight.

Lift-off patterning with PMMA

Figs 12.15 and 12.16 show metal patterns produced by evaporating about 2 nm of PtPd onto developed PMMA patterns and then lifting off the metal on top of the resist by dissolving the resist with acetone (Hatzakis, 1967). The metal produces a "shadow" image, which provides higher contrast in the TEM images than the original resist patterns and allows smaller features to be examined.

Care has to be taken in interpreting these results. In one case, shown in Fig. 12.17, the lines remaining on the substrate suggested that the measurements on the resist patterns had been pessimistic. Some metal reached the substrate in the location of the 32 Å line, showing that it had developed

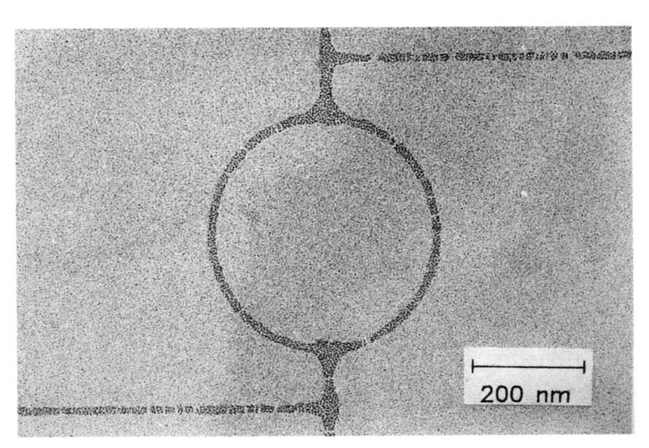

Fig. 12.15 Au ring patten fabricated by PMMA lift-off process. Hair line breaks across the wires cold be due to single molecules of PMMA bridging the lines. The PMMA resist in this case was slightly under-exposed (Broers, 1989).

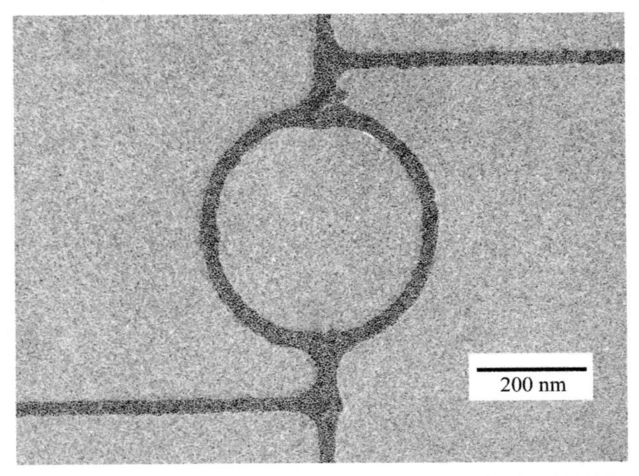

Fig. 12.16 Same pattern as that used in Fig. 15.16 but with the PMMA more heavily exposed. Lines are now continuous to the extent that the gold film is continuous (Broers, 1989).

through to the substrate, at least in some locations. In the site of the line written with a width of 64 Å, there was an almost continuous metal line. However, it turned out that the test pattern used was one where the exposure dose was about twice the minimum required to open large shapes, and the width of the metal lines was almost twice the written linewidth. In the site of the line written with a width of 64 Å, there was an almost

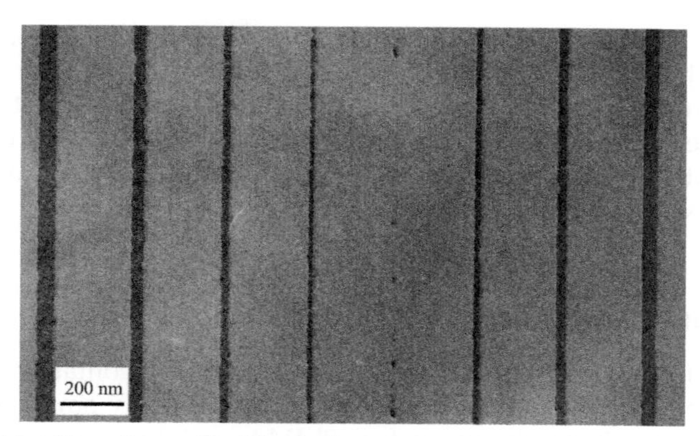

Fig. 12.17 Metal lines formed by lift-off. The written dimensions of the lines were those of the pattern used to measure the resolution of PMMA However, the resist is over-exposed so that the metal lines are wider than their 'written width. For example, the second narrowest line with a written width of 64A is 130A wide.

continuous metal line, but careful measurement showed that the metal line had a width of about 130 Å. In the site of the 94 Å line, there was a continuous metal line, but its width was 180 Å, not 94 Å.

Another fascinating observation was that the patterns obtained when the resist was exposed at a dose slightly below that required for complete removal of the resist on development contain 'hair-line' breaks. See, for example, Fig. 12.15. Fig. 12.16 shows the pattern exposed at a slightly heavier dose where the metal pattern is continuous to the extent that the metal layer itself is continuous. It is possible that the breaks shown in Fig. 12.15 are due to single PMMA molecules bridging the lines. They appear in all pattern areas and are predominantly oriented normally to the lines. This would be expected because it is the shortest path across the lines, and therefore, molecules oriented this way would have the least probability of being exposed. The breaks have a width of 2–3 nm. If this explanation is correct, it suggests that at least some of the molecules in the resist are stretched out.

After exposure, the smaller fragments are dissolved away in the developer, and broken molecules left protruding into the exposed area must "curl up" against the edge of the lines to produce the clean edges observed in the developed patterns. These explanations are obviously highly speculative and need further verification, but the lift-off shadowing method's resolution is high enough to allow the observation of molecular-scale phenomena.

These nano-scale breaks had not been observed in our earlier experiments using the high-resolution probe at IBM. This may have been because we examined the samples in STEM mode rather than TEM mode.

Inevitably, these samples were contaminated with hydrocarbons as they had gone through a series of processes, and the higher contamination rate in STEM mode obscured the finest details on the surface. In retrospect, it might have been better to use low-loss surface imaging where it is the high energy scattered electrons that produce the contrast in the surface image, and thus is not affected by contamination.

Contamination-resist

The hydrocarbon molecules that comprise the contamination that forms when electrons strike a surface are presumably small compared to the molecules in a polymer such as PMMA, and this explains why they can be used to produce smaller structures. There are, however, significant disadvantages to using it as a resist, which include its relative insensitivity and the difficulty of removing it after it has been used. It had been used since the early1960s when Mollenstedt used ions, and Fabian Pease and I used electrons to form what we would now call nano-structures. Fabian Pease formed a free-standing P with the electron beam of his SEM.

The exposure distribution for contamination resist cannot be measured in the same way as it can with resists such as PMMA. This is because the rate of buildup of contamination depends on the hydrocarbon supply in the immediate vicinity of the beam and this varies with the exposure rate and with the pattern density. In the 4000 EX, most of the contamination was carried in on the sample and was not deposited in situ. When this is the case, there is a steady decrease in the writing rate in any given region, making it impossible to use the type of resist pattern used to measure the RCF for PMMA. Instead, the width of the exposure distribution can be estimated from the linewidth obtained with ion-milled metal patterns. For 20nm thick AuPd layers, the minimum linewidth is between 5 and 15nm. This suggests that the resist has a width of about 15nm. This is similar to the width of the distribution in PMMA, and it seems likely that the mechanisms that delocalize the exposure are the same.

Contamination patterns written with the 350kV electron beam of the JEOL 4000EX have very high aspect ratios (50:1) because of the reduced lateral scattering, and some remarkably well-defined nanostructures have been made. Fig. 12.18 shows a ~20nm thick AuPd ring produced by ion milling through a contamination-resist mask. The diameter of the ring is 160nm, and the linewidth is about 12nm. The new ring is compared with one of the rings used to study the Aharonov-Bohm oscillations in Fig. 12.21. Linewidth variations are less than ±2 nm. Figure 12.19 shows a gold–palladium wire that has been reduced in width by continuing to etch the sample after the metal had been removed from the unprotected areas. The wire's width varies from about 4 to 8 nm, but it remains

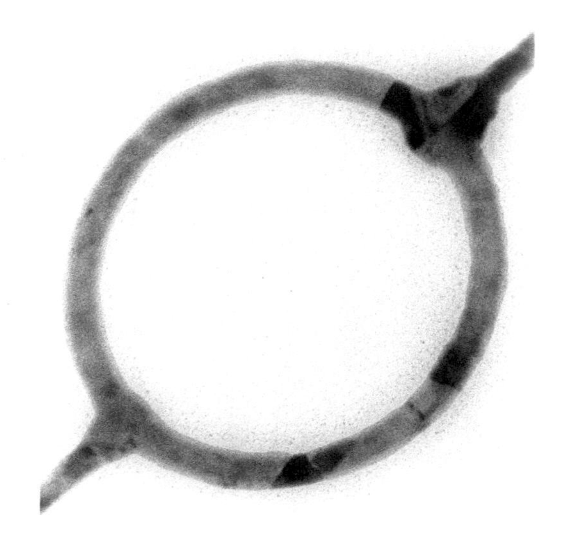

Fig. 12.18 160 nm Diameter gold ring (Broers, 1989).

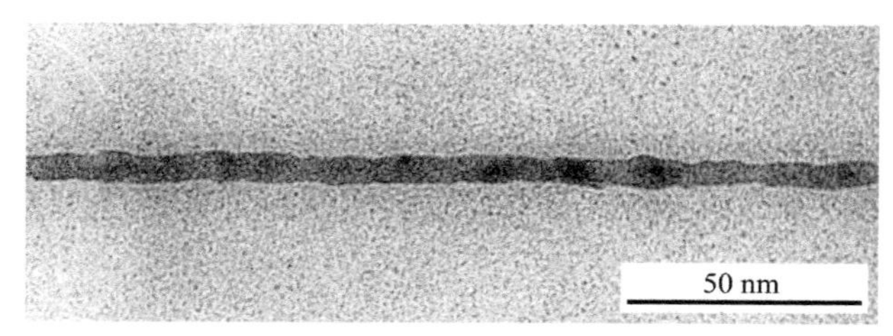

Fig. 12.19 5 nm Wide gold–palladium wire on Si_3N_4 membrane fabricated with contamination resist and ion milling (Broers, 1989).

continuous. The variation in width is a function of the thickness of the metal and of the contamination mask.

Methods for manipulating resist processes to make sub-10 nm structures

There are a number of ways to get around the 30 times degradation between the written linewidth and the developed resist linewidth if the ability to create any shape is sacrificed. The most obvious is to over-expose or under-expose a resist pattern or to over-etch or under-etch in transferring a resist pattern. The 4–8 nm wire shown in Fig. 12.20 was formed by continuing to ion etch the sample until the contamination resist was almost removed. Of course, this process could not have been used to make an array of 5 nm lines and spaces. Another

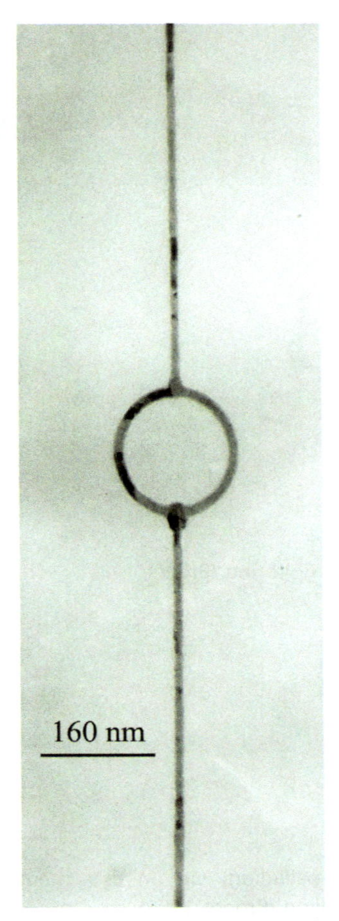

Fig. 12.20 Gold ring shown in Fig. 12.19 at lower magnification (Broers, 1989).

method is to arrange for the size of the small structure to be determined by the thickness of a thin film. This has been done with "Fin" transistors mentioned in Chapter 11. Simple masks for creating narrow wires can also be made by using a sandwich of thin film that adsorbs radiation and one that is transparent to the radiation. This was done in making the mask for x-ray exposure that created the 30nm wire shown in Fig. 11.17. A succession of layers might be used to make a line space array with a limited number of lines.

A further method is to deposit a thin film of material at a glancing angle onto the surface of a developed resist pattern. The material deposited onto the walls of the opening will narrow the opening in the resist,

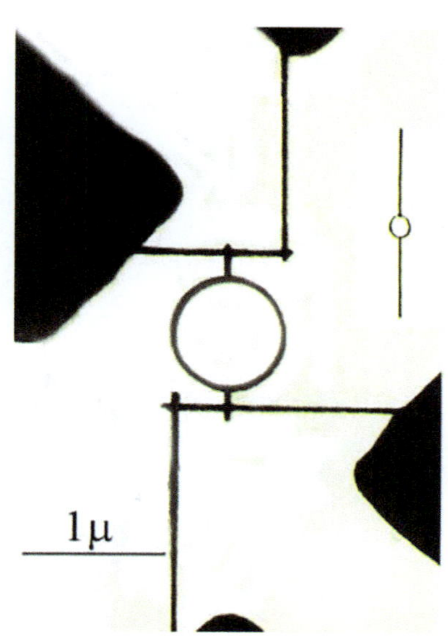

Fig. 12.21 Comparison of the new Ring structure (upper right) with one of the samples that exhibited Aharonov-Bom oscillations described in Chapter 10.

and if the sample is spun around a vertical axis during deposition, all openings could be equally narrowed.

Fig. 12.22 shows an example of a gap in a silicon nitride membrane that has been controllably closed to 3 nm. This dimension could not have been obtained by resist exposure. The protrusions on either side of the initially 50nm wide gap were produced by scanning an electron beam across the gap and allowing contamination to build up on either side. The process can be monitored with a resolution of better than 0.5 nm by switching between the writing mode and the projection mode. In the writing mode, the beam is focused on the sample, and a highly magnified image of it, as it scans out the pattern, can be observed on the projection screen at the bottom of the TEM. In the projection mode, the current in one of the condenser lenses above the sample is decreased so that a larger sample area is illuminated. The currents in the objective lens and the projector lenses remain the same, so an in-focus image of this area appears on the screen at the bottom of the TEM. The hydrocarbon molecules that migrate across the surface to form the contamination are trapped by the outer annulus of the illuminated area, and contamination stops near the gap. This allows the dimensions to be monitored without being further reduced. By choosing the appropriate sequence of processes, it is possible to use the resulting resist pattern to make a lateral wire with a gap equal to the width of the gap.

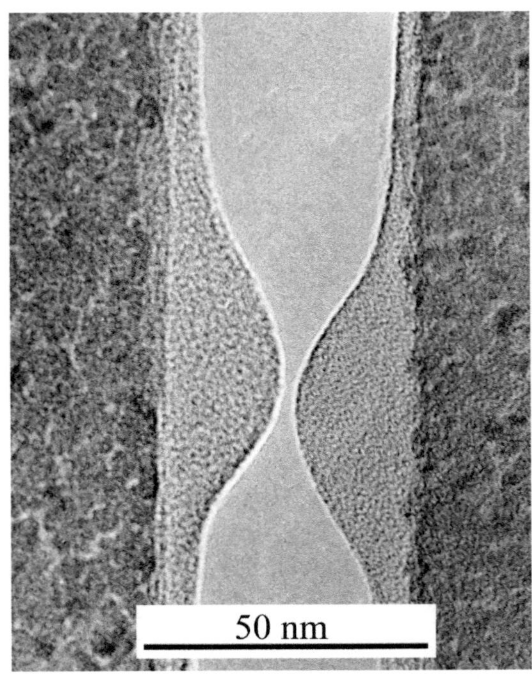

Fig. 12.22 Gap in SiN membrane closed with contamination writing to a width of 3 nm (Broers, 1989).

Discussion

The experiments made with the 4000EX confirmed that there is a fundamental limit to the width of an electron beam exposed resist line of about 15nm. This width is the same at 350kV as it is at lower accelerating voltages and is also about the same in contamination resist as it is in liquid spin-coated resist, although in practice, it is easier to produce narrower isolated lines with contamination resist. The limit is evidently set by the range of the interactions between the electrons and the resist molecule and is not a function of the resolution of the electron optical systems. There is evidence that, with PMMA, the resist molecules are stretched out in the resist rather than coiled up, and for the molecular weight range of 20,000 to 500,000 we used in our experiments, this indicates that the narrowest lines are smaller than the length of the molecules. The lines must, therefore, develop out by the developer removing partial molecules.

All of our experiments with the 4000EX were made on thin membranes where the influence of backscattered electrons is negligible. For many practical devices, thin membranes cannot be used, and backscattered electrons must be considered. As the accelerating voltage increases, the area from which the backscattered electrons emerge increases, and by 350 kV, it is >100 μm in diameter. The actual width of the distribution at 350 kV was not measured. For voltages above about 50 kV, where the width is about 8 μm, the area is so large compared with the dimensions of the structures to be made that the backscattered electrons can be ignored. They merely create a faint background fog to the exposure. This would not be the case for a device requiring a dense array of features over an area larger than the backscatter range, but at present most devices are small compared to the backscatter range, particularly at 350 kV.

A more significant problem with bulk samples is that it is more difficult to examine the structures and to position the electron beam with respect to previously fabricated structures in order to make multilayer devices or to make contact to them. Backscattered or secondary electrons must be collected, and a scanning image formed. In general, scanning images do not have the resolution of transmission images, and contamination rates are higher, but they can nevertheless be used to achieve resolution down to about 2 nm and to position the beam with the same accuracy. This allows dimensions of 15–20 nm to be usefully employed. It is possible to place electron detectors inside the pole pieces of immersion lenses of the type used in the 4000 EX, and JEOL supplied us with a secondary electron detector that detected electrons above the objective lens. This was very effective and provided images with a resolution that was limited only by the secondary electron interaction range in the sample.

The limits identified by our experiments prevented dimensions below about 10 nm from being achieved in useful devices, but our interest in exploring methods that offered this capability remained, and, as mentioned at the end of Chapter 9, one of the main aims of our research in Cambridge was to find ways to make devices and structures with dimensions smaller than 10 nm. At IBM, we had failed to do this with the direct sublimation of materials such as NaCl or with the direct patterning of Langmuir Blodgett films. The process we had not tried was the etch rate enhancement of SiO_2 created by electron bombardment, discovered by O'Keeffe and Handy in 1967.

Direct exposure of SiO_2 on bulk substrates

O'Keefe and Handy showed in 1967 (O'Keeffe and Handy, 1968) that it was possible to increase the chemical etch rate of SiO_2 in p-etch (15:10:300 HF:HNO_3) by three to four times by bombarding it with 1–15 KeV electrons. The patterned layers of SiO_2 formed in this manner could potentially be used both as diffusion masks and to delineate metalized regions for contacts, opening up the possibility of making semiconductor devices without using a photoresist. O'Keefe and Handy had not demonstrated that it was capable of nanometer resolution, but that was not their intention. They made windows 0.6 × 5 μm. The major disadvantage was that an exposure dose of 1 C/cm^2 was needed, which was four orders of magnitude higher than that required for resists.

Despite this, Andrew Hoole, David Allee, Joe Ryan, Xiaodan Pan and I in Cambridge, and Cory Umbach in IBM decided to explore this technique to see if it offered both higher resolution than PMMA and the ability to be useful in making semiconductor devices. As this research gathered pace, I became increasingly involved with administrative responsibilities as Head of the Department of Engineering and Master of Churchill College, and my colleagues conducted most of the research described in the rest of this chapter.

Initially, we had a problem repeating O'Keeffe and Handy's results because the contamination that built up on the surface of the oxide around the highly intense electron beam blocked the wet etch. Even a trace of contamination seemed to block the etch. We overcame this problem with two methods. With the first, the oxide was exposed through a sacrificial layer (Allee and Broers, 1990). The contaminated sacrificial layer was then stripped either mechanically or chemically before etching the SiO_2 in p-etch, aqueous HF, or aqueous NaOH. Both PMMA and aluminum were used successfully as sacrificial layers. The second used oxygen reactive ion etch (RIE) to remove the contamination directly from the oxide surface, eliminating the need for a sacrificial layer (Allee et al., 1991). The latter approach improved resolution by minimizing the forward scattering of the electron beam.

Experiments

The early experiments were carried out on bulk substrates using a sacrificial layer to remove the contamination. The oxide was thermally grown on p-type bulk silicon in wet O_2 at 900 °C (Allee and Broers, 1990). The thickness was measured with an ellipsometer to be 253 nm. A 100 nm layer of PMMA was spun on as a sacrificial layer. Scattering of the 300 keV electrons in this layer, and in the SiO_2, was estimated using a Monte Carlo program to be 12 nm.

The samples were exposed with a relatively high beam current of about 100 picoamps because of the high doses required for this technique. The beam diameter was about 2 nm. After exposure, the PMMA was removed ultrasonically in acetone. Some polymerized PMMA remained in the region of the exposures and was removed by rubbing with a q-tip, leaving the oxide uniformly clean and smooth, as observed at a magnification of 2000X in an optical microscope. The exposed areas could not be distinguished from the unexposed areas in the optical microscope.

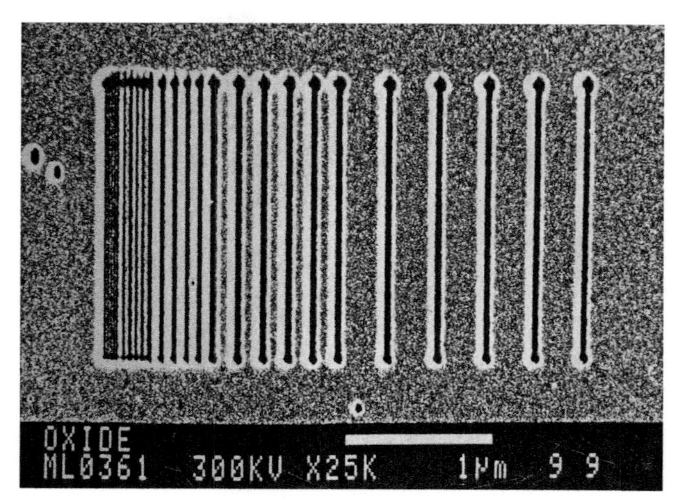

Fig. 12.23 Secondary electron scanning micrograph of an array of lines written at 300 kV in SiO_2 at a line exposure rate of 7.5 µC/cm on a bulk substrate with periods from 340 to 21 nm decreasing in factors of two. Samples are shadowed from the right with gold (Allee and Broers, 1990). *(Reprinted from Broers A.N., Pan, X, Allee, D.R., Umbach, C.P. 1992a. Nanolithography and direct exposure of SiO_2 layers molecular electronics science and technology, Aviram Ari (Ed.) AIP Conference Proceedings 262, (Vol. 262. pp. 151–162). American Institute of Physics; Broers A.N., Hoole, A.C.F., Pan, X.D. 1992b. Electron beam methods for fabricating Sub-20 nanometer structures Internatl. Symposium on New Phenomena in Mesoscopic Structures. Hawaii with the permission of AIP Publishing).*

The exposed and cleaned sample was then etched either in buffered HF or p-etch. The etch time was calibrated to remove 100 nm of the field oxide. The exposed areas etched at a faster rate resulting in trenches in the oxide. These were visible in the optical microscope and high-resolution scanning electron micrographs. The samples were then shadowed at 45° with Au.

Figs 12.23 and 12.24 show an array of lines in SiO_2 with decreasing periods decreasing in factors of two from 340 nm. The line dose for all the lines was 7.5 µC/cm. The smallest period, 21 nm, is discernable and is close to the limit we expected due to the forward scattering in the oxide. The actual width of these trenches is not clear, however, because of the fundamental resolution limit of the secondary electron scanning images and the grain size of the Au coating, both of which are several nanometers.

Nevertheless, these results were encouraging as an array of 105 Å lines with a period of 210 Å would not have been resolved with PMMA.

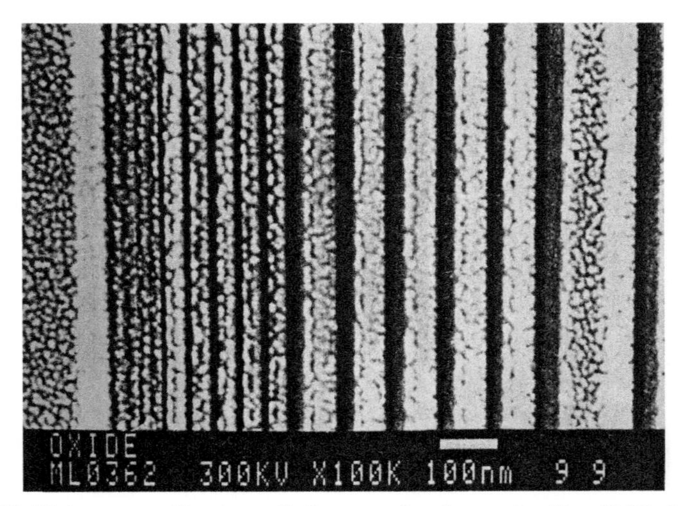

Fig. 12.24 Higher magnification of the sample shown in Fig. 12.23 (Allee and Broers, 1990). *(Reprinted from Broers A.N., Pan, X, Allee, D.R., Umbach, C.P. 1992a. Nanolithography and direct exposure of SiO_2 layers molecular electronics science and technology, Aviram Ari (Ed.) AIP Conference Proceedings 262, (Vol. 262. pp. 151–162). American Institute of Physics; Broers A.N., Hoole, A.C.F., Pan, X.D. 1992b. Electron beam methods for fabricating Sub-20 nanometer structures Internatl. Symposium on New Phenomena in Mesoscopic Structures. Hawaii, with the permission of AIP Publishing).*

Direct exposure of SiO_2 on membrane substrates

Following these initial experiments with bulk substrates, we measured the resolution of the direct exposure method using Si_3N_4/SiO_2 bilayer membrane substrates. This eliminated exposure due to back-scattered electrons and allowed us to examine the patterned samples using transmission electron microscopy. As discussed in Chapter 6, we did not use thin SiO_2 membranes because they typically exhibit buckling due to compressive stress in thermally grown SiO_2. We used Si_3N_4/SiO_2 bilayer membranes.

Again, the patterns were exposed at 300 kV in the 4000 EX using a beam current of several hundred picoamps and a beam diameter of 3 nm as measured on the phosphor screen at the bottom of the microscope. Exposures were also made at 50 kV using the high-resolution electron probe in the IBM Thomas J Watson Research Center using a beam current of 50 picoamps and a beam diameter of 2 to 3 nm.

With the membrane samples, contamination was either removed by using an aluminum sacrificial layer or with O_2 reactive ion etching. When the aluminum sacrificial layer was used, the aluminum sacrificial layer was removed with a standard aluminum etch. The contamination buildup on the other membrane substrates was removed with two 15-min oxygen-reactive ion etches, one for each side of the membrane. To avoid charging of the membrane when reactive ion etching was used, a thin layer of aluminum was deposited on the SI_3N_4 side of the membrane. This was removed with the standard aluminum etch after the exposures.

The exposed samples were then developed with either p-etch (6 min), aqueous HF (room temperature, 48%, 8.7%, and 4.6% for 5, 60, and 180 s, respectively), or aqueous NaOH (75°C, 1.3 M for 15 min). The etch time for the p-etch was calibrated to remove approximately 75 nm or 100 nm of the unexposed oxide. The resulting oxide profiles were shadowed with AuPd at 45°.

With the membrane samples, we were able to examine the etched patterns using transmission electron microscopy with a resolution of a few tenths of a nanometer.

This was to be compared with the bulk samples, for which we had to use secondary electron SEM images with a resolution of several nanometers because of the lateral range of the secondary electrons in the sample.

Fig. 12.25 shows an array of lines in SiO_2 exposed with 300 kV electrons. The sample was shadowed with AuPd along the diagonal. Unfortunately, the structure in the AuPd obscures the details of the trenches, just as similar coatings obscure the details of biological specimens when using high-resolution, low-loss scanning electron microscopy. There are five periods with five lines each. All five are visible, although the smallest period of 12 nm is obscured by the structure of the AuPd coating. It was surprising to us that a period of 12 nm was resolved as the forward scattering of the 300 kV electrons in the 250 nm thick SiO_2 was estimated with a Monte Carlo program to be 12 nm (90% diameter) at the base of the oxide. However, this result showed that the resolution of the process was adequate to produce 6 nm lines on 12 nm centers, clearly better than could have been achieved with PMMA.

Simulation of exposure profiles

Our experiments with shadowing large area exposures confirm O'Keeffe and Handy's observation that the maximum etch rate enhancement is three to four times. This, of course, is much lower than the ratio between exposed and unexposed photoresists and results in trenches that are wider near the surface of the oxide where there has been more lateral etching of the unexposed material. With dense arrays of lines, the sloping sidewalls overlap, decreasing the thickness of the oxide between the lines. A simple calculation gives the sidewall angle Θ as

$$\tan(\theta) = \frac{v_2}{v_1},$$

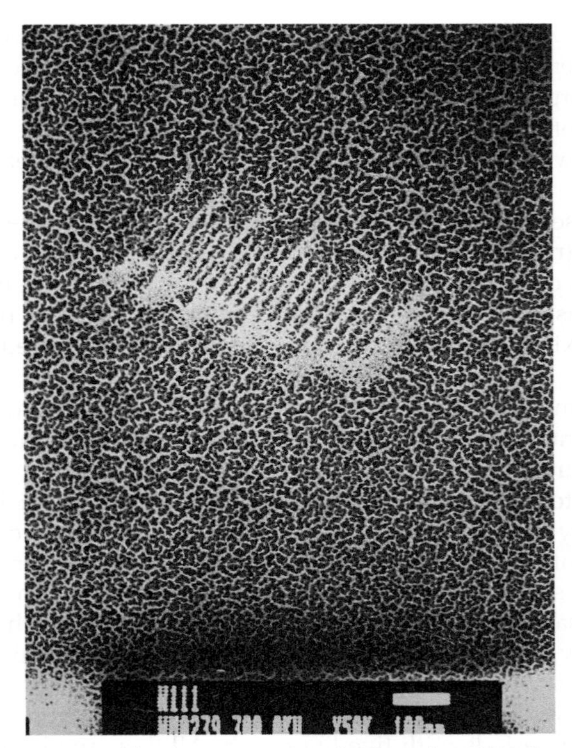

Fig. 12.25 Pattern created by direct exposure of SiO_2 with 300kV electrons. The finest lines with a period of 12nm are resolved but cannot be seen in detail because of the structure in the AU film (Allee et al., 1991) Magnification bar = 100 nm. *(Reprinted from Broers A.N., Pan, X, Allee, D.R., Umbach, C.P. 1992a. Nanolithography and direct exposure of SiO_2 layers molecular electronics science and technology, Aviram Ari (Ed.) AIP Conference Proceedings 262, (Vol. 262. pp. 151–162). American Institute of Physics; Broers A.N., Hoole, A.C.F., Pan, X.D. 1992b. Electron beam methods for fabricating Sub-20 nanometer structures Internatl. Symposium on New Phenomena in Mesoscopic Structures. Hawaii, with the permission of AIP Publishing).*

where v_1 and v_2 are the etch rates of the exposed and unexposed regions, respectively. An etch rate ratio of 3 gives a sidewall angle of 72°. This simple model ignores forward scattering of the electrons. By superposition, when dense arrays of lines are written, the sloping walls overlap, decreasing the amplitude of the oxide between the trenches. This amplitude, h, is given by

$$h = \frac{v_2}{v_1} \frac{(p - w)}{2},$$

where w is the width of an exposed line, and p is the period.

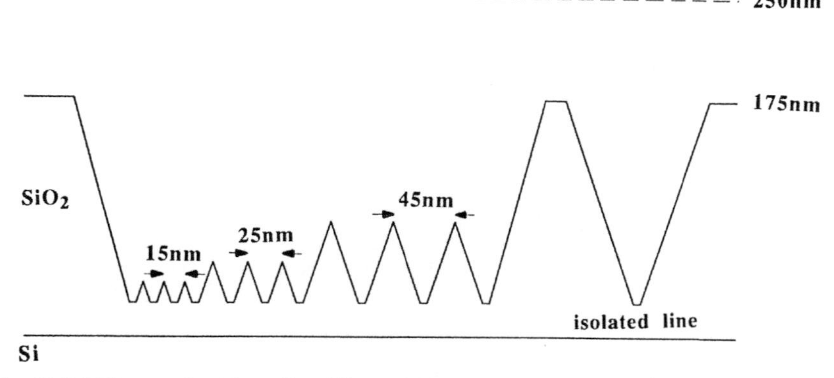

Fig. 12.26 The predicted profile of lines written in SiO_2 assuming the exposed lines are 5 nm wide and have an etch rate 3× faster than the unexposed regions. Forward scattering of the electrons is neglected in this simple model (Allee et al., 1991). The dashed line is he surface of the oxide before development. The relatively small difference in etch rates between the exposed and unexposed regions results in trenches with sloping sidewalls. By superposition, closely spaced lines result in a decreased oxide amplitude. Reprinted from (Broers et al., 1992a,b), with the permission of AIP Publishing.

Fig. 12.26 shows a profile of an isolated line and closely spaced lines with three periods. The etch rate enhancement was assumed to be 3, and the exposed line width was 5 nm.

The depth to which the unexposed oxide had etched was set at 75 nm. The dashed line represents the oxide surface before development. This profile is in qualitative agreement with our experimental results.

Fig. 12.27 shows the profiles obtained when lines are written at accelerating voltages of 300, 200, and 100 kV. The width at the top and bottom of the oxide trench written at 100 kV is wider than those at 200 kV and 300 kV. This is evidence that the reduced electron scattering at higher voltages yields a narrower and deeper exposure profile.

Fig. 12.28 shows a T-gate formed on top of a trench using PMMA lift-off. The gate's mushroom shape is ideal for reducing the resistance of the gate electrode. The width at the bottom of the T is only about 10 nm wide, while the bulk of the gate is 450 nm wide. The structure is mechanically stable because some oxide remains at the base of the T.

Finally, Xiaodan Pan repeated the SiO_2 trench experiments using heating of the sample to prevent contamination rather than removing it with a sacrificial layer (Pan and Broers, 1993). The membrane sample was heated in situ to about 200 C in a JEOL EM-SHH40 heating stage while writing the test patterns. This was successful, and the samples could be

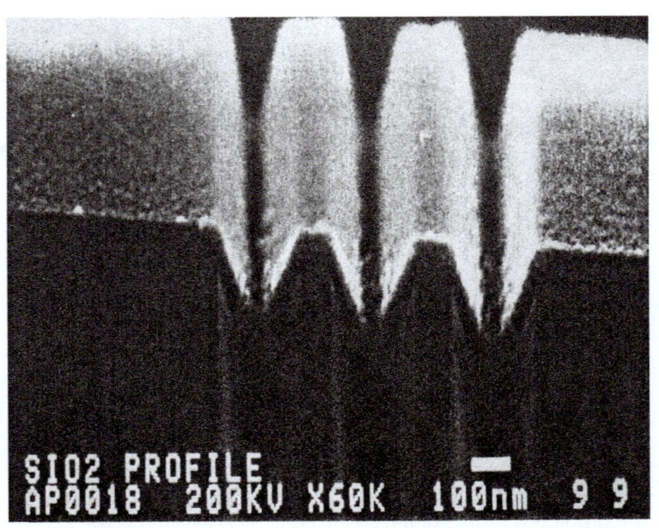

Fig. 12.27 Profiles obtained by Hoole and Pan for electron irradiated SiO_2 trenches written at 300, 200 and 100 kV. *(Used with permission of Elsevier Science and Technology Journals from Broers A.N., Pan, X, Allee, D.R., Umbach, C.P. 1992a. Nanolithography and direct exposure of SiO_2 layers molecular electronics science and technology, Aviram Ari (Ed.) AIP Conference Proceedings 262, (Vol. 262. pp. 151–162). American Institute of Physics; Broers A.N., Hoole, A.C.F., Pan, X.D. 1992b. Electron beam methods for fabricating Sub-20 nanometer structures Internatl. Symposium on New Phenomena in Mesoscopic Structures. Hawaii permission conveyed through Copyright Clearance Center, Inc.).*

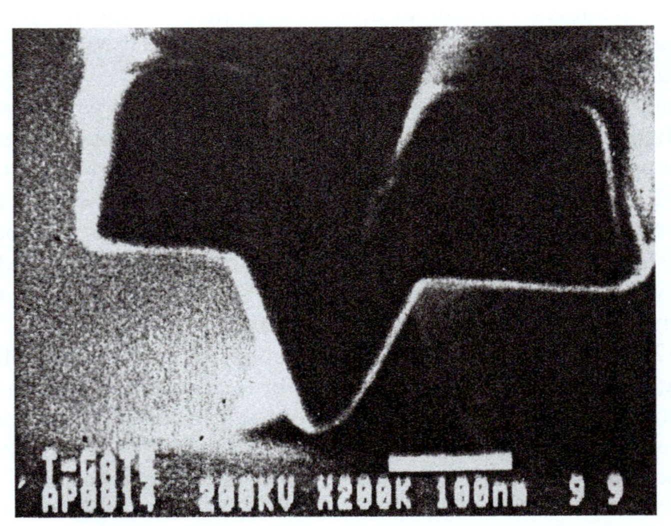

Fig. 12.28 T-gate structure fabricated by Hoole and Pan using PMMA lift-off over an electron-irradiated SiO_2 trench. *(Used with permission of Elsevier Science and Technology Journals from Broers A.N., Pan, X, Allee, D.R., Umbach, C.P. 1992a. Nanolithography and direct exposure of SiO_2 layers molecular electronics science and technology, Aviram Ari (Ed.) AIP Conference Proceedings 262, (Vol. 262. pp. 151–162). American Institute of Physics; Broers A.N., Hoole, A.C.F., Pan, X.D. 1992b. Electron beam methods for fabricating Sub-20 nanometer structures Internatl. Symposium on New Phenomena in Mesoscopic Structures. Hawaii permission conveyed through Copyright Clearance Center, Inc.).*

A

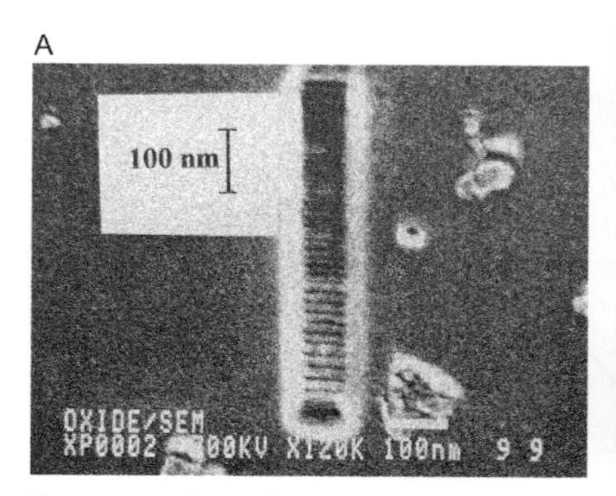

B

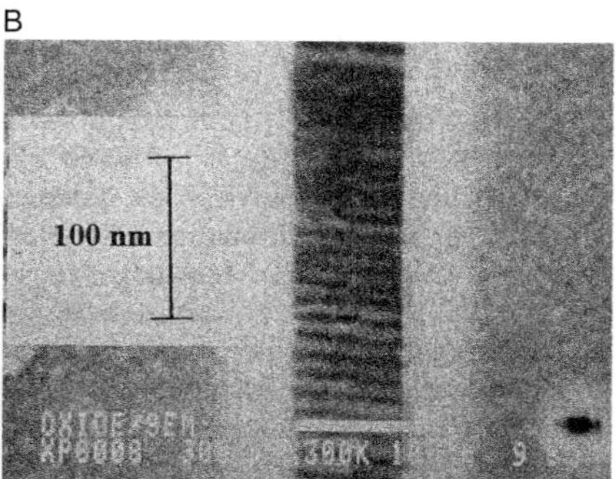

Fig. 12.29 Secondary electron SEM of trenches in SiO_2 on periods of 20.5, 17.2, 13.6, 12.2, 10.8, 9.4 and 8 nm on a Si_3N_4/SiO_2 membrane sample. The sample was shadowed with ~5 nm of titanium at 65° to the surface normal and then viewed from the oxide side. The trenches on the third smallest Period—10.8 nm—are visible at both low magnification and high magnification. *(Reprinted from Pan X., Broers, A.N. 1993. Improved electron beam pattern writing in SiO2 with the use of a sample heating stage Appl. Phys. Lett., 63, 1441–1442. with the permission of AIP Publishing).*

developed immediately after exposure. Using this process, 10.8 nm periods in oxide were demonstrated despite problems with electrical noise perturbing the electron beam so that its diameter was effectively 4 nm and the beam was scattering in the 250 nm thick SiO_2 layer (Fig. 12.29).

Ryan and Hoole have shown that modeling of the etching of the oxide after irradiation under the conditions used by (Pan and Broers, 1993), see Fig. 12.30, shows that the σ of the Gaussian spread function within the oxide must be less than 4 nm to achieve the 11 nm spacing. This is less than half of that measured for PMMA. Heating of the sample proved to be the simplest way to prevent contamination.

Replication of SiO_2 trenches into polycrystalline and single-crystal silicon

To explore whether patterned SiO_2 layers could be used in the fabrication of semiconductor devices we collaborated with Tang and Wilkinson at the University of Glasgow. They used a RIE 80 plasma etcher, with $SiCl_4$, to explore the transfer of patterned SiO_2 into single-crystal silicon and polycrystalline silicon. In these preliminary experiments, the gas flow rate of the etcher was set at 2.5 sccm, the chamber pressure at 10^{-2} Torr, and the self-bias voltage of the substrate at −135 V (corresponding to a power of 35 W). It had been reported that under similar etching conditions, directional etching could be achieved, and the selectivity of $SiCl_4$ RIE of oxide over polycrystalline silicon, as well as over single-crystal silicon, was

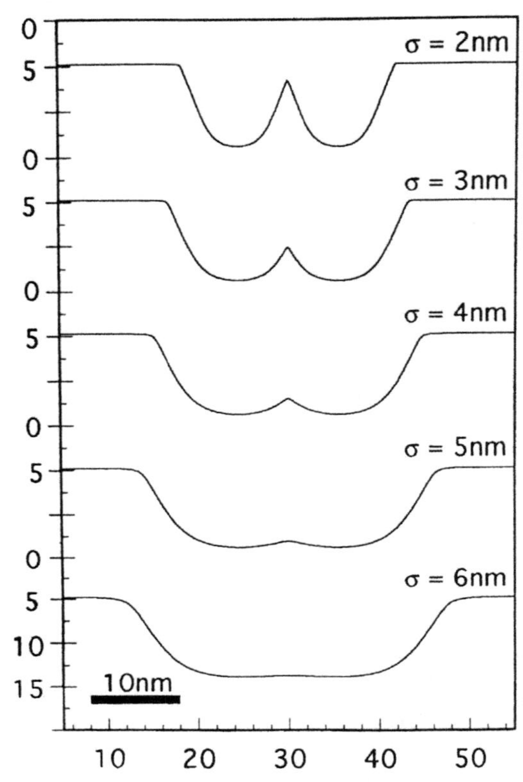

Fig. 12.30 Simulations by Ryan and Hoole of the profiles in SiO_2 after irradiation with 300 kV electrons for a line spacing of 11 nm as used by Pan(ref). They show that the sigmas of the exposure dose profile must had a sigma of less than 4 nm for the period of 11 nm to be resolved. *(Used with permission of Elsevier Science and Technology Journals, from Broers A.N., Hoole, A.C.F., Ryan, J.M. 1996. Electron beam lithography—Resolution limits, Microelectron. Eng. 32, 131–142 permission conveyed through Copyright Clearance Center, Inc.).*

1:10–1:13 (Yang and Wilkinson, 1991). Grooves replicated using this method should have steeper side walls than the original patterns in the oxide. The etching time was chosen to be sufficient to etch 200 nm deep trenches into both polycrystalline silicon and single-crystal silicon. After the RIE, the patterns are all visible in the optical microscope and in high-resolution scanning electron micrographs (Fig. 12.31). It was not possible to measure the depth of the trenches directly, so we were not able to confirm whether the etch depth of 200 nm, which would have been obtained on unasked silicon, was realized in the narrow openings.

To make sure that grooves had been produced in the underlying materials, we stripped off the oxide masking layer in buffered HFF (BHF). The stripping could also have created undercuts of ~1 nm in both materials

Fig. 12.31 Replicated trenches in single-crystal silicon before removal of the patterned oxide mask. *(Reprinted from Pan, X., Allee, D.R., Broers, A.N., Tang, Y.S., Wilkinson, C.W., 1991. Nanometer scale pattern replication using electron beam direct patterned SiO_2 as the etching mask. Appl. Phys. Lett. 59, 3157–3158 with the permission of AIP Publishing).*

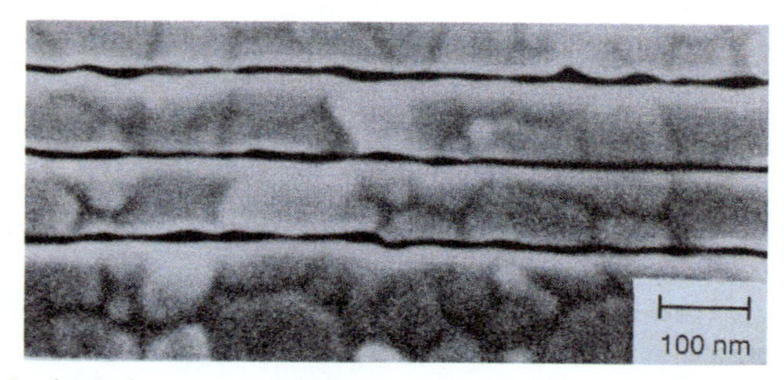

Fig. 12.32 Replicated 10 nm trenches in n-type polysilicon; the oxide masking-layer has been stripped off in BHF. *(Reprinted from Pan, X., Allee, D.R., Broers, A.N., Tang, Y.S., Wilkinson, C.W., 1991. Nanometer scale pattern replication using electron beam direct patterned SiO_2 as the etching mask. Appl. Phys. Lett. 59, 3157–3158 with the permission of AIP Publishing).*

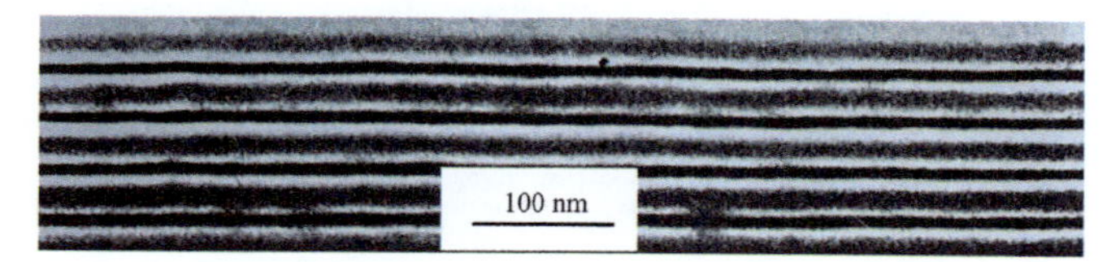

Fig. 12.33 Replicated 10 nm trenches in single-crystal silicon: this is the same sample shown in Fig. 12.30 after the oxide masking layer has been stripped off.

because the etching rates of single-crystal silicon and polycrystalline silicon in BHF are 0.5 nm/min and ~1.2 nm/min. Despite this, fine trenches with widths as small as 10 nm were replicated into both polycrystalline silicon (Fig. 12.32) and single-crystal silicon (Fig. 12.33, which is part of Fig. 12.31 taken after stripping off the oxide). The trenches in polycrystalline silicon were not uniformly etched due to inhomogeneities of the material, such as grain boundaries and defects caused by dopant implantation. The oxide exposure dose for the smallest grooves in both materials is approximately 2×10^{-3} C/m.

Use of deposited SiO_2 as a resist (Pan et al., 1991)

All of our early experiments with direct exposure of SiO_2 used thermally grown SiO_2. Attempts to expose deposited oxides did not produce satisfactory results. The process was usually negative, rather than positive and the resolution was much poorer (see Fig. 12.34). However, Xiaodan Pan discovered that the lithographic properties of the deposited oxide were critically changed by implanting the deposited oxide with 50keV oxygen ions at a dose of $1 \times 10^{16}/cm^2$ and annealing the sample at 600°C for 30min. The process became positive and the p-etch rate of the irradiated LPCVD oxide was reduced by a factor of 5–7 times until it was close to the etch rate of thermal oxide.

Test patterns were written in the deposited and irradiated oxide with a 300kV electron beam with a beam diameter of about 2 nm and a beam current of 10^{-10}A. The resolution was limited by forward scattering in the oxide layer, which spread the beam of about 12nm at the bottom of the oxide layer but it was still possible to produce linewidths of about 10 nm. The lowest electron dose that produced trenches in the oxide was 1.8μC/cm (see Fig. 12.35).

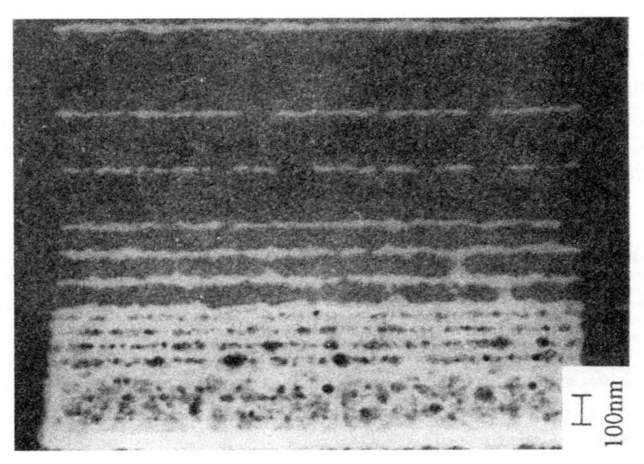

Fig. 12.34 Poor resolution produced with LPCVD SiO_2 film as deposited. Writing process is negative, i.e., etch rate reduced by electron beam exposure. *(Reprinted from Pan, X., Allee, D.R., Broers, A.N., Tang, Y.S., Wilkinson, C.W., 1991. Nanometer scale pattern replication using electron beam direct patterned SiO_2 as the etching mask. Appl. Phys. Lett. 59, 3157–3158, with the permission of AIP Publishing).*

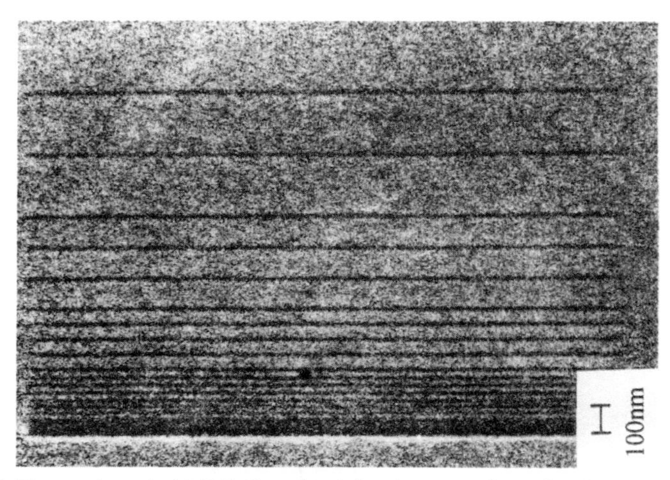

Fig. 12.35 Lines written in LPCVD film after it has been implanted with oxygen ions and annealed. Writing mode is positive and trenches are about 10 nm wide. *(Reprinted from Pan, X., Allee, D.R., Broers, A.N., Tang, Y.S., Wilkinson, C.W., 1991. Nanometer scale pattern replication using electron beam direct patterned SiO_2 as the etching mask. Appl. Phys. Lett. 59, 3157–3158, with the permission of AIP Publishing).*

In these experiments, a PMMA sacrificial layer was used to remove the contamination built up during exposure rather than removing it with plasma-assisted oxygen chemical etching or preventing it from forming by heating the sample, as already described. After stripping off the PMMA layer, the sample was cleaned with a q-tip which left the oxide optically smooth and clean with no visual difference between the irradiated and non-irradiated regions. However, when examined in a high-resolution secondary electron SEM the exposed regions showed a lower secondary electron yield than the exposed regions (see Fig. 12.36). A similar effect has been found in the electron beam irradiated thermal oxide. A possible explanation is that the exposing electrons ionize some of the substrate atoms, leaving a residual positive charge that suppresses the secondary emission of electrons.

The exposed LPCVVD oxide was then developed in *p* etch, where the irradiated regions etched 3–4 times faster than the unirradiated regions. The trenches produced in the deposited oxide are shown in Fig. 12.37. The dose was the same for all the lines at 7.5 μC/cm, and the oxide was developed for 2 min. There are four different periods decreasing by factors of two from 120 nm. The smallest period has 10 nm trenches on a 15 nm period. This is comparable to the resolution of the thermally grown oxide.

The fact that the O+ implantation greatly improved the lithographic properties of the deposited oxide and subsequent annealing may help in understanding how electron bombardment increases the oxide's etch rate.

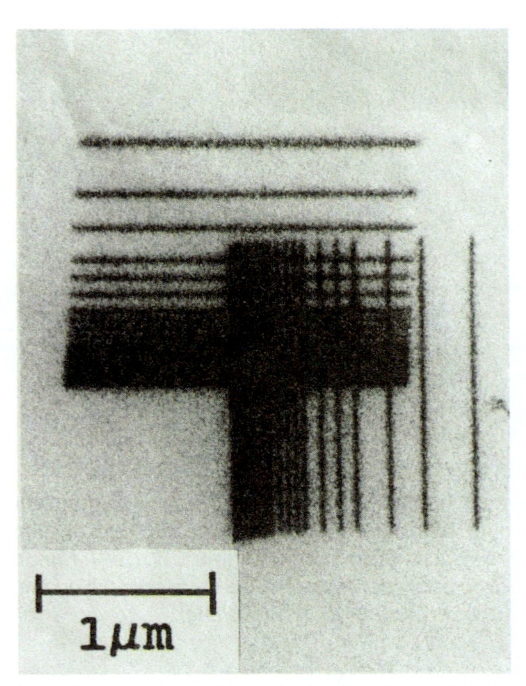

Fig. 12.36 Secondary electron SEM micrograph of SiO_2 sample after exposure with electrons but before etching. *(Reprinted from Pan, X., Allee, D.R., Broers, A.N., Tang, Y. S., Wilkinson, C.W., 1991. Nanometer scale pattern replication using electron beam direct patterned SiO_2 as the etching mask. Appl. Phys. Lett. 59, 3157–3158, with the permission of AIP Publishing).*

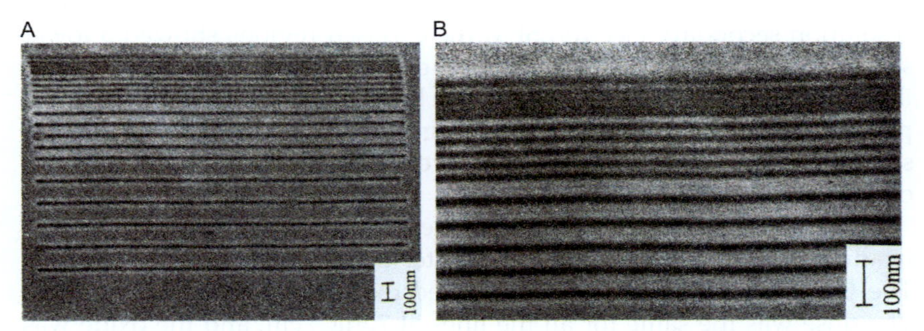

Fig. 12.37 Two micrographs of an array of lines exposed in the modified LPCVD oxide with decreasing periods. The smallest period of 15 nm, is visible in both micrographs. *(Reprinted from Pan, X., Allee, D.R., Broers, A.N., Tang, Y.S., Wilkinson, C.W., 1991. Nanometer scale pattern replication using electron beam direct patterned SiO_2 as the etching mask. Appl. Phys. Lett. 59, 3157–3158, with the permission of AIP Publishing).*

Mechanism of the direct exposure of SiO_2

The mechanism of this electron beam exposure of SiO_2 is not clearly understood. In light of O'Keefe and Handy's experiments at 1–15kV, which is well below the nuclear damage threshold (Vasilov et al., 1975), it is unlikely that processes associated with direct atom displacement dominate the exposure mechanism(s). The enhancement of the SiO_2 etch rate in HF-based etches may be due to the effects of ionizing radiation. It is known that electron beam irradiation of amorphous SiO_2 can cause oxygen vacancies, peroxy radicals, non-bridging oxygen (Beale Fowler, 1983), and neutral traps that can subsequently fill with either electrons or holes (Aitkin, 1980). Defect creation in amorphous SiO_2 has often been explained in terms of breaking strained Si-O-Si bonds (Beale Fowler, 1983) or more recently, in terms of interatomic Auger processes (Knotek, 1983). Of particular importance is whether the minimum energy required for an exposure event is above the energy of the secondary electrons (several tens of eV).

There is some evidence that the delocalized exposure due to secondary electrons is the source of the resolution limit in PMMA, as I have already discussed at length (Broers, 1981). The mechanisms associated with defect formation due to interatomic Auger processes have energy thresholds of 30 to 40eV (Feidelman and Knotek, 1978). The mechanism of exposure of SiO_2 may be related to the density and stress in the oxide film. The density and stress are known to be sensitive functions of the oxide growth conditions and correlate well with the refractive index (Taft, 1978). Furthermore, the etch rate of the HF-based SiO_2 etches, particularly that of p-etch, is known to be sensitive to oxide density, bond strain, etc. (Pliskin, 1977).

Use of patterned SiO_2 as a template

The deposition of thin films can be influenced by creating artificial topography on a surface. I have observed this with metals with high mobility on surfaces, such as lead. A simple experiment with patterned SiO_2 was also successful in producing fine metal wires. An array of trenches of decreasing width were formed in the SiO_2 layer and metal was deposited on the sample. An angled dry-etch process was then used to remove the metal from the surface leaving fine wires in the shadowed base of the trenches. The narrowest wires were about 10 nm wide (Fig. 12.38).

Device applications

Although we were not able to establish the resolution limits for the direct exposure of the SiO_2 process, it seems clear that it can be used to produce center-to-center spacings in dense arrays of about 10 nm, which is 3 to 4 times smaller than can be produced with PMMA. Even smaller spacing

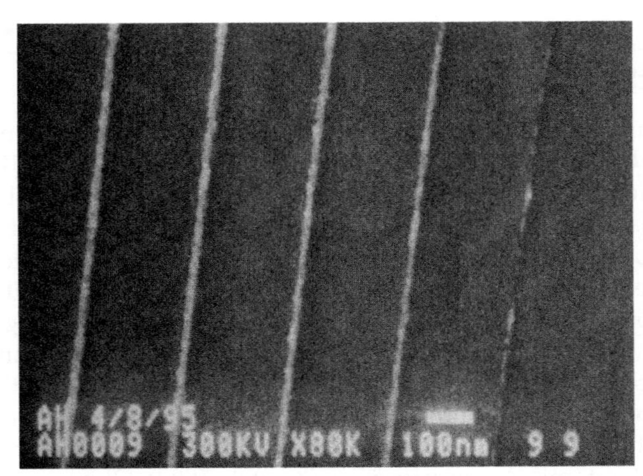

Fig. 12.38 Micrograph of the fine metal wires fabricated using a patterned SiO_2 template obtained by Hoole and Ryan. *(Used with permission of Elsevier Science and Technology Journals, from Broers A.N., Hoole, A.C.F., 1996. Ryan, J.M. Electron beam lithography—Resolution limits, Microelectron. Eng. 32, 131–142 permission conveyed through Copyright Clearance Center, Inc.).*

may be achievable using thinner layers of SiO_2 and smaller electron beams, both of which can readily be produced.

It was also important that the patterned layers could be used to transfer features on this size scale into silicon, making it potentially useful in the fabrication of silicon devices and circuits, but perhaps most importantly, nanometer-scale patterned SiO_2 could itself be used to fabricate several quantum and conventional field-effect transistors with dimensions significantly smaller than had previously been possible. An FET with a corrugated gate oxide forms a lateral superlattice (Fig. 12.39) where the gate metal in the thinner regions of the oxide depletes the two-dimensional electron gas, forming an array of electrostatic barriers, for example (Warren et al., 1985). If only two trenches are formed, a lateral resonant tunneling FET is formed, for example (Chou et al., 1989). The disadvantage of the short scattering lengths in the Si/SiO_2 relative to the AlGaAs/GaAs system is partially offset by the smaller periods possible and the large band gap of SiO_2 (9 eV). This large band gap enables the gate insulator to be much thinner without getting appreciable tunneling from the gate to the 2D gas. Since the spatial harmonics of the gate potential decay exponentially with distance into the device, the electrostatic barriers at the interface are significantly larger. Both the larger electrostatic barriers and the smaller period possible with SiO_2 increase the separation of the quasibound levels and strengthen quantum effects. Other quantum devices, such as Aharonov Bohm rings described in Chapter 10, where

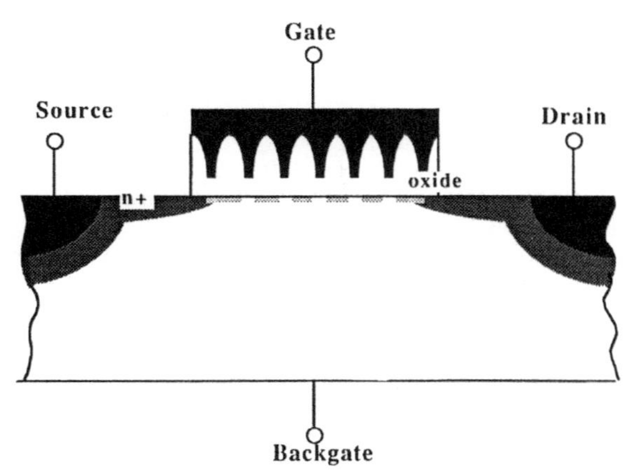

Fig. 12.39 A drawing of a lateral surface superlattice in a MOSFET structure incorporating a nanometer scale patterned SiO_2. If only two trenches in the oxide are present, a lateral resonant tunneling FET is formed. A single trench forms a depletion mode MOSFET with a sub-10 nm gate length. *(Reprinted with permission from Allee D.R., Umbach, C.P., Broers, A.N. 1991. Direct Nanometer Scale Patterning of SiO2 with Electron Beam Irradiation Proc. 35th Internat. Symposium on Electron Ion and Photon Beams, May 28–31. J. Vac. Sci. Technol. B9 (6), 2838–2841. Copyright 1991, American Vacuum Society).*

electron wires are induced under the thin portions of the oxide, are also possible. It should also be noted that a single trench in the oxide could be used to form a depletion mode metal-oxide-semiconductor field effect transistor MOSFET) with sub-10 nm gate length.

References

Aitkin, J., 1980. Noncryst. Sol. 40, 31.
Allee, D.R., Broers, A.N., 1990. Direct nanometer scale patterning of SiO2 with electron beam irradiation through a sacrificial layer. Appl. Phys. Lett. 57 (21), 2271–2273.
Allee, D.R., Umbach, C.P., Broers, A.N., 1991. Direct Nanometer Scale Patterning of SiO2 with Electron Beam Irradiation. In: Proc. 35th Internat. Symposium on Electron Ion and Photon Beams, May 28–31, pp. 2838–2841. J. Vac. Sci. Technol. B9 (6).
Beale Fowler, W., 1983. Semicond. Insul. 5, 583.
Broers, A.N., 1981. Resolution limits of PMMA resist for electron beam exposure. J. Electrochem. Soc. 128 (1), 166–170.
Broers, A.N., 1989. Resolution Limits for Electron Beam Lithography and Methods for Avoiding These Limits. Nanostructure Physics and Fabrication. Copyright Elsevier (Academic Press, N.Y.).
Broers, A.N., Timbs, A.E., Koch, R., 1989. Nanolithography at 350 kV in a TEM. Microelectronic Engineering 89. North-Holland, pp. 187–190.
Broers, A.N., Pan, X., Allee, D.R., Umbach, C.P., 1992a. Nanolithography and direct exposure of SiO2 layers molecular electronics science and technology.

In: Ari, A. (Ed.), AIP Conference Proceedings 262. Vol. 262. American Institute of Physics, pp. 151–162.
Broers, A.N., Hoole, A.C.F., Pan, X.D., 1992b. Electron beam methods for fabricating Sub-20 nanometer structures Internatl. In: Symposium on New Phenomena in Mesoscopic Structures. - Hawaii.
Chou, S.Y., Allee, D.R., Pease, R.F.W., Harris Jr., J.S., 1989. Observation of electron resonant tunneling in a lateral dual-gate resonant tunneling field-effect transistor. Appl. Phys. Lett. 55, 76.
Feidelman, P.J., Knotek, M.L., 1978. Phys. Rev. B 18 (12), 6531–6539.
Hatzakis, M., 1967. Semiconductor metalization process. IBM Tech. Disclosure Bull 10 (4).
Knotek, M.L., 1983. Semicond. Insul. 5, 361.
O'Keeffe, T.W., Handy, R.M., 1968. Solid State Electronics. vol. 11 London Pergamon Press, pp. 21–266.
Pan, X., Allee, D.R., Broers, A.N., Tang, Y.S., Wilkinson, C.W., 1991. Nanometer scale pattern replication using electron beam direct patterned SiO_2 as the etching mask. Appl. Phys. Lett 59, 3157–3158.
Pan, X., Broers, A.N., 1993. Improved electron beam pattern writing in SiO_2 with the use of a sample heating stage. Appl. Phys. Lett. 63, 1441–1442.
Pliskin, W.A.J., 1977. Comparison of properties of dielectric films deposited by various methods. Vac. Sci. Technol. 14 (5), 1064–1081.
Taft, E.A.J., 1978. Electrochem. Soc 125, 968.
Vasilov, V.S., Kiv, A.E., Niyazova, O.R., 1975. "Sub-threshold" defect formation in natural diamond. Phys. Sol. (a) 32, 11.
Warren, A.C., Antoniadis, D.A., Melngailis, J., Smith, H.I., 1985. Surface superlattice formation in silicon inversion layers using 0.2-micron period grating-gate electrodes. IEEE Electron Dev. Lett. 6, 294.
Yang, Y.S., Wilkinson, C.D.W., 1991. Reactive Ion etching of polycrystalline-silicon using $SiCl_4$. Appl. Phys. Lett. 58, 2898.

CHAPTER THIRTEEN

Nanolithography for modulation doped field effect transistors

Contents

Introduction

The first 5 years after I returned to Cambridge were spent gathering our small research group together, building the nanostructure laboratory, and establishing the capabilities of the JEOL 4000 E electron microscope. This went well. The microscope met, or exceeded, all of its specifications, and we found that the resolution of the standard electron beam fabrication processes at electron energies of 300–400 keV was similar to that at lower energies. However, the higher voltage allowed us to use thicker resists and thicker and more robust membrane substrates without incurring a loss of resolution due to electron scattering in the resist. The steep-sided contamination resist patterns produced at the higher voltage also allowed us to pattern thicker metal layers and make more resilient devices.

It was now time to fabricate some practical devices, and we enthusiastically joined a European consortium led by IMEC, the Inter-University Microelectronics Centre in Belgium, to work on the fabrication and scaling of HEMTs (High-Electron Mobility Transistors). We also collaborated with Fabian Pease, who had been working on HEMTs with David Allee at Stanford University.

Advances in Imaging and Electron Physics, Volume 231
ISSN 1076-5670
https://doi.org/10.1016/B978-0-443-31462-9.00013-7

Our initial plan was to explore using the direct exposure of SiO_2 process, as described in Chapter 12, to create smaller and faster transistors. However, we soon realized that this would require extensive changes to the transistor fabrication process. In light of this, we made a strategic decision to concentrate on making smaller gates with established processes. The dimensions we could already achieve with PMMA, and with contamination resist, were considerably smaller than those used to produce HEMTs in 1989.

IMEC was founded by Baron Professor Roger Van Overstraeten in 1984 at Leuven University in Belgium. By 1989, it already had an outstanding reputation for partnering with leading semiconductor companies and had its own state-of-the-art fabrication facility for making high-performance transistors.

Since then, under the leadership of Gilbert Declerck and Luc Van den Hove, IMEC has gone from strength to strength. Today, in 2024, with over 5500 researchers and an annual budget of about $1 billion, it is one of the strongest research organizations in the semiconductor world. Their contributions have covered the spectrum from fundamental science to the practicalities of semiconductor manufacturing. For example, they have partnered with ASML to develop the EUV lithography technology that is used to make the world's most advanced chips.

Material has been reprinted in this chapter from (Van Hove et al., 1993) with permission from the American Vacuum Society, from (Allee et al., 1991) with permission from the IEEE, and from (Alle et al., 1990) with permission of Springer Nature.

The scaling of high electron mobility transistors HEMTs

The lithography challenges of making discrete HEMTs are very different from those of making silicon chips, where the patterns to be written are astronomically complex, and high throughput is essential. The formation of the substrate that contains the two-dimensional electron gas channel for the HEMT is very complex and relies on state-of-the-art atomic layer deposition, but the geometry of the source, drain, and gate is relatively straightforward from a lithographic point of view. The aim is to achieve extremely short electron transit times by using very short gate lengths and, at the same, minimize the gate resistance and the other device parasitics that limit device speed. Electron beam lithography is ideal for achieving these small dimensions and has been used to make most of the highest-performance HEMTs. The need to keep gate resistance to a few ohms means that very thick gates, T-gates, or multi-gates are needed. High-temperature superconductors have also been tried, but as discussed in this chapter, their use does not compete with the use of normally conducting multi-gates.

The consortium was financially supported by ESPRIT BRA 3042. It had five members: IMEC was developing fabrication processes suitable for ultra-submicron HFETs with the exception of the gate electrodes, CNRS-Laboratoire de Microstructures et de Microelectronique (L2M) was investigating and optimizing the influence of the epitaxial layer structure of AlGaAs/InGaAsHFETs for gate lengths down to 0.1 μm, CNRS-Institute d'Electronique Fondamentale (IEF) was developing the design modeling and electrical characterization tools including Monte Carlo simulations and high-frequency s-parameter characterizations, and The Laboratoire d'Optique Appliquée (LOA) was using time-resolved optical spectroscopy to study the carrier dynamics. We were responsible for the fabrication of the nanometer scale gate electrodes.

Fabian Pease and David Allee from Stanford University had been working on high-performance HEMTs for some time. David joined our group in Cambridge in October 1989 as a research associate. Xiaodan Pan and Andrew Hoole were already working with me in Cambridge. As time went on, I became increasingly involved in administration as Head of the Electrical Division of the Engineering Department, and in 1993 as Head of the Engineering Department, and David, Xiaodan, and Andrew performed most of the research reported here on HEMTs.

The consortium was to study the physical limits of carrier transit time and hence device switching speed as heterojunction field effect transistors were scaled down from submicron to ultra-submicron dimensions (0.2–0.02 μm gate lengths). It was also to use Monte Carlo simulations, high-frequency s-parameter characterizations, and ultra-fast laser spectroscopy to explore the conditions where the carriers might exceed the saturation velocity due to near ballistic transport. The ultimate goal was to reach carrier transit times in the 1 ps regime and to maximize the corresponding device switching times by simultaneously reducing the device parasitics. AlGaAs/GaAsHFETs were used for these studies for gate lengths down to 0.1 μm, and pseudomorphic AlGaAs/InGaAs HFETs for dimensions below 0.1 μm. The results of this basic research action were to provide guidelines for those developing the next generation of HFETs for information technology.

Normally conducting gates

Maintaining a low gate resistance as the gate length is reduced below 0.1 μm is particularly difficult because unrealistically large gate aspect ratios are needed. Gate resistances for typical sub-0.1 μm gate lengths fabricated with PMMA lift-off were several hundred ohms (Allee et al., 1988), which was two orders of magnitude too high for high performance. As a result, the fastest FETs at the time had larger gate lengths between 0.1 and 0.15 μm (Lester et al., 1988).

Fig. 13.1 shows the configuration for the MODFET devices we considered. We first analyzed various approaches to reduce the gate resistance for normally conducting gate metals. We then considered superconducting gate electrodes. There were several possible techniques to reduce the gate resistance to a few ohms. The most obvious, though often impractical, approach was to increase the thickness of the gate metallization to form a high aspect ratio rectangular electrode (Fig. 13.2, Structure A) Another approach was to fabricate a mushroom gate (Fig. 13.2, Structure B) with a very small footprint (Chisholm et al., 1989; Writzel and Doane, 1986). Both of these structures were limited primarily by the mechanical stability of the electrodes and secondarily by the parasitic sidewall gate to source capacitance.

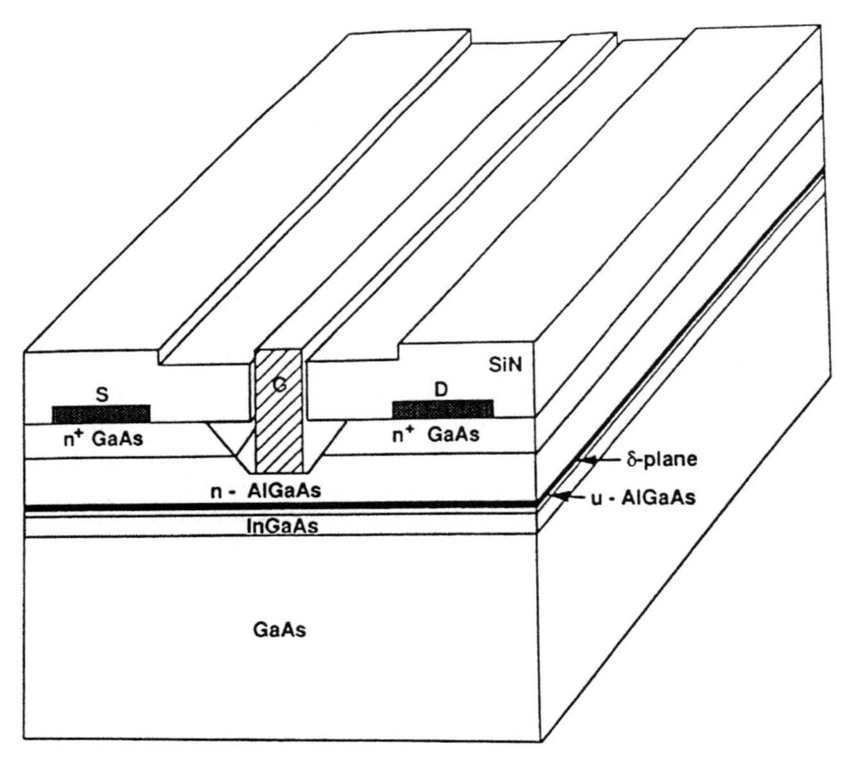

Fig. 13.1 Schematic diagram of MODFET without the interconnection level. *(Reprinted with permission from Van Hove, M., Zou, G., De Raedt, W., Jansen, Ph., Jonckheere, R., Van Rossum, M., Hoole, A., Allee, D.R., Broers, A.N., Crozat, P., Jin, Yan, Aniel, F., Adde, R., 1993. Scaling behavior of delta-doped AlGaAs/InGaAs high electron mobility transistors with gatelengths down to 60 nm and Source-Drain gaps to 230 nm. J. Vac. Sci. Technol. B 11: American Vacuum Society, pp. 1203–1208. Copyright 1993, American Vacuum Society).*

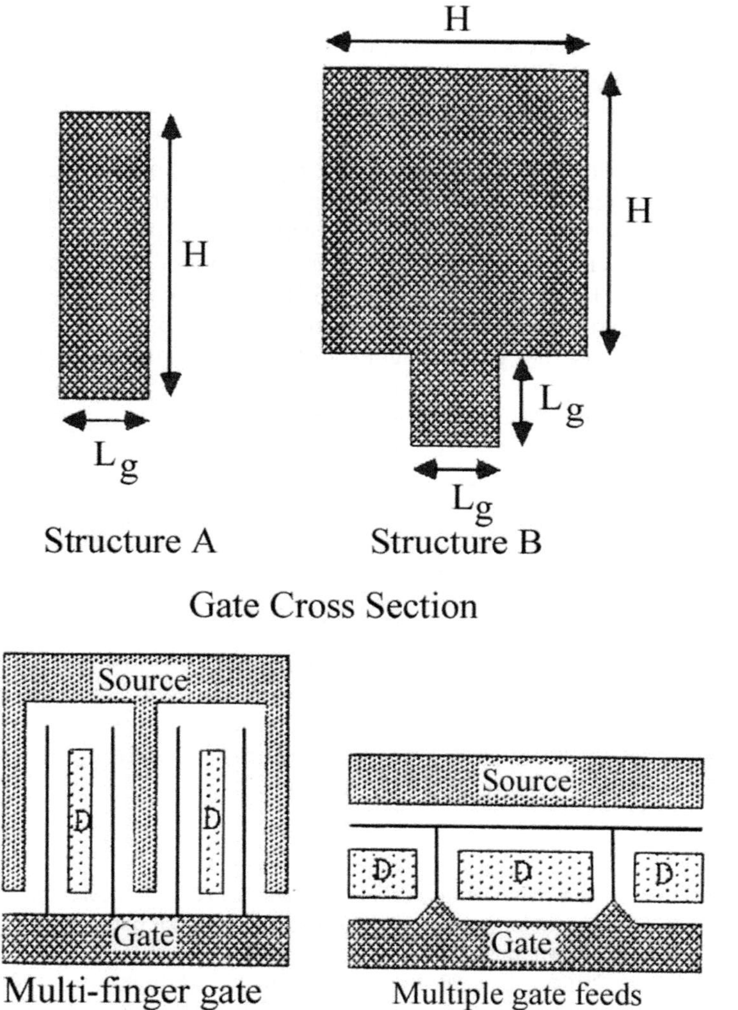

Fig. 13.2 Structures A and B are approximate cross-sections of two common gate structures. A is a simple rectangle and B is a T-gate. L_g is the gate length, and H/L is the gate aspect ratio. The multi-finger gate and the multiple gate feeds configuration can be used to reduce gate resistance by n^2 by breaking the gate into n equivalent segments. *(Reprinted with permission from Limits of Nanogate Fabrication Allee, D.R., Broers, A.N., Pease, R.F.W., 1991. Limits of nano-gate fabrication. Proc. IEEE. 79, 1093. © 1991 IEEE).*

A third approach was to break the gate electrode into several segments (Fig. 13.2 Multi-finger gate). Each segment was placed at the interstice between an interdigitated source and drain, forming a multi-fingered gate. While maintaining the same total device width, the gate resistance was reduced by a factor of n^2 where n was the number of segments. One factor of n arose from each gate segment being shorter, and the second factor of n was from the number of segments. Alternatively, n equivalent gate segments could be formed with multiple gate feeds as also shown in Fig. 13.2.

The dimensional parameters H and L_g are defined for the two gate structures in Fig. 13.2. L_g is the gate length. H is the thickness of the gate metallization for the rectangular gate (Structure A) and the vertical and horizontal dimensions of the upper portion of the T-gate (Structure B). The values of H and the aspect ratio H/L_g necessary to achieve a total DC gate resistance of 10 Ω are shown in Table 13.1 as a function of gate length for a rectangular gate, a T-gate, and 10 rectangular gates in parallel. The calculations assume the gate metal is gold at room temperature with a bulk resistivity of 2.04 μΩcm. The actual resistivity and magnitude of H would be somewhat higher because of the extremely small dimensions of these structures. The total device width is assumed to be 100 μm as this was the width used in most commercial quarter-micron microwave HFETs. A single rectangular gate is implausible for gatelengths below 0.1 μm because the aspect ratio would have to be greater than 20:1. For the T-gate, H must be about 400 nm and becomes independent of the gate length below 0.1 μm because the connecting stem of the mushroom becomes negligibly small. These dimensions are just plausible for 0.1 μm gate lengths where the aspect ratio must be 4:1, but not for the smaller gatelengths. The only technique that appears promising for gate lengths down to 20 nm is the 10-finger multi-gate option, although the aspect ratio will still have to be 5:1 at 20 nm.

Table 13.1 Shows the size of H (upper number) and aspect ratio (lower number) necessary to achieve a total resistance of 10 Ω as a function of gate length L_g for the various structures shown in Fig. 13.2.

Gatelength	**20 nm**	**40 nm**	**60 nm**	**80 nm**	**100 nm**
Rectangle	10 μm 500	4.8 μm 120	3.4 μm 57	2.6 μm 33	2 μm 20
T-gate	420 nm 21	420 nm 11	420 nm 6.7	420 nm 5.3	420 nm 4.2
Multi-gate	100 nm 5	50 nm 1.3	35 nm 0.58	26 nm 0.33	15 nm 0.15

The gate metal is assumed to be gold at 300 K (σ_n = 2.04 μΩcm). The total device width is 100 μm. The multi-gate assumes 10 simple rectangular gates in parallel; each gate is 10 μm wide to maintain the total device width of 100.

Data from Allee, D.R., Broers, A.N., Pease, R.F.W., 1991. Limits of nano-gate fabrication. Proc. IEEE. 79, 1093, fig. 14.

Table 13.2 Shows the size of H (upper number) and aspect ratio (lower number) necessary to achieve a total resistance of 10 Ω as a function of gate length L_g for the various structures shown in Fig. 13.2.

Gatelength	20 nm	40 nm	60 nm	80 nm	100 nm
Rectangle	2.3 μm 115	1.2 μm 30	0.8 μm 13	0.7 μm 8.8	0.5 μm 5
T-gate	220 nm 11	220 nm 5.5	220 nm 3.7	220 nm 2.8	220 nm 2.2
Multi-gate	25 nm 1.2	12 nm 0.3	8 nm 0.13	6 nm 0.08	5 nm 0.05
High-temperature superconductor	220 nm 11	100 nm 2.5	64 nm 1.6	52 nm 0.65	42 nm 0.42

The total device width is 100 μm. For the rectangle, the T-gate, and the multi-gate cases the metal is assumed to be gold at 77 K ($\sigma_n = 0.5\,\mu\Omega$cm). The total width of the gate is 100 μm in all cases. The superconductor is a simple rectangle and is made of YBaCuO at 77 K. The superconducting material properties are assumed to be: $T_c = 90$ K, ρ_n (100K) = 150 μΩcm, $\lambda_L(0\text{ K}) = 140$ nm.
Data from Allee, D.R., Broers, A.N., Pease, R.F.W., 1991. Limits of nano-gate fabrication. Proc. IEEE. 79, 1093, fig. 15.

These calculations were repeated assuming a device temperature of 77°K where gold has a resistivity of 0.5 μΩcm (Table 13.2). A single rectangular gate becomes viable at 0.1 μm but the aspect ratio climbs to 8:8 for an 80 nm gatelength. The T-gate might be possible down to a gatelength of 40 nm where the aspect ratio would be 5.5. Again, the parallel gate segments option is the most promising with an aspect ratio of only 1.2:1 for a gatelength of 20 nm.

Although DC resistance has been calculated, the data in Table 13.1 are valid up to the frequency at which the current is localized at the conductor surface due to the skin effect. The most limiting case is the T-gate where uniform current is assumed in the large upper portion. The assumption will break down when the skin depth is approximately equal to H/2. Using the well-known formula for skin effect depth (Ramo et al., 1984).

$$\lambda_s = \frac{1}{\sqrt{\frac{1}{2}\omega\mu_0\sigma_n}}$$

where ω is the frequency in rad/s. μ_o is the permeability of free space, and σ_n is the conductivity, the current in the T-gate will be uniform for frequencies up to 129 GHz at 77 K.

Superconducting gates

High T_c superconductors (Wu et al., 1987) had been discovered 4 years before we were conducting this study, and although there were unsolved problems in growing these materials on III–V substrates, we decided to see if it would be possible to reduce the gate resistance with a superconducting gate electrode operating at liquid nitrogen temperature. However, the gate resistance would strictly only be zero at DC, and we were interested in frequencies deep into the millimeter wave band (30–300 GHz) beyond the f_{max} of longer gate length HFETs. To understand this situation, it was best to use the London two-fluid model (Gittleman and Rosenblum, 1964) in which the superconductor is modeled as a parallel combination of a superconducting channel and a normally conducting channel (Fig. 13.3). The superconducting channel would have no resistance but would have inductance due to the inertia of the Cooper pairs. The normal channel would have resistance and inductance due to the inertia of the electrons. The inertial or kinetic resistance of the normal channel is negligible compared to the resistance and is usually ignored. At DC, the inductance of the superconducting channel shorts out the normal channel resistance, resulting in a perfectly conducting wire. As the frequency increases, the reactance of the superconducting channel increases, causing current to flow in the normal channel and resulting in a net series resistance for the superconductor.

To determine the magnitude of this resistance at microwave and millimeter frequencies, we assumed that we had a gate electrode made of c-axis orientated $YBa_2Cu_3O_{7-\delta}$, the most thoroughly studied high T_c superconductor. As a thin film, it has a critical temperature T_c of 90 K, a c-axis coherence length ξ of 1.2 nm, an energy gap at 0 K of 54 meV, a normal state resistance at 100 K of 150 $\mu\Omega$ cm, and a DC critical current density at 77 K greater than $2 \times 10^6 A/cm^2$ (Kapitulnik and Char, 1989). In bulk form, the London penetration depth at 0 K λ_L (0), has been measured to be 140 nm (Cava et al., 1987).

There are several assumptions in the London two-fluid model. The normal electron mean free path and the coherence length of the superconducting electrons must be less than the London penetration depth in order to use the local theory, $\mathbf{J} = \sigma \mathbf{E}$. Since the penetration depth is proportional to $\{1 - t_r^4\}^{-1/2}$ where t_r is the temperature normalized to T_c. λ_L at

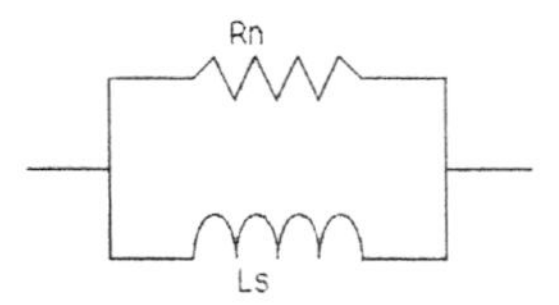

Fig. 13.3 The London fluid model of a superconductor. A superconducting inductance, L_S, whose physical origin is the inertia of the Cooper pairs is in parallel with a normal channel resistance, R_n. *(Reprinted with permission from Limits of Nanogate Fabrication Allee, D.R., Broers, A.N., Pease, R.F.W., 1991. Limits of nano-gate fabrication. Proc. IEEE. 79, 1093. © 1991 IEEE).*

77K is 205nm. This is much larger than the coherence length and also larger than the mean free path of the normal electrons if we assume sufficiently "dirty" material. Furthermore, the cross-section dimensions of the gate electrode must be equal to or less than twice the penetration depth to assume uniform current density. This is also true since the largest gate length we consider is 100nm. Finally, the operating frequency must be less than the energy gap frequency at which the electric field will break the Cooper pairs. Since the energy gap is proportional to $(1-t_r)$, the energy gap at 77K is 20.7 meV, corresponding to a frequency of 5.02THz. The largest frequency we will consider is 300GHz, the upper limit of the millimeter wave band. Under these conditions, the London two-field model is applicable.

The total current through the superconductor can be written as $_2$.

$$\mathbf{J} = (\sigma_n + \sigma_s)\mathbf{E}$$

where σ_n and σ_s are the normal channel and superconducting channel conductivities, respectively. J is the current density, and E is the electric field. These conductivities can be conveniently expressed in terms of measured parameters as follows:

$$\sigma_n = \sigma_{n0} t_4^4 \quad \sigma_s = \frac{1-r_r^4}{j\omega\mu_0\lambda_L^2(0)}$$

σ_{no} is the normal channel conductivity just above T_c: j is the square root of negative 1. The remaining variables maintain their previous definitions. The normal channel resistance, R_n, and the superconducting inductance L_s are

$$R_n = \frac{Z}{L_g H \sigma_{n0} t^4} \quad L_S = \frac{Z\mu_0\lambda_L^2(0)}{L_g H\left(1-t_r^4\right)}$$

where Z is the device width. A rectangular cross-section length, L_g, and height, H, is assumed (structure I). The equivalent series resistance, R_{series}, is

$$R_{series} = \frac{R_n\omega^2 L_s^2}{R_n^2 + \omega^2 L_S^2}$$

The end-to-end gate resistance for a 100μm wide device as a function of frequency for $H=L_g=20$, 50 and 100nm are plotted (Fig. 13.4). While the gate resistances are all less than 1 Ω at 10GHz, they reach 1000Ω for $L_g=20$nm at 300GHz.

At 100GHz, the values of H and the corresponding aspect ratio required for a total gate resistance of 10 Ω as a function of gate length are also given for a superconducting gate electrode in Table 13.2. Significantly, H for a superconducting electrode at 100GHz is still an order of magnitude larger than the H required for the multi-gate case.

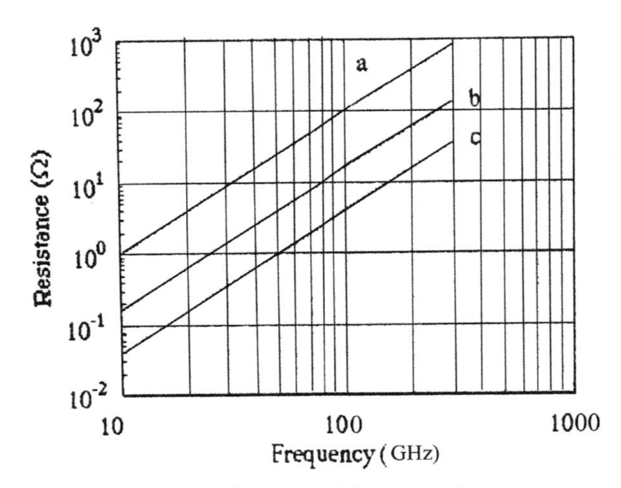

Fig. 13.4 The gate resistance as a function of frequency for superconducting YBaCuO at 77 K. The three gate electrodes shown all have square cross-sections with (a) $L_g = H = 20$ nm, (b) $L_g = H = 50$ nm and (c) $L_g = H = 100$ nm. *(Reprinted with permission from Limits of Nanogate Fabrication Allee, D.R., Broers, A.N., Pease, R.F.W., 1991. Limits of nano-gate fabrication. Proc. IEEE. 79, 1093. © 1991 IEEE).*

While the superconducting gate electrode could be fabricated in a parallel structure as well, there isn't much motivation to do so because of the sufficiently low aspect ratio of the normally conducting gate electrodes in parallel. This is particularly true given the current processing difficulties associated with high T_c superconductors: the difficulty of depositing films on GaAs substrates and the high temperature anneal that is incompatible with GaAs.

For completeness, the current carrying capabilities of high T_c superconductors at millimeter wave frequencies must also be considered. To our knowledge, this measurement has not yet been made. Kwon has estimated the AC critical current to be 50 mA/μm based on the low critical magnetic field of the high T_c superconductors (Kwon, 1986). This current density would be sufficient for small signal applications.

In conclusion, the most promising approach to reducing the gate resistance for sub-0.1 μm gate lengths was to use multi-fingered gates or multiple gate taps with normally conducting gate metal, and gate aspect ratios should be as large as possible.

Experimental results

The most significant advantage of the high beam voltage of the JEOL 4000EX in fabricating low-resistance nanometer-scale gate electrodes was the reduction in forward scattering of the electrons in the resist.

Thicker resists could be used while still maintaining linewidths close to the resolution limit. Thicker resists allowed larger gate aspect ratios either through lift-off of an evaporated gate metal or through electroplating on a pre-deposited plating base. The latter approach was being extensively investigated for the fabrication of X-ray lithography masks at the time (Windbracke et al., 1986). To our knowledge, plating had not been applied to the fabrication of nanometer gates, presumably because of the risk of contaminants from the plating solution. Plating might have had significant advantages over lift-off. The tearing of metal at the gate edges during lift-off would be avoided possibly resulting in greater edge uniformity and a reduced likelihood of discontinuities. The thin plating base could have been removed after the resist with a short chemical etch. A third approach was to use contamination resist. At the time, we were pursuing all three approaches.

By May 1990, several nanometer-scale test structures had been fabricated on bulk substrates, both GaAs and Si, using PMMA and lift-off. Au lines down to 16nm had been fabricated on GaAs (Fig. 13.5). The edge definition on the lines defined by contamination resist was markedly better than the lift-off lines (Fig. 13.6). An array of 25nm AuPd lines on a period of 65nm were fabricated with contamination resist on a silicon sample prepared by IMEC (Figs. 13.6 and 13.7).

A 35nm Au gate electrode was fabricated in the source-drain gap of a device structure prepared at IMEC (Fig. 13.8). The source was broken into

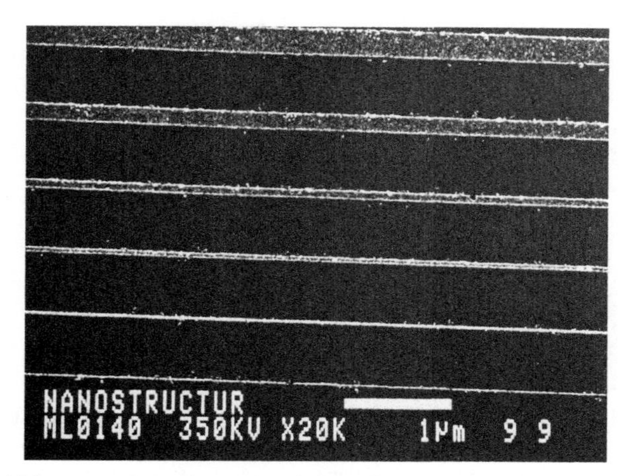

Fig. 13.5 Gold lines on GaAs formed with lift-off. Linewidths from top to bottom are 270, 170, 94, 56, 31, and 16nm. *(Reprinted from Alle, D.R., Broers, A.N., Van Rossum, M., Borghs, S., Launois, HH., Etienne, B., Adde, R., Castagne, R., Antonetti, A., Hulin, D., 1990. Performance and Physical Limits of Heterostructure Field Effect Tranbsistors" (ESPRIT Basic Research Action 3042) [Report]: Springer Nature, with permission of Springer Nature).*

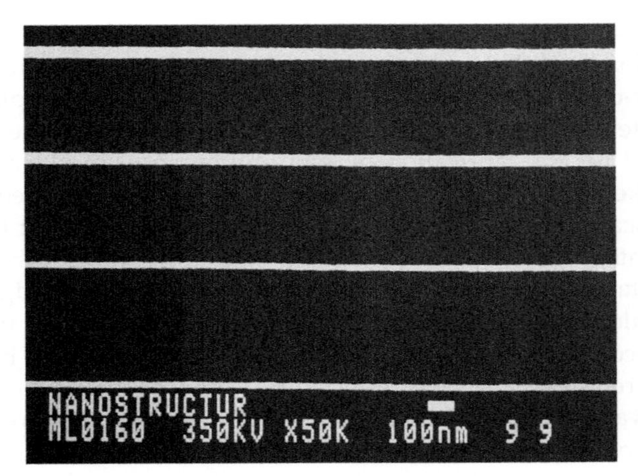

Fig. 13.6 AuPd lines formed by contamination resist and ion milling. The two upper lines are 50 nm; the two lower lines are 25 nm. Note the excellent edge uniformity. *(Reprinted from Alle, D.R., Broers, A.N., Van Rossum, M., Borghs, S., Launois, HH., Etienne, B., Adde, R., Castagne, R., Antonetti, A., Hulin, D., 1990. Performance and Physical Limits of Heterostructure Field Effect Tranbsistors" (ESPRIT Basic Research Action 3042) [Report]: Springer Nature, with permission of Springer Nature).*

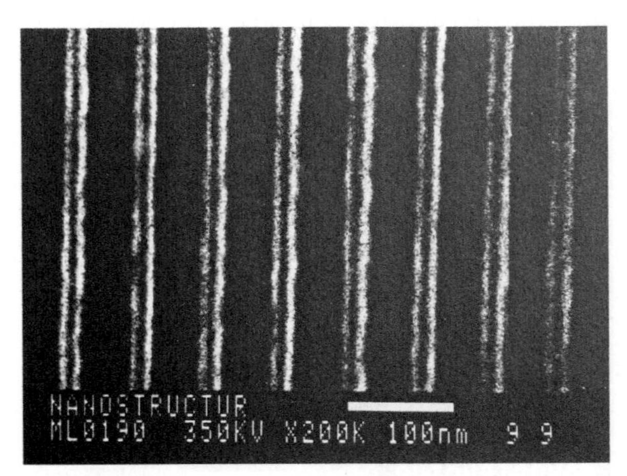

Fig. 13.7 AuPd lines on silicon substrate formed with contamination resist and ion milling. Lines are 25 nm on 65 nm period. *(Reprinted from Alle, D.R., Broers, A.N., Van Rossum, M., Borghs, S., Launois, HH., Etienne, B., Adde, R., Castagne, R., Antonetti, A., Hulin, D., 1990. Performance and Physical Limits of Heterostructure Field Effect Tranbsistors" (ESPRIT Basic Research Action 3042) [Report]: Springer Nature, with permission of Springer Nature).*

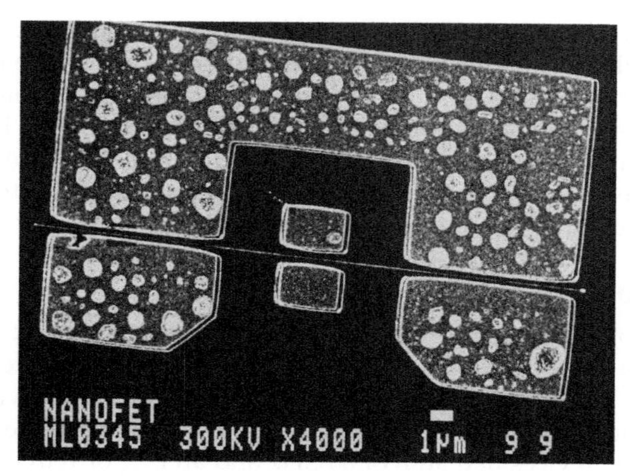

Fig. 13.8 35 nm gate electrode aligned in the source drain gap of a MODFET device structure. *(Reprinted from Alle, D.R., Broers, A.N., Van Rossum, M., Borghs, S., Launois, HH., Etienne, B., Adde, R., Castagne, R., Antonetti, A., Hulin, D., 1990. Performance and Physical Limits of Heterostructure Field Effect Tranbsistors" (ESPRIT Basic Research Action 3042) [Report]: Springer Nature, with permission of Springer Nature).*

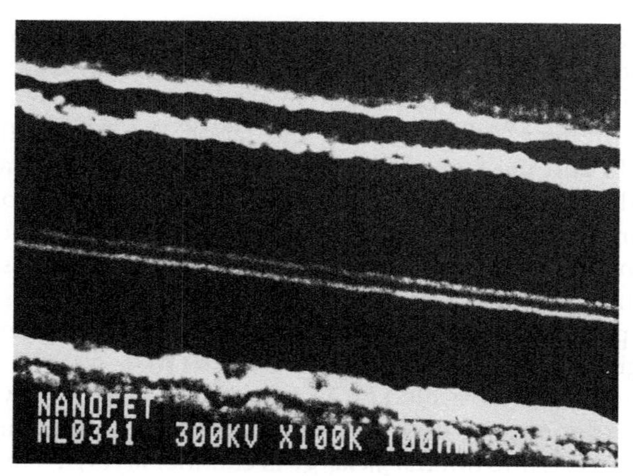

Fig. 13.9 High magnification view of 35 nm gate shown in Fig. 13.8. *(Reprinted from Alle, D.R., Broers, A.N., Van Rossum, M., Borghs, S., Launois, HH., Etienne, B., Adde, R., Castagne, R., Antonetti, A., Hulin, D., 1990. Performance and Physical Limits of Heterostructure Field Effect Tranbsistors" (ESPRIT Basic Research Action 3042) [Report]: Springer Nature, with permission of Springer Nature).*

two segments: a central gate contact (not shown) divided the gate into two equivalent segments and hence reduced the gate resistance by a factor of four. Fig. 13.9 shows a higher magnification image of the 35 nm gate.

By the end of the first 9 months of the collaboration, we had analyzed the various techniques needed to reduce gate resistance for sub 0.1 μm gatelengths. The most promising approach was to use multi-fingered gates or multiple gate feeds with a normally conducting gate metal, and gate aspect ratios as large as it was practical to fabricate. Nanometer-scale structures had been fabricated on bulk substrates with PMMA lift-off, and with ion milling and contamination resist down to 16 nm. A 35 nm by 25 μm gate electrode was fabricated in the 0.4 μm source to drain gap of a device structure.

Device fabrication

By the end of the first year of the NANOFET project, both bulk and planar doped AlGaAs/GaAs HFETs were fabricated at L2M with gate lengths of 0.2 μm. The transconductances were 370 and 625 mS/mm with gm/gd (transconductance/output conductance) ratios of 13 and 16, respectively. Both of these devices had an f_t of 60 GHz and an f_{max} of 100 GHz. 0.15 μm planar doped gate lengths devices were also fabricated both with a double and single gate recess. The single gate recess device had a transconductance of 612 mS/mm, a g_m/g_d of 8.3, f_t of 81 GHz, and f_{max} of 102 GHz. As expected, the double gate recess devices were superior, having a transconductance of 694 mS/mm, a g_m/g_d of 13, f_t of 112 GHz, and f_{max} of 132 GHz.

At IMEC, pseudomorphic AlGaAs/InGaAs layer growth conditions were being optimized for bulk and delta doping. For bulk doping, room temperature mobilities of 28,000 cm^2/Vs and sheet carrier densities of $1.95 \times 10^{12}/cm^2$ were achieved. For the fabrication of sub-0.1 μm gate length HFETs and the corresponding reduction of source-drain spacing to a few tenths of a micron, the ohmic contact edges definition became critical. The common alloyed AuGeNiAu contacts were inadequate. IMEC had developed InAs grown ohmic contacts with contact resistances less than 0.1 Ohm-mm. The contact metal was optimized to be PdGeTiPt, which also provided high contrast in the scanning electron microscope, making alignment easy.

The devices were fabricated on an epitaxial layer grown by molecular epitaxy. In order of growth, the layer structure consists of an undoped GaAs buffer layer, a 13 nm $In_{0.20}Ga_{0.80}As$ channel layer, a 5 nm undoped $Al_{0.25}Ga_{0.75}As$ spaced layer, a Si sheet layer delta-doped at $5 \times 10^{12}/cm^2$, a 30 nm $Al_{0.25}Ga_{0.75}As$ layer doped at $5 \times 10^{17}/cm^3$ and a 40 nm cap layer doped at $5 \times 10^{18}/cm^3$. Under identical growth conditions, a calibration layer was grown without the highly doped GaAs cap layer. High layer quality was indicated by Hall mobilities, and sheet carrier densities measured by the Van de Pauw method on this reference layer: $\mu_H = 5800\ cm^2/Vs$ and $ns = 1.6 \times 10^{12}/cm^2$ at 300 K, $\mu H = 26{,}000\ cm^2/Vs$ and $n_s = 1.55 \times 10^{12}/cm^2$ at 77 K and $\mu_H = 30{,}400\ cm^2/Vs$ and

$n_s = 1.55 \times 10^{12}/cm^2$ at 4 K. At 4 K the same carrier density was derived from Stubnikov de Haas oscillations, indicating a single conducting path in the two-dimensional electron gas.

The device fabrication process that would incorporate the high-resolution gates grown in Cambridge started with an alignment level defined by optical lithography and Ti/Au metal lift-off. The ultrasmall ohmic gaps were defined by electron beam lithography exposure at an acceleration voltage of 20 kV and lift-off of the evaporated metal (120 nm AuGe/15 nm Ni/60 nm Au). A 50 nm thick silicon nitride layer was deposited by plasma-enhanced chemical vapor deposition, and the ohmic metal was nitride cap alloyed at 400 °C for 30 s. The nitride-capped alloy ensures an ohmic contact with low contact resistance and an extremely smooth morphology that was necessary for accurate alignment of the subsequent gate exposure in the small ohmic gaps (ref). Other advantages of using a nitride layer were the improved gate metal lift-off yield and the possibility of using it as a supporting layer in a T-gate fabrication process. A possible drawback was its higher dielectric constant compared to air, leading to higher parasitic gate capacitances (Nummila et al., 1991).

The wafer was scribed into 5 mm square pieces for gate exposure in the lithography system in Cambridge. A 450 nm thick bilayer of PMMA was spun on the wafer, consisting of a layer of low molecular weight at the bottom and a layer of high molecular weight at the top. Each gate was exposed individually due to the limited field of view in the TEM. Before each exposure, the system was focused, and the rotation corrected. The thick resist scheme was necessary to improve the gate metal lift-off yield. Thinner layers of resist did not provide enough coverage, especially around the mesa and ohmic edges, which had a total step height of over 400 nm. A drawback of the thicker resist scheme was that the limit of PMMA resolution could not be approached. The smallest device that could be successfully lifted off had a gate length of 40 nm, to be compared with a linewidth of 16 nm that could be achieved by exposure in the same lithographic system using planar substrates and thinner resists (Allee et al., 1991).

The exposed silicon nitride was removed in a CF4 plasma recess etching in a $H_2SO_4/H_2O_2/H_2O$ (1/8/1000) solution. Lateral under-etching problems, which were related to the bad adhesion of PMMA to GaAs, were solved by the nitride-assisted process. This lateral etching reduced both the device performance by seriously increasing the source resistance and the device yield (Lamarre and Mctaggart, 1990). Before the wet gate recess etching, the sample was exposed to an oxygen plasma for 10s to improve the etching uniformity. The nanogates were formed by e-gun evaporation and lift-off of Ti/Au (25/45) metal layers. The fabrication

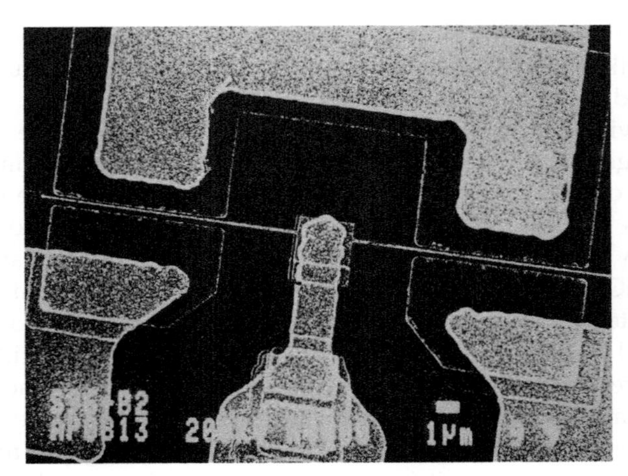

Fig. 13.10 SEM micrograph of a 85 nm gatelength device with a 430 nm source-drain gap. *(Reprinted with permission from Van Hove, M., Zou, G., De Raedt, W., Jansen, Ph., Jonckheere, R., Van Rossum, M., Hoole, A., Allee, D.R., Broers, A.N., Crozat, P., Jin, Yan, Aniel, F., Adde, R., 1993. Scaling behavior of delta-doped AlGaAs/InGaAs high electron mobility transistors with gatelengths down to 60 nm and Source-Drain gaps to 230 nm. J. Vac. Sci. Technol. B 11: American Vacuum Society, pp. 1203–1208. Copyright 1993, American Vacuum Society).*

process was completed with a macroscopic interconnection level defined by optical lithography and Ti/Au lift-off. A schematic cross-section of the device without the interconnection level is shown in Fig. 13.1. An SEM overview micrograph of an 85 nm gatelength device with a 430 nm source-drain gap is shown in Fig. 13.10. The devices had a small width ($2 \times 7.5\ \mu m$) because of the limited field of view in the TEM. Fig. 13.11 shows a high magnification view of the gate region of a 40 nm gatelength device. The nitride cap and the wet gate recessed region can be seen.

We processed devices with gatelengths ranging down from 250 to 60 nm in source to drain gaps of nominally 760, 430, and 230 nm on the pseudomorphic layer structure described above. We used different gate recess depths so that we could study the influence of the aspect ratio on the performance of the device.

DC characteristics

The observed threshold voltages (V_{th}) vs gatelength (L_G) for gate recess times of 60, 80, 90, and 120 s are shown in Fig. 13.12. Since the AlGaAs layer is pulse-doped (Moll et al., 1988), ignoring the weak uniformly doping as a first approximation, the threshold voltage shifts ΔV_{th} are related with differences in gate-to-channel separations Δd as

$$\Delta V_{th} = \frac{eN_{sh}}{\varepsilon} \Delta d$$

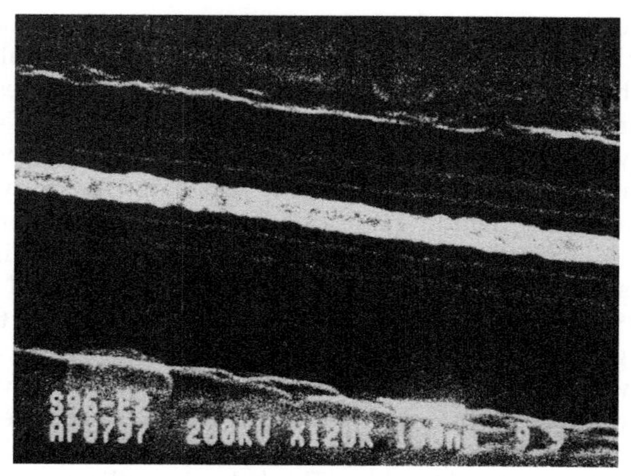

Fig. 13.11 SEM micrograph of a 40 nm gatelength device. *(Reprinted with permission from Van Hove, M., Zou, G., De Raedt, W., Jansen, Ph., Jonckheere, R., Van Rossum, M., Hoole, A., Allee, D.R., Broers, A.N., Crozat, P., Jin, Yan, Aniel, F., Adde, R., 1993. Scaling behavior of delta-doped AlGaAs/InGaAs high electron mobility transistors with gatelengths down to 60 nm and Source-Drain gaps to 230 nm. J. Vac. Sci. Technol. B 11: American Vacuum Society, pp. 1203–1208. Copyright 1993, American Vacuum Society).*

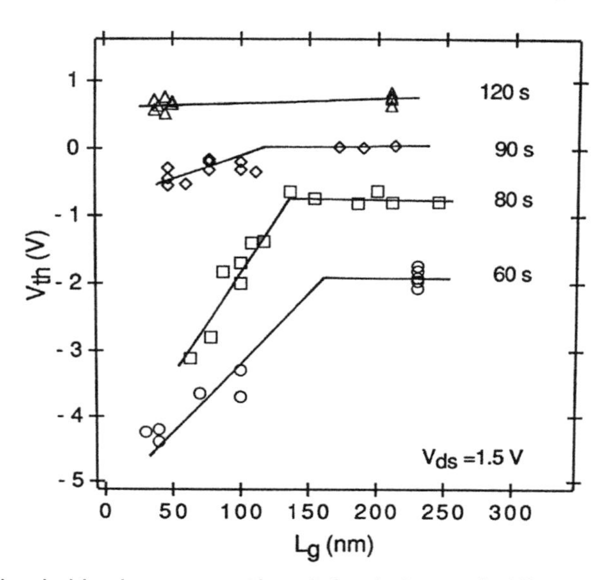

Fig. 13.12 Threshold voltage vs gate length for devices with different recess depths. *(Reprinted with permission from Van Hove, M., Zou, G., De Raedt, W., Jansen, Ph., Jonckheere, R., Van Rossum, M., Hoole, A., Allee, D.R., Broers, A.N., Crozat, P., Jin, Yan, Aniel, F., Adde, R., 1993. Scaling behavior of delta-doped AlGaAs/InGaAs high electron mobility transistors with gatelengths down to 60 nm and Source-Drain gaps to 230 nm. J. Vac. Sci. Technol. B 11: American Vacuum Society, pp. 1203–1208. Copyright 1993, American Vacuum Society).*

with e the electron charge, ϵ the dielectric constant of AlGaAs, and N_{SH} the sheet donor concentration. Following this relation, the Δd values, calculated from the shifts in V_{th} for long gate lengths devices, correspond well to the etch times obtained by assuming a calibrated etch rate of 40 nm/min with a delay of 20s due to surface conditioning. With this assumption, gate-to-channel separations of 8, 28, 35, and 45 nm were calculated in the same way as the data in Table 13.1. The same values were obtained by measuring the width of the nitride undercut from the scanning electron micrographs of the various devices. Since the wet etchant used was isotropic (Shaw, 1981) the etched region in the lateral direction was the same as the vertical gate recessed depth.

Devices with gatelengths shorter than 100 nm show severe v_{th} shifts for shallow gate recesses but only weak shifts for the deeper recesses. For these measurements, the drain was biased at 1.5 V. Smaller v_{th} shifts were observed at 0.2 V source-drain bias. This behavior indicates that the shifts were due to short channel effects and not because of possible nonuniformities of the wet gate recess etching process.

Peak room temperature g_{ms} measured at a bias voltage of 1.5 V for shallow gate recessed devices (gate-to-channel separation of 35 nm) are shown in Fig. 13.13. The different source-drain gaps are indicated. For devices with gatelengths longer than 100 nm, the g_m was very high (>88 mS/mm) and

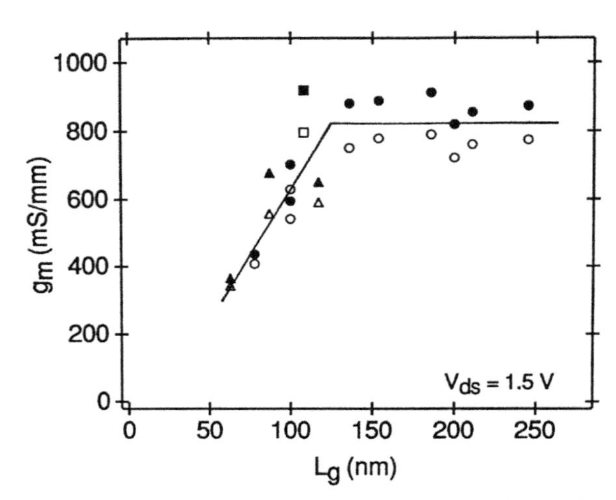

Fig. 13.13 Peak room temperature g_{ms} measured at a bias voltage of 1.5 V for shallow gate recessed devices with a gate-to-channel separation of 35 nm. *(Reprinted with permission from Van Hove, M., Zou, G., De Raedt, W., Jansen, Ph., Jonckheere, R., Van Rossum, M., Hoole, A., Allee, D.R., Broers, A.N., Crozat, P., Jin, Yan, Aniel, F., Adde, R., 1993. Scaling behavior of delta-doped AlGaAs/InGaAs high electron mobility transistors with gatelengths down to 60 nm and Source-Drain gaps to 230 nm. J. Vac. Sci. Technol. B 11: American Vacuum Society, pp. 1203–1208. Copyright 1993, American Vacuum Society).*

independent of gatelength. For shorter gatelengths, a strong decrease in g_m was observed. The decrease was strongly related to the V_{th} shift, as can be seen by comparing Figs. 13.12 and 13.13 and can be ascribed to short channel effects. Monte Carlo calculations (Shaw, 1981) predicted that the smaller parasitic resistance would lead to an increase in the extrinsic maximum g_m. Indeed, the highest value was achieved for a 100 nm gatelength device in a 260 nm wide source-drain gap. However, due to the limited amount of data and the experimental error, this result was inconclusive and needed to be confirmed by a more systematic study. The intrinsic transconductance g_{mi} was obtained from the measured values of g_{mx} using the equation

$$g_{mi} = \frac{g_{mx}}{1 - g_{mx} R_S}$$

The source resistances R_S (average 0.160 Ω mm) were extracted from a high-frequency equivalent circuit model, described in the next paragraph. The R_S values obtained from the sheet resistance $R_{SH} = 160\,\Omega/\text{square}$, the contact resistance $Rc = 0.07\,\Omega\,\text{mm}$ measured on a transmission line contact pattern, and the source-gate spacing were approximately 0.045 Ω lower. The difference can be explained by the parasitic resistance caused by the open sidewalls of the recess trench and the higher contact resistance to the two-dimensional electron gas in a gate-recessed structure.

The transconductance and full transistor curves for a 60 nm gatelength device in a 430 nm source-drain are shown in Figs. 13.14 and 13.15, respectively. The extrinsic peak g_m was 340 mS/mm. The device showed perfect

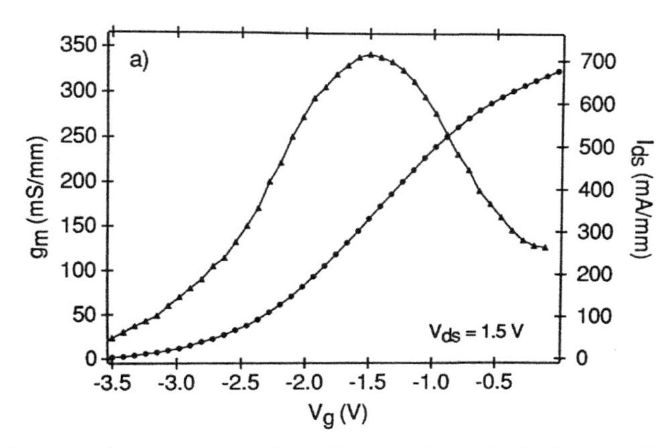

Fig. 13.14 Transconductance curve for a 60 nm gatelength device in a 430 nm source drain gap. The extrinsic peak g_m was 340 mS/mm. *(Reprinted with permission from Van Hove, M., Zou, G., De Raedt, W., Jansen, Ph., Jonckheere, R., Van Rossum, M., Hoole, A., Allee, D.R., Broers, A.N., Crozat, P., Jin, Yan, Aniel, F., Adde, R., 1993. Scaling behavior of delta-doped AlGaAs/InGaAs high electron mobility transistors with gatelengths down to 60 nm and Source-Drain gaps to 230 nm. J. Vac. Sci. Technol. B 11: American Vacuum Society, pp. 1203–1208. Copyright 1993, American Vacuum Society).*

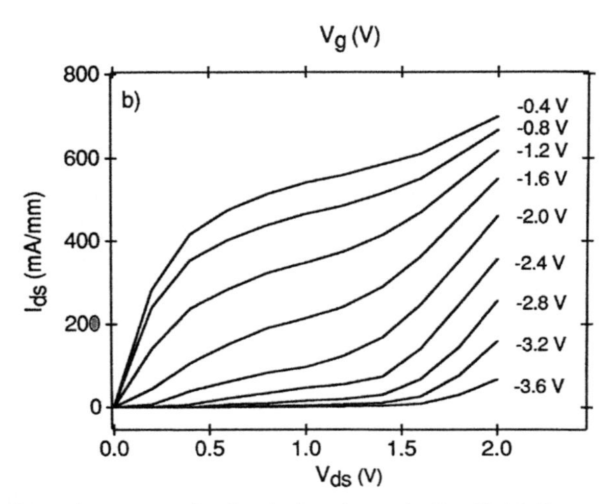

Fig. 13.15 Full transistor curve for the device shown in Fig. 12.14. Due to the very small gate-drain distance (<100 nm) drain breakdown occurred for bias voltgaes higher than about 1 V. *(Reprinted with permission from Van Hove, M., Zou, G., De Raedt, W., Jansen, Ph., Jonckheere, R., Van Rossum, M., Hoole, A., Allee, D.R., Broers, A.N., Crozat, P., Jin, Yan, Aniel, F., Adde, R., 1993. Scaling behavior of delta-doped AlGaAs/InGaAs high electron mobility transistors with gatelengths down to 60 nm and Source-Drain gaps to 230 nm. J. Vac. Sci. Technol. B 11: American Vacuum Society, pp. 1203–1208. Copyright 1993, American Vacuum Society).*

pinch-off behavior. Due to the very small gate-drain distance (<100 nm for this device) drain breakdown occurred for bias voltages higher than about 1 V. The higher output conductance and the negative v_{th} shift compared to longer gatelength devices (Figs. 13.16 and 13.17) indicated significant carrier injection into the buffer layer.

The g_m of ultrashort gatelength devices can be significantly improved by using a deeper gate recess, hereby increasing the aspect ratio and reducing short channel effects. Fig. 13.18 shows the g_m curve (peak g_m600 mS/mm) for a 60 nm gatelength device with a gate-channel separation of 28 nm. These data illustrate that accurate control of the recess depth is extremely important to decrease the short channel effects in very small gatelength HEMTs. However, a comparison of Figs. 13.16 and 13.18 shows that, even for devices with equal V_{th}s, the maximum g_m was not higher for the 60 nm gatelength device.

There had been speculation that ballistic transport and velocity overshoot would occur at very short gatelengths. These would result in higher effective saturation velocities and, hence, higher transconductances. However, our data showed no evidence for these effects. Monte Carlo simulations (Dollfus et al., 1992) predicted an increase of the "ballistic peak" from 46% of the total electron concentration at the end of the gate for the

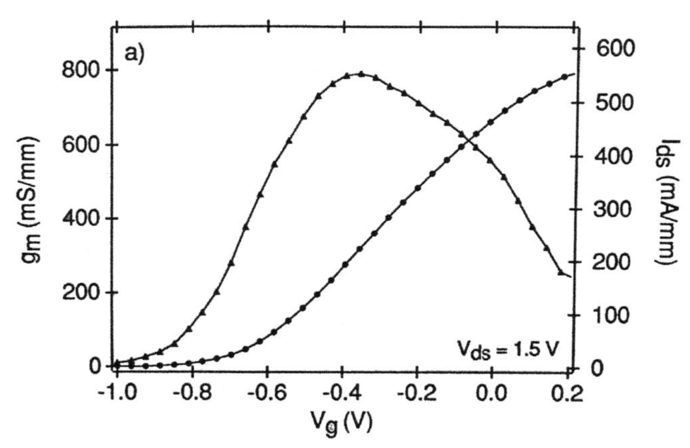

Fig. 13.16 I_{ds} and g_m vs V_g for a device with a longer gatelength than the device used for Figs. 13.14 and 13.15. *(Reprinted with permission from Van Hove, M., Zou, G., De Raedt, W., Jansen, Ph., Jonckheere, R., Van Rossum, M., Hoole, A., Allee, D.R., Broers, A.N., Crozat, P., Jin, Yan, Aniel, F., Adde, R., 1993. Scaling behavior of delta-doped AlGaAs/InGaAs high electron mobility transistors with gatelengths down to 60 nm and Source-Drain gaps to 230 nm. J. Vac. Sci. Technol. B 11: American Vacuum Society, pp. 1203–1208. Copyright 1993, American Vacuum Society).*

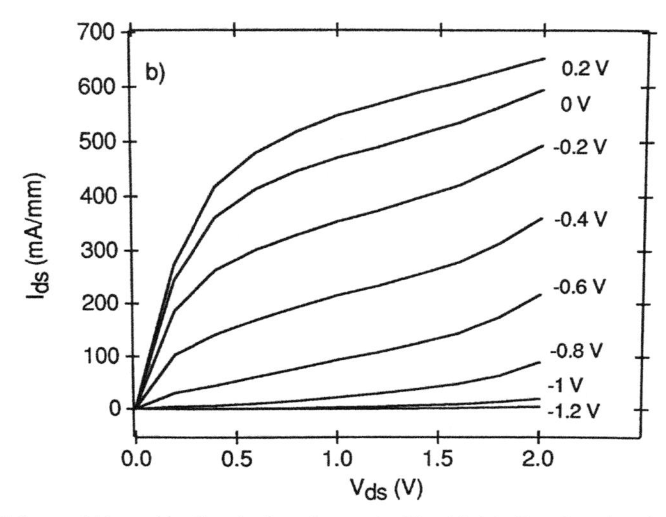

Fig. 13.17 I_{ds} and V_G vs V_{ds} for device shown in Fig. 13.16. *(Reprinted with permission from Van Hove, M., Zou, G., De Raedt, W., Jansen, Ph., Jonckheere, R., Van Rossum, M., Hoole, A., Allee, D.R., Broers, A.N., Crozat, P., Jin, Yan, Aniel, F., Adde, R., 1993. Scaling behavior of delta-doped AlGaAs/InGaAs high electron mobility transistors with gatelengths down to 60 nm and Source-Drain gaps to 230 nm. J. Vac. Sci. Technol. B 11: American Vacuum Society, pp. 1203–1208. Copyright 1993, American Vacuum Society).*

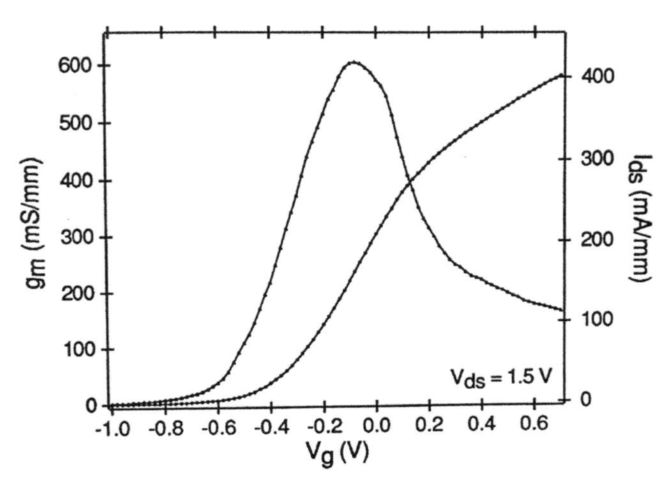

Fig. 13.18 g_m curve for a 60 nm gatelength device with a deeper recess than the device whose characteristics are shown in Fig. 16.14. The deeper recess resulted in a gate-to-channel separation of 28 nm. This increased the g_m from 340 mS/mm to about 600 mS/mm, but this was still lower than the peak g_m of 800 mS/mm for the device whose characteristics are shown in Fig. 13.16. *(Reprinted with permission from Van Hove, M., Zou, G., De Raedt, W., Jansen, Ph., Jonckheere, R., Van Rossum, M., Hoole, A., Allee, D.R., Broers, A.N., Crozat, P., Jin, Yan, Aniel, F., Adde, R., 1993. Scaling behavior of delta-doped AlGaAs/InGaAs high electron mobility transistors with gatelengths down to 60 nm and Source-Drain gaps to 230 nm. J. Vac. Sci. Technol. B 11: American Vacuum Society, pp. 1203–1208. Copyright 1993, American Vacuum Society).*

150 nm gatelength, to 87% for the 50 nm gatelength. A maximum drift velocity overshoot as high as 9×10^7 cm/s was calculated for the 50 nm gatelength device compared to 8×10^7 cm/s for a 150 nm gatelength device. It is shown that this should result in very short transit times, i.e., very high transition frequencies, but this is not sufficient to improve transconductances. The calculations are in qualitative agreement with our results.

Microwave characteristics

The high-frequency characteristics of the devices were determined from S-parameter measurements made with an HP8510B network analyzer and Tektronix coplanar microwave probes over a frequency range from 45 MHz to 18 GHz. Due to the very small width of the devices (15 μm) the current levels were rather low ($\leq$10 mA). Our results showed for the first time parameter levels for devices with such a small gatelength and low current level. The parasitics in the 100 μm square access pads were considerable. However, the accuracy of the capacitance extraction made it possible to make a comparative study of the high-frequency performance of devices with different gatelengths fabricated with the same design.

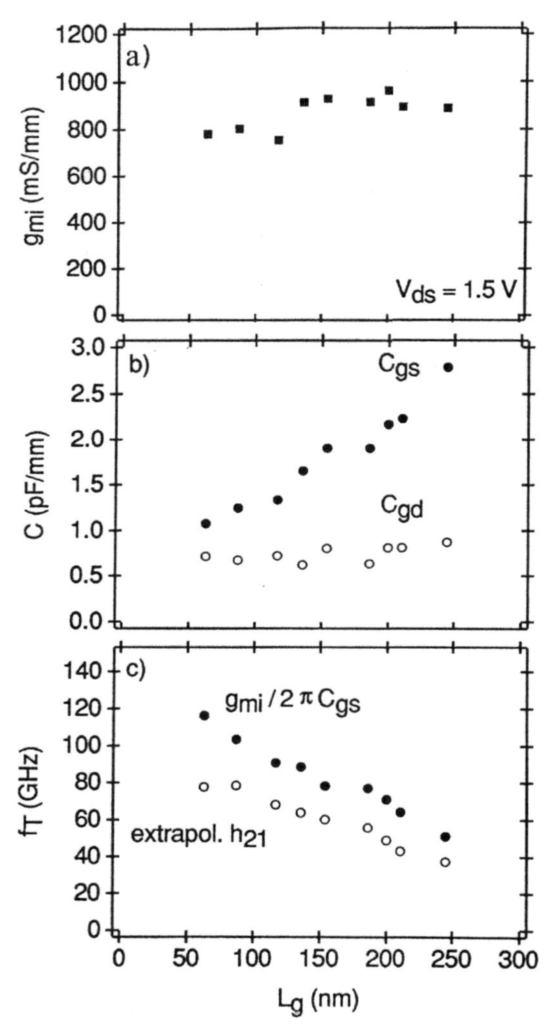

Fig. 13.19 HF parameters; HF g_m, gate-source and gate-drain capacitances C_{gs} and C_{gd} and f_T for devices with gatelengths that ranged from 250 to 60 nm.

The HF parameters were extracted from the measured S-parameters using an equivalent circuit model (Berroth and Bosch, 1991). The extracted HF g_m, the gate-source and gate-drain capacitances C_{gs} and C_{gd}, and f_T values are presented in Fig. 13.19A–C, respectively. Good agreement between DC and HF g_m was observed for the longer gatelength devices, but the HF g_m was substantially increased for the shorter gatelengths. Since this was correlated with less negative V_{th} shifts, the difference between HF and DC performance is believed to be due to less important short-channel effects at high frequencies. Although not very well understood at the time, these effects were remarkable and, to our best knowledge, had not been observed and explained earlier.

Extrinsic cut-off frequencies F_{t1} were extrapolated from h_{21} measurements and intrinsic values f_{T2} were calculated from $f_{T2}=g_{mi}/2\pi C_{gs}$. The rather low f_T values compared to state-of-the-art data were explained by the higher C_{gs} values since, in our design, the gate contact finger was made on top of the active material to avoid the step of the narrow gate over the mesa. However, the general trend showed that due to the reduction of the gate capacitance, there was still an improvement in the high-frequency performance of the devices when the gatelength was reduced from 250 to 60 nm. This meant that in the future, it should be possible to reduce gate resistance by using T-gates or multi-gates and achieve both higher g_ms and f_ts. Wider devices would also allow more accurate parasitic capacitance extraction.

Summary and conclusions

We successfully studied pseudomorphic delta-dopes AlGaAs/InGaAs HEMTs with gatelengths as small as 60 nm in ultra-submicron source-drain gaps.

Different aspect ratios were studied. For large gate-to-drain separations, important short channel effects were observed, but they were less important for deeper gate recesses. For gatelengths longer than 0.1 μm, the transconductance was found to be independent of gatelength, and the transconductance of shorter gatelength devices was decreased by short channel effects. In qualitative agreement with Monte Carlo calculations, no evidence for velocity overshoot was observed. When comparing DC and HF peak transconductance for the devices with gatelengths shorter than 0.1 μm, appreciably higher values were measured at high frequencies, indicating less important short channel effects. The important conclusion to be drawn from our work was that due to decreased gate capacitances, the current cut-off frequency continuously increased as the gatelength was reduced into the sub-0.1 μm regime, showing that further reductions in gatelength would be beneficial for the high-frequency performance.

References

Alle, D.R., Broers, A.N., Van Rossum, M., Borghs, S., Launois, H.H., Etienne, B., Adde, R., Castagne, R., Antonetti, A., Hulin, D., 1990. Performance and Physical Limits of Heterostructure Field Effect Tranbsistors. (ESPRIT Basic Research Action 3042) [Report], Springer Nature.

Allee, D.R., de la Houssaye, P.R., Schlom, D.G., Harris Jr., J.S., Pease, R.F.W., 1988. Sub-0.1μm gate length GaAs MESFETs and MODFETs. J. Vac. Sci. Technol. A B6 (1), 328–332.

Allee, D.R., Broers, A.N., Pease, R.F.W., 1991. Limits of nano-gate fabrication. Proc. IEEE 79, 1093.

Berroth, M., Bosch, R., 1991. High frequency equivalent circuit of GaAs FETs for large signal applications. IEEE Trans. Microwave Theory Tech. 39, 224.

Cava, R.J., Batlogg, B., van Dover, R.B., Murphy, D.W., Sunshine, S., Siegrist, T., Remeika, J.P., Rietman, E.A., Zahurak, S., Espinosa, G.P., 1987. Bulk superconductivity at 91K in single-phase oxygen-deficient perovskite $Ba_2YCu_3O_9$-δ. Phys. Rev. Lett. 58 (16), 1676–1679.

Chisholm, A., Sainson, S., Feuillade, M., Clei, A., 1989. 0.15μm e-beam T-shaped gates for GaAs FETs. Microcircuit Eng. Abstracts, 36.

Dollfus, P., Bru, C., Galdin, S., Hesto, P., 1992. Influence of short channel effects on the microwave performance of AlGaAs/InGaAs HENTs using Monte Carlo simulations. In: ESSDERC Sept. 14–17. Leuven.

Gittleman, J.I., Rosenblum, B., 1964. Microwave properties of superconductors. Proc. IEEE 52, 1138–1147.

Kapitulnik, A., Char, K., 1989. Measurement on thin film high-Tc superconductors. IBM J. Res. Devel. 33 (3), 252–261.

Kwon, O.K., 1986. Chip-to-chip interconnections for very high speed system level integration, Ph.D. dissertation. Stanford University, CA.

Lamarre, P., Mctaggart, R., 1990. A positive resist adhesion promoter for PMMA on GaAs MESFETs. IEEE Trans. Electron Devices ED-37, 2406.

Lester, L.F., Smith, P.M., Ho, P., Chao, P.C., Tiberio, R.C., Duh, K.H.G., Wold, E.D., 1988. 0.15 um gate length double recess pseudomorphic HEMT with fmax of 350 Ghz. IEDM Technical Digest., 172–175.

Moll, N., Hueschen, M., Fischer-Colbrie, A., 1988. Pulse-doped AlGaAas/InGaAs pseudomorphic MODFETs. IEEE Trans. Electron Devices ED-325, 879.

Nummila, K., Tong, M., Ketterson, A.A., Adesida, I., 1991. Fabrication of sub-100nm T gates with SiN passivation layer. J. Vac. Sci. Technol. A B9, 2870.

Ramo, S., Whinnery, J.R., van Duzer, T., 1984. Fields and Waves in Communication Electronics. Wiley.

Shaw, D.W., 1981. Localized GaAs etching with acidic hydrogen peroxide solutions. J. Electrochem. Soc. 128, 874.

Van Hove, M., Zou, G., De Raedt, W., Jansen, P., Jonckheere, R., Van Rossum, M., Hoole, A., Allee, D.R., Broers, A.N., Crozat, P., Jin, Y., Aniel, F., Adde, R., 1993. Scaling behavior of delta-doped AlGaAs/InGaAs high electron mobility transistors with gatelengths down to 60 nm and Source-Drain gaps to 230 nm. J. Vac. Sci. Technol. B 11, 1203–1208. American Vacuum Society.

Windbracke, W., Betz, H., Huber, H.L., Pilz, W., Pongratz, S., 1986. Critical dimension control in X-ray masks with electroplated gold absorbers. Microelectron. Eng. 5 (1–4), 73–80.

Writzel, C.E., Doane, D.A., 1986. A review of GaAs MESFET gate electrode fabrication technologies. J. Electrochem. Soc., 133. 409Ch.

Wu, M.K., Ashburn, J.R., Tong, C.T., Hor, P.H., Meng, R.I., Gao, L., Huang, Z.J., Chu, C.W., 1987. Superconductivity at 93K in a new mixed-phase Y-Ba-Cu-O compound system at ambient pressure. Phys. Rev. Lett. 58, 908.

Epilogue

I stopped my research in 1996 when I became Vice Chancellor of the University of Cambridge. I realized that this would be more than a full-time job, and that is how it turned out.

When my 7-year contract as Vice Chancellor ended, Mary and I moved to London, and I became a member of the House of Lords in 2004. I was President of the Royal Academy of Engineering from 2001 until 2006, which was an exceptional privilege, and held a number of industrial directorships so there was no time to get back to research. However, I retained my interest in developments in microscopy and lithography.

Progress with semiconductor lithography was rapid but evolutionary during this time. Optical projection continued to dominate mainstream lithography for manufacturing chips, and scanning electron beam lithography was used to make the masks for the optical cameras and to explore the characteristics of future devices and circuits. In the early 1990s, it was thought it would be impossible to extend optical lithography below about 0.25 μm

EUV lithography

What was not known was that ASML in The Netherlands would combine lasers that produced high-intensity 13.5 nm wavelength EUV light with precision multilayer mirror lenses tuned for this wavelength to build high-throughput EUV lithography exposure systems theoretically capable of producing dimensions down to and beyond the capability of today's photoresists.

Initially, these EUV systems had a numerical aperture of 0.35, but the latest systems have an NA of 0.55. The Rayleigh criterion, as defined as the wavelength divided by the NA multiplied by a constant K, which is about 0.5 for lithography, suggests that such systems would be able to make structures with dimensions approaching the wavelength of 13.5 nm. Lower values of K might even be possible with immersion imaging and multi-imaging. Optimists point out that the theoretical limit for $K1$ is 0.25, which would suggest dimensions below 10 nm, although at present, there is no evidence that there are resists with this resolution. The minimum metal pitch in the Semiconductor Technology Roadmap for 2024 is 21 nm.

The EUV exposure systems are hugely complex and expensive. The exposure system is about the same size as a compact electron synchrotron storage ring, about 7 m × 2.5 m, and the laser illumination shipping the system is about the same size and is located on a floor beneath the exposure system. The overall system costs about $250 million, and delivering it internationally takes three Boeing 747s.

Nonetheless, their exposure speed is so high that EUV lithography has continually outperformed rival technologies, including proximity printing with X-rays generated by a compact electron synchrotron storage ring that was thought to be the most likely successor to optical projection in the early 1990s. A mask-to-wafer gap of less than 10 μm would have been needed for X-ray proximity printing for sub-0.1 μm dimensions, as pointed out in Chapter 11, which was considered impractical. The ALF project at IBM East Fishkill was abandoned around the turn of the century despite encouraging results with 0.25 μm dimensions.

Electron beam lithography

Progress with electron optics and electron microscopy continued with a steady increase in the use of thermal and Schottky field emission electron sources and improved focusing, deflection, and projection optics. A remarkable advance was made in transmission electron microscopy, where techniques for correcting spherical aberration were finally perfected, and the $C_S^{1/4}\lambda^{3/4}$ barrier that had prevented electron microscopes following optical microscopes in attaining resolution approaching the wavelength of the electrons was broken. Resolution well below 1 Å has been achieved.

Advances in examining surfaces with scanning electron microscopes were less revolutionary, but the increased application of high-brightness sources made the ultimate resolution of secondary electron surface microscopy more widely available at an acceptable cost.

Progress with electron beam lithography systems concentrated on increasing throughput in an attempt to compete with optical lithography in manufacturing. Major examples were the SCALPEL and PREVAIL electron beam projection systems built at Bell Labs and IBM, described at the end of Chapter 11. They offered higher throughput than the variable-shaped beam scanning systems and were capable of linewidths down to about 0.1 μm. However, throughput was still low compared with the EUV systems, their thin substrate multilayer mask was more complex, and their resolution and overlay performance were marginal for 0.1 μm dimensions. It was also clear by 2000 that dimensions below 0.1 μm were going to be needed, and significant changes would have been needed for these electron beam projectors were to match the throughput and resolution of the EUV systems. Support for them faded away at the turn of the century when most major semiconductor companies opted for UV and EUV systems.

Multibeam electron beam masks writers

Despite the large increase in the writing speed of scanning electron beam systems achieved by using variable-shaped beams as described in Chapter 11, the fastest mask writers today are taking times of 18 h or more to write a single reticle, stretching the limits of system reliability and increasing cost and turnaround time. To try to shorten exposure times, interest has returned to multibeam systems. These had been investigated in the 1990s when building very small electron beam columns with field emission cathodes became possible. For example, Chang, Kern and Muray (Chang, 1992) developed a system based on a concept they called scanning tunneling microscope aligned field emission (SAFE). This used an array of field emission cathode microcolumns, one or more per chip, operating at a relatively low voltage of 1 kV to minimize the proximity effect. Each column had its own focusing and deflection system. Kratschmer et al. (1995) also showed that a microcolumn using a Schottky field emission tip can produce a 10 nm diameter, 1 nA, electron beam. Chang predicted that SAFE could produce a throughput of 100–200 m wafers per hour for linewidths below 250 nm.

Twenty years later, IMS Nanofabrication, Nuflare Technology, and DNP have built multibeam systems that operate completely differently. They use an array of hundreds of thousands of 10–20 nm diameter electron beams formed by flood illuminating a thin metal plate with an array of apertures. This is followed by a second array of apertures, each with its own blanking system. The array of beams is then scanned with a curvilinear lens and deflection system similar to the systems developed for the electron beam projectors to complete the exposure of a reticle that is about 10 cm square. In the multibeam case, the throughput seems to become limited by the time taken to deliver the vast data volume (tens of terabytes) needed to describe a modern chip, rather than the writing rate of the electron beams. The total electron beam current delivered to the wafer is about the same as in the variable-shaped beam systems, about 0.5 μA. The exposure time is about 10 h, even with hundreds of thousands of beams, but this exposure time is constant and does not depend on the complexity of the pattern.

Looking back on the development of semiconductor chip technology, it becomes clear that the ability to, in effect, store vast quantities of information on a physical mask and replicate it in seconds has been essential in realizing the incredible reductions in cost that have driven the semiconductor industry. It also explains why the attempts to reach acceptable costs with scanning electron beam lithography failed.

It also makes it unlikely that direct exposure methods for patterning devices that avoid the need for masks are going to succeed in manufacturing.

Thermal field and Schottky emission cathodes

Most high-resolution round beam electron probes used for microscopy, microanalysis, and microfabrication now use thermal field emission (TFE) or Schottky field emission (SFE) cathodes. Cold field emission cathodes of the type used, for example, by Albert Crewe back in the 1960s and 1970s (Crewe, 1970; Crewe and Wall, 1970), are seldom used because vacuum levels of 10^{-10} mmHg are required for stable emission, and these finely pointed cathodes are susceptible to damage through microdischarges. TFE cathodes operate at about 1800 °C where the noise and beam instabilities are reduced and vacuum levels of about 10^{-8} mmHg can be tolerated. Vacuum pumps and seals that produce these pressures have also become more available and less expensive. Electron brightnesses above 10^8 A/cm^2 sr. at 50 kV are available.

SFE emitters are coated with low work-function materials such as ZrO_2, and the thermionic emission is combined with field emission.

Thermal emitters, mainly lanthanum hexaboride, are still used when high beam currents are needed and where vacuum pressures are above about 10^{-8} mmHg. These conditions are often encountered in microfabrication and nanofabrication systems, especially variable beam size and projection systems. Polished and pointed single crystal LaB_6 cathodes operating with Schottky enhancement can produce brightness above 2×10^7 A/cm^2 sr at 45 kV.

Limited-area cathodes

Another way to obtain high brightness with thermal cathodes may be to limit the size of the emitting area so that a Wehnelt electrode is no longer needed to determine the size of the emitting area. This increases the electric field at the cathode surface minimizing space-charge divergence and increasing brightness.

The method might also be used to produce an array of electron beams, removing the need for vacuum levels below 10^{-8} mmHg.

Chris Maloney, Shanhong Xia, Hiroshi Nakamura, Les Peters, and I in Cambridge worked with Eric Munro and Xieqing Zhu in the Blackett Laboratory at Imperial College to simulate the performances of these limited-area cathodes (Broers, 1991).

Several types of cathode were modeled. Two were of particular interest. They were rod cathodes with a conical tip with an included angle of

140°, a tip radius of 15 μm, and an emitting area of 10 μm in diameter. They had a work function of 2.4 eV, approximately that of LaB_6. One operated in a conventional triode configuration where the emission area is set by the intersection of the zero potential line with the cathode surface. The second was operated in the limited-area mode without a Wehnelt electrode but with an emitting area limited to 10 μm diameter. The brightness for the first cathode was 3.6×10^6 A/cm^2/sr at 25 kV, and the second limited-area cathode was 2×10^7 A/cm^2 sr.

Such cathodes are difficult but not impossible to make. They may be made by masking all but the apex of the cathode or by coating the apex of the cathode with a small area of low-work function material. Alternatively, a small conical Wehnelt electrode with a very small hole could be used. Maloney made limited measurements, but the full potential of the cathodes is yet to be realized.

Ultimate resolution of optical, electron beam, and X-ray lithographies

It has now become clear that the ultimate resolution of conventional resist-based lithographies is set by the range of the interaction between the radiation and the molecules in the resist and by the molecular scale mechanisms of development. This delocalization limit has been measured to be between 10 nm and 20 nm for electrons as described in Chapters 9 and 12. With X-rays, it has been estimated from the minimum features found in X-ray micrographs made by contact printing specimens into resist. Low-loss scanning electron microscopy with a resolution of about 2 nm was used to show that the smallest features in the resist pattern were between 10 nm and 20 nm. An example of such a contact print is given in Chapter 11.

With practical lithography systems, many other factors have to be considered. For systems using visible and ultraviolet light with wavelengths larger than 20 nm, resolution is limited by diffraction and lens aberrations.

With electron exposure, electron scattering in the resist and backscattering from the substrate have a major influence on image contrast. High current may also be needed to reach adequate throughput, and this may lead to the beam size, or beam definition being greater than the delocalization limit.

For X-ray proximity printing, resolution is determined by the minimum acceptable gap between the mask and the resist. This is set at 10 μm, to avoid damage between mask and wafer, and diffraction between the mask and the resist limits dimensions to about 0.1 μm. This is an order of magnitude larger than the delocalization limit.

Fig. 1 shows one way to compare the resolution of the different exposure methods. It shows the exposure dose received in the center of a small

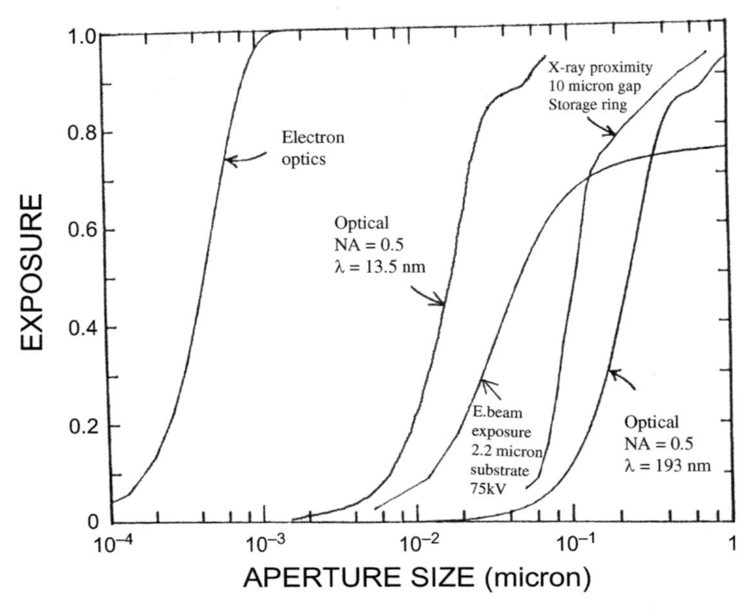

Fig. 1 Exposure dose received at the center of a square aperture as a function of aperture size. The exposure dose is normalized to the background dose. The practical resolution limit is given when the dose at the center of the aperture has fallen to approximately 50% of the background level.

square aperture as a function of the size of the square. The exposure dose is normalized to the exposure dose received in the center of an infinitely large shape. When the exposure flux falls to 50%, it can be said that the practical resolution limit has been reached. At this level, there will be a variation of 2:1 in the flux received by the largest and the smallest shapes. Beyond 2:1, the resist process can no longer be relied upon to handle the variation.

The data for electron beam exposure use experimentally determined distributions rather than the usual, less accurate models that are based on adding in quadrature a series of Gaussian distributions. The X-ray curve is derived from calculations made by Alan D. Wilson when working in the Advanced Lithography Facility in IBM for line exposures. The aperture value is obtained by squaring the line value.

Several conclusions can be drawn from Fig. 1. First, the resolution of X-ray proximity printing with a 10 μm gap between mask and wafer was only a factor of 2 better than the deep-UV ($\lambda = 250$ nm, NA $= 0.5$) step-and-repeat camera lens. However, the effective depth of focus for the X-ray case was much better than that for the deep UV lens. This was the situation in the late 1990s, and it sustained interest in X-ray proximity printing. However, this was of no avail as the evolutionary route of progress through further reductions in wavelength with optical projection offered larger advantages and avoided the need for a 1:1 mask with its extreme dimensional stability requirements. The 13.5 nm, 0.5 NA EUV case offers 10 times higher resolution than the 10 μm gap X-ray proximity case.

Fig. 1 also shows the 100-fold difference between the ultimate resolution of electron optical systems and the real-life case where electron scattering in the resist and backscattering from the substrate and the delocalization limit are taken into account. The leveling off at larger dimensions in the electron beam case is due to electron backscattering from the substrate.

Fig. 2 shows the contrast obtained in an array of lines and spaces for electron beam and optical lithography, as defined in Chapter 9. In particular, it shows the case of electron beam lithography, where a thin substrate and a thin resist are used. It should be noted that the resolution of electron optics ignoring the delocalization limits shown in Fig. 1 falls off the scale of Fig. 2. It is a hundred times smaller than the delocalization limit.

The EUV case is for the optical system and does not include a delocalization limit. This would move the curve to the right.

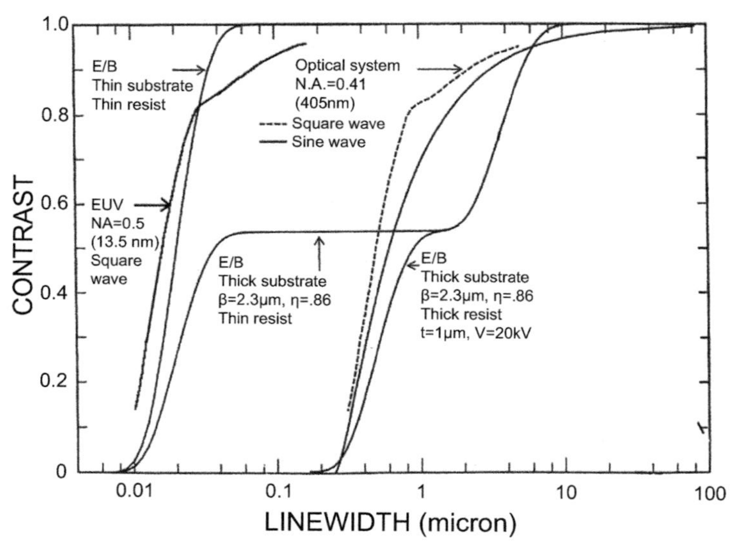

Fig. 2 Contrast for an array of lines and spaces vs linewidth for optical and electron beam lithography. The resolution of the EUV case coincides with the ultimate delocalization limit measured for electron beam exposure.

The most important observation is that the resolution of the EUV system is already at the delocalization limit as measured by electron beam exposure for available resists. Hence, improvements in the resolution of the exposing system may not yield smaller structures unless new resists or processes are found. The reduction in the size of electron beams from about 5 nm to less than 1 nm does not yield smaller structures using conventional resist processes. The smallest pitch in an infinite array of lines and spaces is about 20 nm. Perhaps this explains why the minimum metal pitch in the International Semiconductor Roadmap remains at 20 nm in 2024 and only falls to 16–12 nm in 2036.

Unless new resists or processes can be found with a smaller delocalization limit, further progress in the performance of semiconductor chips will have to rely on device and circuit innovations rather than miniaturization. The challenge is to find resist processes with higher resolution. It is possible to place atoms with scanning tunneling microscopes and therefore manipulate materials at the atomic level (Eigler and Schweizer, 1990), but the challenge of using such techniques to produce devices with the complexity of a modern chip is astronomical. The latest EUV lithography cameras print an image that contains the equivalent of more than 10^{14} pixels in about 10 s.

References

Broers, A., Shanhong, X., Maloney, C., Zhu, X., Munro, E., 1991. High brightness limited area cathodes. J. Vac. Sci. Technol. B 9 (6), 2929–2933.

Chang, T.H.P., Kern, D.P., Muray, L.P., 1992. Arrayed miniature electron beam columns for high throughput sub-100 nm lithography. J. Vac. Sci. Technol. B 10 (6), 2743–2748.

Crewe, A.V., 1970. The current state of scanning electron microscopy. Q. Rev. Biophys. 3, 137–175.

Crewe, A.V., Wall, J., 1970. Contrast in a High-Resolution STEM. Optik 30, 461.

Eigler, D.M., Schweizer, E.K., 1990. Positioning single atoms with a scanning tunnelling microscope. Nature 344, 524–526.

Kratschmer, E., Kim, H.S., Thomson, M.G.R., Lee, K.Y., Rishton, S.A., Yu, M.L., Chang, T.H.P., 1995. An electron-beam micro-column with improved resolution, beam current, and stability. J. Vac. Sci. Technol. B 13 (6).

Index

Note: Page numbers followed by "*f*" indicate figures and "*t*" indicate tables.

N

O

P

Printed and bound by CPI Group (UK) Ltd, Croydon, CR0 4YY
09/12/2024
01803132-0001